Recent Evolution and Seismicity of the Mediterranean Region

NATO ASI Series

Advanced Science Institutes Series

A Series presenting the results of activities sponsored by the NATO Science Committee, which aims at the dissemination of advanced scientific and technological knowledge, with a view to strengthening links between scientific communities.

The Series is published by an international board of publishers in conjunction with the NATO Scientific Affairs Division

A	Life Sciences	Plenum Publishing Corporation
B	Physics	London and New York
C	Mathematical and Physical Sciences	Kluwer Academic Publishers Dordrecht, Boston and London
D	Behavioural and Social Sciences	
E	Applied Sciences	
F	Computer and Systems Sciences	Springer-Verlag
G	Ecological Sciences	Berlin, Heidelberg, New York, London,
H	Cell Biology	Paris and Tokyo
I	Global Environmental Change	

NATO-PCO-DATA BASE

The electronic index to the NATO ASI Series provides full bibliographical references (with keywords and/or abstracts) to more than 30000 contributions from international scientists published in all sections of the NATO ASI Series.
Access to the NATO-PCO-DATA BASE is possible in two ways:

– via online FILE 128 (NATO-PCO-DATA BASE) hosted by ESRIN,
Via Galileo Galilei, I-00044 Frascati, Italy.

– via CD-ROM "NATO-PCO-DATA BASE" with user-friendly retrieval software in English, French and German (© WTV GmbH and DATAWARE Technologies Inc. 1989).

The CD-ROM can be ordered through any member of the Board of Publishers or through NATO-PCO, Overijse, Belgium.

Series C: Mathematical and Physical Sciences - Vol. 402

Recent Evolution and Seismicity of the Mediterranean Region

edited by

E. Boschi
Istituto Nazionale di Geofisica,
Rome, Italy

E. Mantovani
Dipartimento di Scienze della Terra,
Università di Siena,
Siena, Italy

and

A. Morelli
Istituto Nazionale di Geofisica,
Rome, Italy

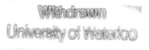

Kluwer Academic Publishers

Dordrecht / Boston / London

Published in cooperation with NATO Scientific Affairs Division

Proceedings of the NATO Advanced Research Workshop on
Recent Evolution and Seismicity of the Mediterranean Region
Erice, Italy
September 18-27, 1992

Library of Congress Cataloging-in-Publication Data

```
Recent evolution and seismicity of the Mediterranean region / edited
   by E. Boschi, E. Mantovani, A. Morelli.
       p.   cm. -- (NATO ASI series. Series C, Mathematical and
   physical sciences ; vol. 402)
     ISBN 0-7923-2325-4 (acid-free)
     1. Geology, Structural--Mediterranean Region--Congresses.
   2. Geodynamics--Congresses.  3. Earthquakes--Mediterranean Region-
   -Congresses.    I. Boschi, E.  II. Mantovani, E.  III. Morelli, A.
   IV. Series: NATO ASI series. Series C, Mathematical and physical
   sciences ; no. 402.
   QE633.M43R43  1993
                                                      93-19532
```

ISBN 0-7923-2325-4

Published by Kluwer Academic Publishers,
P.O. Box 17, 3300 AA Dordrecht, The Netherlands.

Kluwer Academic Publishers incorporates the publishing programmes of
D. Reidel, Martinus Nijhoff, Dr W. Junk and MTP Press.

Sold and distributed in the U.S.A. and Canada
by Kluwer Academic Publishers,
101 Philip Drive, Norwell, MA 02061, U.S.A.

In all other countries, sold and distributed
by Kluwer Academic Publishers Group,
P.O. Box 322, 3300 AH Dordrecht, The Netherlands.

Printed on acid-free paper

All Rights Reserved
© 1993 Kluwer Academic Publishers
No part of the material protected by this copyright notice may be reproduced or
utilized in any form or by any means, electronic or mechanical, including photo-
copying, recording or by any information storage and retrieval system, without written
permission from the copyright owner.

Printed in the Netherlands

TABLE OF CONTENTS

Introduction	vii
List of Participants	xiii
A.M.C. Şengör, *Some current problems on the tectonic evolution of the Mediterranean during the Cainozoic*	1
J. Jackson, *Rates of active deformation in the Eastern Mediterranean*	53
E. Mantovani, D. Albarello, D. Babbucci and C. Tamburelli, *Post Tortonian deformation pattern in the Central Mediterranean: a result of extrusion tectonics driven by the Africa-Eurasia convergence*	65
D. Albarello, E. Mantovani, D. Babbucci and C. Tamburelli, *Africa-Eurasia kinematics in the Mediterranean: an alternative hypothesis*	105
M. Barazangi, D. Seber, T. Chaimov, J. Best, R. Litak, D. Al-Saad and T. Sawaf, *Tectonic evolution of the Northern Arabian Plate in Western Syria*	117
J.L. Mercier, D. Sorel, S. Lalechos and B. Keraudren, *The tectonic regimes along the convergent border of the Aegean Arc from the late Miocene to the present; southern Peloponnesus as an example*	141
E. Patacca, R. Sartori and P. Scandone, *Tyrrhenian basin and Apennines. Kinematic evolution and related dynamic constraints*	161
A. Argnani, *Neogene basins in the Strait of Sicily (Central Mediterranean): tectonic settings and geodynamic implications*	173
A. Morelli, S. Mazza, N.A. Pino and E. Boschi, *Seismological studies of upper mantle structure below the Mediterranean with a regional seismograph network*	189
D. Giardini, B. Palombo and E. Boschi, *The determination of earthquake size and source geometry in the Mediterranean Sea*	213
G.A. Papadopoulos, *Tectonic and seismic processes of various space and time scales in the Greek area*	239

D. Slejko, *A review of the Eastern Alps - Northern Dinarides seismotectonics*   251

A. Udias and E. Buforn, *Regional stresses in the Mediterranean region derived from focal mechanisms of earthquakes*   261

J. Drakopoulos and G.N. Stavrakakis, *Source process of some large earthquakes in Greece and its tectonic implication*   269

M. Cocco, G. Selvaggi, M. Di Bona and A. Basili, *Recent seismic activity and earthquake occurrence along the Apennines*   295

G.B. Cimini and A. Amato, *P-wave teleseismic tomography contribution to the delineation of the upper mantle structure of Italy*   313

P. Favali, R. Funiciello and F. Salvini, *Geological and seismological evidence of strike-slip displacement along the E-W Adriatic-Central Apennines belt*   333

M. Dragoni, *Stable fault sliding and earthquake nucleation*   347

E. Márton, *Paleomagnetism in the Mediterraneum from Spain to the Aegean: a review of data relevant to Cenozoic movements*   367

G. Scalera, P. Favali and F. Florindo, *Use of the Paleomagnetic databases for geodynamical studies: some examples from the Mediterranean Region*   403

Introduction

The Mediterranean is one of the most studied regions of the world. In spite of this, a considerable spread of opinions exists about the geodynamic evolution and the present tectonic setting of this zone. The difficulty in recognizing the driving mechanisms of deformation is due to a large extent to the complex distribution in space and time of tectonic events, to the high number of parameters involved in this problem and to the scarce possibility of carrying out quantitative estimates of the deformation implied by the various geodynamic hypotheses. However, we think that a great deal of the present ambiguity could be removed if there were more frequent and open discussions among the scientists who are working on this problem.

The meeting of ERICE was organized to provide an opportunity in this sense. In making this effort, we were prompted by the conviction that each step towards the understanding of the Mediterranean evolution is of basic importance both for its scientific consequences and for the possible implications for society. It is well known, for instance, that the knowledge of ongoing tectonic processes in a given region and of their connection with seismic activity may lead to the recognition of middle-long term precursors of strong earthquakes. The few cases of tentative earthquake prediction in the world occurred where information on large scale seismotectonic behavior was available. This led to identify the zones prone to dangerous shocks, where observations of short-term earthquake precursors were then concentrated.

The choice of participants to the ERICE meeting was made so as to set up a multidisciplinary group of experts in many fields of earth sciences, able to discuss in detail all types of possible implications of the proposed geodynamic hypotheses. Presentations and discussions were mainly focused on some crucial problems, with particular reference to the Africa-Eurasia relative motion in the Mediterranean and to geodynamic processes and structural-seismotectonic features in the peri-Adriatic and eastern Mediterranean regions.

As concerns the first problem, the main object of debate was the direction of the Africa- Eurasia convergence. So far, this information has been determined through the analysis of kinematic indicators in the North Atlantic. By these data it is possible to estimate the Africa-North America and Eurasia-North America relative motions from which the Africa- Eurasia motion can be determined. The results obtained by this type of approach suggest that Africa has moved N to NW with respect to Eurasia during the last 5-10 My. This hypothesis, however, encounters some difficulties in explaining the post-Tortonian deformation pattern in the Central Mediterranean. Furthermore, it is not easily reconcilable with reconstructions of previous (pre-Tortonian) Africa-Eurasia kinematics, which provide a NEward convergence between these two blocks. This implies that around 9-10 My the drifting trend of Africa must have undergone a change of roughly 90°, passing from NE to NW.

The above difficulties led some authors to check the reliability of the assumptions which were used in previous analyses of North Atlantic data. One crucial assumption is the internal coherence of the Africa and Eurasia blocks. This hypothesis cannot easily be reconciled with the occurrence of significant deformation and seismic activity in the region comprised between the Rhine Graben system and the Azores Gibraltar belt. If this intraplate deformation is not negligible, the kinematic indicators observed in the Atlantic ridges just north of the Azores and along the Azores-Gibraltar belt cannot be straightforwardly used to constrain the Eurasia-North America and Africa-Eurasia relative motions. This may have important consequences, especially if one considers that the other kinematic data in the North Atlantic can be reconciled with Africa-Eurasia relative motions significantly different from the ones previously obtained. This would mean, for instance, that a NE to NNEward direction of the Africa-Eurasia convergence, which seems to best account for the post-Tortonian deformation pattern in the Central Mediterranean, is not necessarily in contrast with North Atlantic data (see, Albarello et al.).

However, if one decides to explore this last kinematic hypothesis, there remains the problem of understanding the tectonic mechanisms through which part of the Atlantic spreading is absorbed by deformation in the western European region and the adjacent oceanic area. Moderate seismicity affects the Rhine Graben system and the Pyrenees, while stronger activity is observed in the southern part of Iberia and in the oceanic zone offshore of Portugal and Morocco. Is this deformation sufficient to absorb part of the spreading which develops in the Mid Atlantic ridge lying just north of the Azores? Finding an answer to this question may have important consequences for understanding Africa-Eurasia kinematics in the Mediterranean area, while we are waiting for the time when geodetic measurements will help to solve this basic problem.

Another argument which attracted attention and triggered lively discussion is the formation of the Tyrrhenian basin and the structural evolution of the other peri-Adriatic regions (see, Argnani, Mantovani et al., Mercier et al., Patacca et al., Sengor, Slejko). The main difficulty in explaining these tectonic events is due to the fact that the framework of evidence does not fit satisfactorily with any simple known geodynamic process. Some years ago, the formation of the Tyrrhenian basin was generally explained as a consequence of oceanization processes connected with mantle plumes or as back-arc spreading in response to the active subduction of the Ionian lithosphere beneath the Calabrian Arc. The difficulties that these models encountered in accounting for some major features led many authors to look for alternative interpretations. This search has been considerably influenced by the fact that a much better knowledge is now available on the structural-tectonic evolution of the Tyrrhenian-Apennines system. In particular, an important constraint on the geodynamic evolution of the whole central Mediterranean region is that crustal stretching in the Tyrrhenian basin and major deformations in the adjacent thrust belt-foredeep Apenninic system have occurred in three main phases, well differentiated in space and time. Another piece of evidence which may have very important implications for the understanding of the rheological behavior of the lithosphere-asthenosphere system in the region considered, is that a large part of the continental Adriatic foreland has disappeared, since the upper Miocene, beneath the Apenninic chain (see, Patacca et al.).

During the meeting, the discussion about the driving mechanism of the Central Mediterranean deformation pattern mainly concerned two kinds of mechanism. One hypothesizes that the Tyrrhenian opening was caused by the gravitational sinking of the Adriatic-Ionian subducted margin, which was responsible for the progressive E-SEward migration of the rift-thrust belt-foredeep system in the Tyrrhenian-Apenninic region (see, Patacca et al.).

The other model (see, Mantovani et al.) suggests that the retreat of the Adriatic-Ionian foreland and the outward migration of the Apenninic belt were the effects of the compressional regime which affected the central Mediterranean region in connection with a NNE to NEward convergence between Africa and Eurasia. The great amount of light crustal material belonging to the Alpine and Apenninic orogenic belts, which occupied the future Tyrrhenian region around the late Miocene, reacted to this compressional framework through a lateral extrusion towards the Adriatic foreland. This collisional pattern caused the scraping off of the Adriatic Mesozoic platform and the accumulation of thrust units along the external front of the Apennines. Once decoupled from its buoyant crustal material, the Adriatic lithosphere offered much less resistance to sinking and underwent downward bending in response to two possible dynamic causes: negative buoyancy with respect to the surrounding mantle and horizontal compressional forces connected with the Africa-Adriatic convergence. Afterwards, some role in driving the roll-back mechanism might have also been played by the pull of the downgoing slab.

As concerns the eastern Mediterranean area, the debate mainly centered on whether the S-N extension in the Aegean zone has been closely connected with the westward extrusion of Anatolia from the Arabia-Eurasia collisional border (see, Jackson, Papadopoulos, Şengör). This geodynamic hypothesis seems to be the most likely, on the basis of the available evidence but some outstanding problems still exist. In particular, it appears opportune to check the geological information which indicate that the Aegean extension started before the westward migration of Anatolia.

Considerable ambiguity still surrounds the starting time of the ongoing subduction process beneath the Hellenic trench. Some authors suggested a recent activation of this mechanism, 5-6 My ago, whereas other geodynamic reconstructions hypothesize a Miocenic starting time, 13-16 My ago. The huge amount of subducted material beneath the Aegean zone identified by tomographic investigations could imply an even longer duration of the subduction process.

Another peculiar feature whose geodynamic implications should be better understood is the noticeable difference between the convergence rate (at least 6 cm/yr) predicted along the Africa-Aegean boundary (Hellenic trench) by current kinematic models and the seismic energy release, which can only account for a relative motion slower than 1 cm/yr (see, Jackson). This discrepancy has been tentatively explained as due to the occurrence of aseismic slip in the Hellenic trench, but other explanations should be explored. In particular, it could be useful to verify the reliability of the assumption that all extension in the Aegean region is taken up along the Hellenic trench. For example, a Northward drifting of the Balkan system with respect to stable Eurasia, driven by the combined pushes of the Adriatic and Anatolian blocks, might play a non-negligible role in determining extension in the North Aegean trough.

More generally, it becomes ever clearer that understanding the tectonic processes which are taking place in the Mediterranean region can give very important insights into the structural and tectonic evolution of continental collision zones. Paleomagnetic measurements would represent precious indications on the past block kinematics, to be compared with other pieces of evidence. Unfortunately, however, the studies and interpretations of this type of data in some Mediterranean zones have often provided different and even contrasting results. A review of the presently available observations, accompanied by a description of the main possible causes of uncertainty, is included in this volume (see Marton, Scalera et al.).

A fundamental contribution to the growth of knowledge about the structure and kinematics of the complex Mediterranean area comes from seismological studies. The area is characterized by strong seismic activity which we need to understand better in order to improve our view of the active tectonics. At the same time, the occurrence of earthquakes also provides invaluable information about the structure of the area. A general scheme of the regional stress pattern can be delineated by using available source mechanisms for the largest earthquakes (see Udias et al.). More particular studies are carried out to better characterize several key areas, such as the Hellenic subduction trench (see Jackson), or the northern border of the Adriatic (see Slejko), Apenninic structures (see Cocco et al.), and the Arabian plate (see Barazangi et al.).

Source mechanisms computed from the distribution of polarity of first motion may be affected by errors due to unknown calibration of stations if bulletin reports are used. The discussion stressed the importance of referring to the original waveform data. Data availability then becomes an important issue. For strong events, well recorded worldwide, long period analyses allow very reliable reconstruction of the source mechanisms, whereas broadband studies can image in detail the time evolution of the earthquake process (see Drakopoulos et al.). In the Mediterranean region, however, most of the interesting earthquakes are of moderate size, a fact which calls for the presence of local stations and modification of the computation algorithms in order to deal with events at regional distances (see Giardini et al.). A detailed experimental knowledge of source processes is also needed to test theoretical models of fault slip and earthquake nucleation (see Dragoni).

For these reasons the installation of high-quality, very-broad band instrumentation in this area is particularly important. The meeting was also an occasion to discuss results and potentiality of the seismograph network MEDNET installed as a joint project between the Istituto Nazionale di Geofisica and World Laboratory in countries of the Mediterranean (see Morelli et al.).

A fundamental design goal followed to plan MEDNET was the study of Mediterranean upper mantle structure. Attractive tomographic images of the upper mantle were shown, deriving from the analysis of travel time data. Important features like the descending slab under the Aegean, and the remnant subduction system under the Apennines, could be recognized (see Cimini et al.). However, the limitations of these images were also discussed. The complicated propagation of seismic waves at regional distances, with strong deviation of ray paths due to discontinuities, makes it in fact difficult to locate precisely the source of travel time anomalies. This is another instance in which digital, broad-band data — such as those recorded by MEDNET stations — are essential to break the uncertainty tied to studies based on bulletin reports (see Morelli et al.). Waveform inversion yields precise information on the radial structure of the upper mantle and its lateral variations.

We express our thanks to all participants to the ERICE meeting for having contributed in creating a very lively and stimulating, although always friendly, atmosphere. Further important contributions to the discussion would have certainly come from J. Dewey, P. Scandone and L. Beccaluva who, however, were forced to withdraw their participation for personal reasons. The works reported in this volume cover most of the topics presented during the meeting. Unfortunately papers concerning some very important studies, like the ones given by W. Spackman, G.Valensise, R.L.M. Vissers, S.Ward and R. Wortel could not be prepared in time to appear in this volume.

Aknowledgements

We are very grateful to Prof. Antonio Zichichi, Director of the Ettore Majorana centre of Erice, for his continual encouragement and support. The Nato Science Division contributed in a substantial way to allow the participation of scientists coming from different Countries. We thank the Regional Government of Sicily, the National Institute of Geophysics (Rome, Italy). We wish also to thank the Staff of the E. Majorana Centre.

We also thank, for their help at various stages in the organization of the School, T. Cavoli, F. Falciani, C. Jannuzzi and, in particular, G. Calcara, who also managed editorial duties for this volume.

ETTORE MAJORANA CENTRE FOR SCIENTIFIC CULTURE

INTERNATIONAL SCHOOL OF SOLID EARTH GEOPHYSICS
8th Course:
Recent Evolution and Seismicity of the Mediterranean Area
Erice, 18-26 September 1992

List of Participants

Dario ALBARELLO
Università di Siena
Via Banchi di Sotto, 55
53100 Siena
Italy

Paola ALBINI
Ist. per la Geofisica della Litosfera
C.N.R.
Via Ampère, 56
20131 Milano
Italy

Alessandro AMATO
Istituto Nazionale di Geofisica
Via di Villa Ricotti, 42
00161 Roma
Italy

Andrea ARGNANI
Ist. per la Geologia Marina
C.N.R
Via Zamboni, 65
40127 Bologna
Italy

Abdulkrim AOUDIA
C.R.A.A.G.
B.P. 63
16340 Bouzareah, Alger
Algeria

Raffaele AZZARO
Ist. Internazionale di Vulcanologia
C.N.R.
Piazza Roma, 2
95100 Catania
Italy

Daniele BABBUCCI
Università di Siena
Dip. di Scienze della Terra
Via Banchi di Sotto, 55
53100 Siena
Italy

Muawia BARAZANGI
Cornell University
Dept. of Geological Sciences
14853 Ithaca
U.S.A.

Alessandro BONACCORSO
Ist. Internazionale di Vulcanologia
C.N.R.
Piazza Roma, 2
95100 Catania
Italy

Maria Paola BOGLIOLO
Istituto Nazionale di Geofisica
Via di Villa Ricotti, 42
00161 Roma
Italy

Enzo BOSCHI
Istituto Nazionale di Geofisica
Via di Villa Ricotti, 42
00161 Roma
Italy

Vittorio BOSI
Università di Roma
Dip. di Scienze della Terra
P.le Aldo Moro, 5
00185 Roma
Italy

Stefano CARBONARA
Università di Roma
Dip. di Scienze della Terra
P.le Aldo Moro, 5
00185 Roma
Italy

Claudio CHIARABBA
Istituto Nazionale di Geofisica
Via di Villa Ricotti, 42
00161 Roma
Italy

Paolo CHIOZZI
Università di Genova
Dip. di Scienze della Terra
Viale Benedetto XV, 5
Genova
Italy

Cenka CHRISTOVA
Geophysical Institute
Dept. of Seismology
Acad. G. Bonchev BL 3
1113 Sofia
Bulgaria

Fabienne COLLIN
Royal Observatory of Belgium
Dept. of Int. Geophysics
Av. Circulaire, 3
1180 Bruxelles
Belgium

G. Battista CIMINI
Istituto Nazionale di Geofisica
Via di Villa Ricotti, 42
00161 Roma
Italy

Massimo COCCO
Istituto Nazionale di Geofisica
Via di Villa Ricotti, 42
00161 Roma
Italy

Rodolfo CONSOLE
Istituto Nazionale di Geofisica
Via di Villa Ricotti, 42
00161 Roma
Italy

Luigi CUCCI
Istituto Nazionale di Geofisica
Via di Villa Ricotti, 42
00161 Roma
Italy

Pascale DEFRAIGNE
Royal Observatory of Belgium
Dept. of Int. Geophysics
Av. Circulaire, 3
1180 Bruxelles
Belgium

Michele DRAGONI
Università di Bologna
Dip. di Fisica
Viale Berti Pichat, 8
40127 Bologna
Italy

John DRAKOPOULOS
National Observatory of Athens
Seismological Institute
Athens
Greece

Elena EVA
Università di Genova
Dip. di Scienze della Terra
Viale Benedetto XV, 5
16132 Genova
Italy

Paola FABRETTI
Istituto di Geologia Marina
C.N.R.
Via Zamboni, 65
40138 Bologna
Italy

Paolo FAVALI
Istituto Nazionale di Geofisica
Via di Villa Ricotti, 42
00161 Roma
Italy

Icilio FINETTI
Università di Trieste
Dip. di Geofisica
Via Università, 7
34123 Trieste
Italy

Fabio FLORINDO
Università di Roma
c/o Istituto Nazionale di Geofisica
Via di Villa Ricotti, 42
00161 Roma
Italy

Peter GERNER
Hungarian Geological Survey
Stefania ut 14
1143 Budapest
Hungary

Domenico GIARDINI
Istituto Nazionale di Geofisica
Via di Villa Ricotti, 42
00161 Roma
Italy

Ingrid HUNSTAD
Istituto Nazionale di Geofisica
Via di Villa Ricotti, 42
00161 Roma
Italy

James JACKSON
University of Cambridge
Dept. of Earth Sciences
Bullard Labs, Madingley Road
Cambridge, CB3 OE2
United Kingdom

Emo MARTON
E.L. Geophysical Institute of Hungary
Dept. of Earth's Physics
Columbus, 17-23
1145 Budapest
Hungary

Antonietta MEGNA
c/o Istituto Nazionale di Geofisica
Via di Villa Ricotti, 42
00161 Roma
Italy

Paul Th. MEIJER
University of Utrecht
Dept. of Theoretical Geophysics
Budapestlaan, 4
3508 TA Utrecht
The Netherlands

Giuliana MELE
Istituto Nazionale di Geofisica
Via di Villa Ricotti, 42
00161 Roma
Italy

Nadia MENIA
C.R.A.A.G.
B.P. 63
16340 Bouzareah, Alger
Algeria

Jacques MERCIER
Université de Paris Sud
Dep. de Geodynamique et
Geophysique Interne
Bat 509
91405 Orsay
France

Giancarlo MONACHESI
Osservatorio Geofisico
Sperimentale
Viale Indipendenza
62100 Macerata
Italy

Andrea MORELLI
Istituto Nazionale di Geofisica
Via di Villa Ricotti, 42
00161 Roma
Italy

Stephen MORICE
University of Cambridge
Dept. of Earth Sciences
Madingley Road
Cambridge CB3 OE2
United Kingdom

Mario MUCCIARELLI
ISMES
Div. Geofisica
Viale Giulio Cesare, 29
24100 Bergamo
Italy

Anna NARDI
Istituto Nazionale di Geofisica
Via di Villa Ricotti, 42
00161 Roma
Italy

Svetlana NIKOLOVA
Geophysical Institute
Dept. of Seismology
Ac. G. Bonchev, BL 3
1113 Sofia
Bulgaria

Barbara PALOMBO
Istituto Nazionale di Geofisica
Via di Villa Ricotti, 42
00161 Roma
Italy

Gerassimos A. PAPADOPOULOS
Earthquake Planning and
Protection Organization
Dept. of Seismotectonics
226 Messogion Ave.
15561 Athens
Greece

Stefania PASTORE
Università di Genova
Dip. di Scienze della Terra
Viale Benedetto XV, 5
16132 Genova
Italy

Gianluca PATRIGNANI
Soc. Acquater S.p.A.
S. Lorenzo in Campo
Pesaro
Italy

Laura PERUZZA
Osservatorio Geofisico
Borgo Grotta - Opicina
34016 Trieste
Italy

Fedora QUATTROCCHI
Istituto Nazionale di Geofisica
Via di Villa Ricotti, 42
00161 Roma
Italy

Simone RASPOLLINI
Università di Siena
Dip. di Scienze della Terra
Banchi di Sotto, 55
53100 Siena
Italy

Rosa ROSINI
c/o Istituto Nazionale di Geofisica
Via di Villa Ricotti, 42
00161 Roma
Italy

Yasmine ROUCHICHE
C.R.A.A.G.
B.P. 63
16340 Bouzareah, Alger
Algeria

Stefano SANTINI
Università di Urbino
Dip. di Fisica
Via Santa Chiara, 27
61029 Urbino
Italy

Renzo SARTORI
Università di Bologna
Dip. Scienze Geologiche
Via Zamboni, 65
40127 Bologna
Italy

Leonardo SAGNOTTI
Istituto Nazionale di Geofisica
Via di Villa Ricotti, 42
00161 Roma
Italy

Claudio SANACORI
Università di Roma
Dip. di Geologia Strutturale
P.le Aldo Moro, 5
00185 Roma
Italy

Giancarlo SCALERA
Istituto Nazionale di Geofisica
Via di Villa Ricotti, 42
00161 Roma
Italy

A. M. C. ŞENGÖR
I.T.Ü. Maden Fakültesi
JeoloJi Bölümü
Ayazaǧa 80626. Istanbul
Turkey

Giulio SELVAGGI
Istituto Nazionale di Geofisica
Via di Villa Ricotti, 42
00161 Roma
Italy

Valeria SINISCALCHI
Istituto Nazionale di Geofisica
Via di Villa Ricotti, 42
00161 Roma
Italy

Chadaram SIVAJI
Indian Institute of Geomagnetism
Dept. of Solid Earth Geophysics
Colaba, Nana bhoy Moos Marg
Bombay 400005
India

Dario SLEJKO
Osservatorio Geofisico
Sperimentale
Dip. Oceanografia e Geofisica
P.O. Box 2011
34016 Trieste
Italy

Wim SPAKMAN
Dept. of Geophysics
Inst. of Earth Sciences
Budapestlaan, 4
3504 TA Utrecht
The Netherlands

Massimiliano STUCCHI
C.N.R
Istituto di Geofisica della Litosfera
Via Ampère, 56
20131 Milano
Italy

B.V. SUBBA RAO
Indian Institute of Geomagnetism
Dept. of Solid Earth Geophysics
Colaba, Nana bhoy Moos Marg
Bombay 400005
India

Pèter SZAFIÀN
Eotvos University
Dept. of Geophysics
Ludovika ter
Budapest H-1083
Hungary

Quintilio TACCETTI
Istituto Nazionale di Geofisica
Via di Villa Ricotti, 42
00161 Roma
Italy

Caterina TAMBURELLI
Università di Siena
Dip. di Scienze della Terra
Via Banchi di Sotto, 55
53100 Siena
Italy

Agustin UDIAS
Universidad Complutense
Dep. de Geofisica
28040 Madrid
Espana

Gianluca VALENSISE
Istituto Nazionale di Geofisica
Via di Villa Ricotti, 42
00161 Roma
Italy

Reinoud VISSERS
Inst. of Earth Sciences
Dept. of Geology
Budapestlaan, 4
3508 TA Utrecht
The Netherlands

Yue WANG
Center for Analysis
and Prediction
SSB
Fuxing Road 63
100036 Beijing
China

Steven WARD
University of California
Dep. of Earth Sciences
High Street
95064 Santa Cruz, CA
USA

Rinus WORTEL
University of Utrecht
Dept. of Geophysics
Budapestlaan, 4
3584 TA Utrecht
The Netherlands

Shochi YOSHIOKA
University of Utrecht
Dept. of Theoretical Geophysics
Budapestlaan, 4
3508 TA Utrecht
The Netherlands

SOME CURRENT PROBLEMS ON THE TECTONIC EVOLUTION OF THE MEDITERRANEAN DURING THE CAINOZOIC

A. M. C. ŞENGÖR,
ITÜ Maden Fakültesi,
Jeoloji Bölümü,
Ayazağa 80626 Istanbul
Turkey

> The bosom of the Mediterranean like a sea received the principal waters of Africa, Asia and Europe; for they were turned towards it and came with their waters to the base of the mountains which surrounded it and formed its banks.
> And the peaks of the Apennines stood up in this sea in the form of islands surrounded by salt water....; and above the plains of Italy where flocks of birds are flying today fishes were once moving in large shoals.
> Since things are far more ancient than letters, it is not to be wondered at if in our days there exits no record of how aforesaid seas extended over so many countries.......But sufficient for us is the testimony of things produced in the salt waters and now found again in the high mountains, sometimes at a distance from the seas.
> LEONARDO DA VINCI (in the Leicester MS)

ABSTRACT. The Cainozoic history of the Mediterranean and its mountainous frame may be conveniently divided into two stages, not everywhere synchronous. The earlier stage was a continuation of the late Mesozoic Alpide convergence, dominated by subduction tectonics and that culminated in the continental collisions from the Betic and the Rif cordilleras and the Pyrenees via the Provence chains and the Tellian Atlas, Sicily, and the Apennine/Corsican collisional system, the Alps, Carpathians, Dinarides, Hellenides, the Balkan ranges and the Anatolian mountain ranges to the Caucasus and the Zagros. During this episode, subduction led to back-arc basin generation in the western Mediterranean, and the resulting tectonics is little different from the southeast Asian and western Pacific systems, despite the commonly and unjustifiably overemphasized differences. Both compressional arcs and collisions led to widespread fore- and hinterland deformation creating complex germanotype orogens such as the Iberian and the Catalan chains, and the High and Saharan Atlas. The second stage corresponds in the western Mediterranean with the continued rapid expansion of back-arc basins accompanied by trenchward jumping of extensional loci creating remnant arcs such as the Corso-Sardinian block, collisions of migratory arcs with continental margins and eventual diminution of orogenic deformation, except where the subduction of the Eastern Mediterranean interfered as in offshore Calabria. In the Eastern Mediterranean, this stage led to the collisional assembly of a tectonic collage in Anatolia and eventually to the collision of Arabia with Eurasia, to the opening of the Aegean Sea, and to the concurrent expulsion of Anatolia westwards from the collision front of the Arabian promontory with Eurasia along the Bitlis suture in southeastern Turkey. Tectonic

1

escape has played a very major rôle in the Cainozoic tectonics of the Mediterranean and shares the responsibility, together with extensional arcs, for the extremely contorted plan view of the Mediterranean Alpides. Wherever "extensional orogenic collapse" has occurred in the Mediterraenan Alpides (Betic/Rif/Alboran Sea system, the Tyrrhenian Sea, the Pannonian Basin, and the Aegean Sea), it invariably has been caused or at least triggered by either extensional arcs (no collapsing orogen extended over a flat subduction zone in the Mediterranean) or by escape (if not part of an extensional arc, orogenic collapse invariably has occurred in the Mediterranean at the leading part of an escaping fragment: e.g., the Alboran Sea, the Pannonian Basin, the Aegean Sea). I conclude that there is no "special" Mediterraean tectonics, as there are no "Mediterranean type" orogens, for there seems to be no aspect of the tectonic evolution of the Mediterranean that cannot be directly compared with some *presently active* orogenic belt and/or basin complex elsewehere in the world. The overall kinematic (but not temporal!) evolution of the Mediterranean, except in the Eastern Mediterranean, seems to be much as Argand portrayed it nearly seventy years ago. I think that this shows the power of the regional geological methods, now much amplified, but not replaced, by geophysical measurements.

1. Introduction

Seventy years ago, with a bold adoption of Wegener's then heretical hypothesis of vast horizontal motions, Emile Argand cut through the Gordion Knot of tectonics, *viz.* the Mediterranean Sea and its mountainous frame, that had previously stumped all attempts at a rational explanation. So difficult did even Argand find them that he felt compelled to emphasize that "*The Mediterranean* and its chains present to the mobilistic theory a difficult field of application and a test that this theory must undergo if it pretends to have more than a passing acceptance. The smallness of the scale allows neither the large statistical approaches by which this theory is successful nor the easy unfolding of the greater deformations that it uses. The type of small-scale complication displayed by these structures, all this *three dimensional puzzle* of which many pieces are deformed to the extent of being almost unrecognizable, and furthermore the very demanding stratigraphy, ready to lock up in its refined chronology each phase of some importance of the movements, require from the theory a calculated and cautious progress. Besides, it could not, without the help of a great number of operational artifices borrowed from observed tectonics, penetrate very far into this problem, which is too close to the lower limit of the scales within its interest" (Argand, 1924, p. 304, italics Argand's).

Argand's sentences still serve well for an introduction, especially to an essay on the current problems concerning the Cainozoic tectonics of the Mediterranean and its frame. The Mediterranean remains the most difficult testing ground for our views on the tectonics of our globe, not only on account of its tremendous complexity, but also because we know so much about it.

Although the outlines of its evolution (see, for example, Dewey et al., 1973, 1989a; Biju-Duval et al., 1977; Hsü, 1977, 1989; Channel et al., 1979; Rögl and Steininger, 1983; Steininger et al., 1985; Dercourt et al., 1986; Rakús et al., 1990; also see the review by Smith and Woodcock, 1982) and present structure and behaviour (e.g., Ketin, 1948; McKenzie, 1972, 1978; Morelli, 1975; Morelli et al., 1975; Dewey and Şengör, 1979; Finetti, 1982; Jackson and McKenzie, 1984; Anderson and Jackson, 1987; Taymaz et al., 1991; Wortel and Spakman, 1992) are generally agreed upon, many important disagreements still stimulate much discussion. These disagreements may be divided into two classes: The first class contains those disagreements about *processes* and apply equally well to analogous regions. The second class is populated by peculiarly Mediterraean problems arising from disagreements about interpretation of its specific aspects. Although perhaps more numerous and in some cases more entrenched by tradition, these second class of problems are no different in nature from those that animate discussions about the geology of any other orogenic region in the world.

The purpose of this contribution is to discuss primarily the first class of problems, not only because of their general appeal, but also because the circum-Mediterranean geology seems ideally suited as a basis for their discussion. A number of these problems have led recently to controversies raging on extremely scanty data collected from elsewhere. It is unlikely that they are going to be resolved unless a reliable database is employed, and it is my view that for some of these problems at least, the geology of the Mediterranean area might be more suitable than any other region as a database both with respect to the extant information and the relative ease with which more may be collected.

In the process of discussing these general problems, I shall inevitably deal with the tectonic evolution of the Mediterranean, especially with its Cainozoic portion, and it is here that I shall also address a portion of the second class of problems, those that are specific to the Mediterranean.

2. Main problems concerning the tectonic evolution of the Mediterranean during the Cainozoic

The following problems I consider the most significant that currently generate much controversy both in the Mediterraean region and elsewhere. I list them below in an order that allows me to deal with the geology of the Mediterraenan in their framework and also in a logical sequence in the following paragraphs:

1) Is there a special "Mediterranean-type" orogeny different from the usual collision, subduction, obduction, or transpression controlled orogenies?

2) How are highly sinuous orogenic belts such as the Mediterranean generated?

3) What are the significance and cause of "tectonic escape" in collisional processes?

4) How does "extensional orogenic collapse" occur, or what is really the rôle of topography in tectonics?

5) How is orogenic deformation distributed in a collisional zone at any one time?

In the following paragraphs I discuss each of these problems in some detail and in the process also review the Cainozoic evolution of the Mediterranean as a basis for the later discussions on its present tectonics and seismicity. But before doing so it is necessary to summarize the orogenic structure and ancestry of the Mediterranean mountain ranges.

3. Orogenic structure of the circum-Mediterranean mountain ranges

The circum-Mediterranean mountain ranges form the westernmost part of the superorogenic complex *Tethysides* that resulted from the compressional obliteration of the Tethyan domain (Şengör, 1989). The Tethysides are formed from two largely superimposed orogenic complexes, namely the older *Cimmerides* (Carboniferous to early Cretaceous) and the younger *Alpides* (early Jurassic to present), which resulted from the disappearance of *Palaeo-* and *Neo-Tethyan* oceans respectively (Şengör, 1984, 1985a, b, 1986, 1987a, 1989; Şengör et al., 1988). In the circum-Mediterranean mountain ranges, the Cimmerides occupy the Alpide basement only around the Black Sea. Suggestions to extend them farther west (e.g. Şengör, 1984; Şengör et al., 1984; Mountrakis, 1985; Mountrakis et al., 1987; Sideris, 1989; Kozur, 1991a, b) have been controversial. It is thus the Alpides that dominate the orogenic structure of the circum-Mediterranean mountain ranges. Depending on whether they formed along former continental margins or within already consolidated continental lithosphere, the Alpides have been divided into two areas: the *alpinotype Alpides* correspond with the orogenic belts sensu stricto, whereas the *germanotype Alpides* represent deformed areas within the fore- and hinterlands (Şengör, 1984, 1985b). Figures 1 and 2 show the distribution of the two orogenic complexes within the entire Tethysides and the suture distribution in them, respectively.

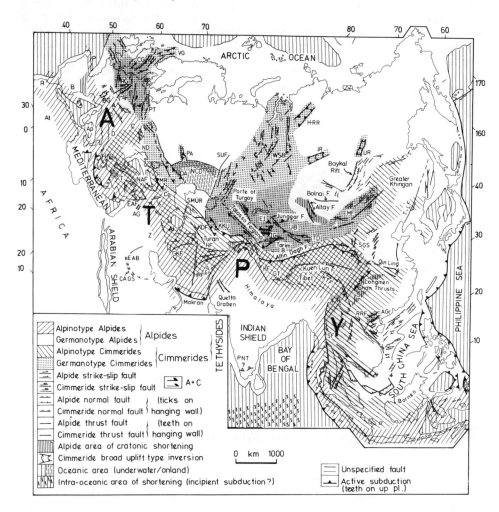

Figure 1. The Tethyside super-orogenic complex showing both the Cimmerides and the Alpides and their associated areas of fore- and hinterland deformation. Key to lettering: Larger bold letters A, T, P and Y are the Alpine, Turkish, Pamir and Yunnan syntaxes respectively. Smaller letters: A = Alps, AG = Akçakale graben, AGr = An Chau graben, Al = Alborz, Ap = Apennines, At = Atlas Mountains (*sensu lato*), B = Betics, BF = Bogdo Fault, BG = Besse graben, C = Carpathians, Ca = Caucasus, CAGS = Central Arabian graben system, CF = Chaman Fault, CG = Central graben, D = Dinarides, DA = Dnyepr-Donetz aulacogen, EAB = East Arabian block,

Şengör (1984) has pointed out that the westernmost parts of the Neo-Tethys (west of the Carpathians) had opened as parts of the central Atlantic Ocean and suggested that they may be termed *Atlantides* as opposed to Tethysides. But this nomenclature would only be useful, where the differences between the Tethys and the Atlantic are discussed. I shall therefore not use it in what follows.

4. Is there a special "Mediterranean-type" orogeny different from the usual collision, subduction, obduction or transpresion controlled orogens?

A glance at Figs. 1 and 2 shows that the circum-Mediterranean Alpides present a much more sinuous map view than the rest of the Tethysides and they enclose deep ocean basins of varied ages. Their divergent directions of tectonic transport (commonly from the enclosed deeps to the surrounding highs! Fig. 3), paucity of west Pacific style island arcs and well-developed magmatic continental margins, existence in them of thin sheets of far-travelled allochthons overriding continental margins of continents across closed oceans and commonly after having lost their original continental basement, and the coeval operation of co-directional extension and shortening in many of them have invited scepticism as to whether their tectonics may be explained by comparison with other active mountain ranges. Most recently, for example, Burchfiel and Royden (1991a) proposed that "Mediterranean-type orogeny" results in back-arc spreading as a consequence of directly density-driven roll-back of the subduction hinge and they used this as an analogue for the fossil Antler orogen of the western United States.

As Argand emphasized in the long quotation from his masterpiece *La tectonique de l'Asie* given above, the Mediterranean is a very complicated region, one that has proved exceptionally difficult to understand, despite the more-than-two-century-old geological research tradition that has studied it. However, it is no more complicated than any other orogenic belt in terms of the procesess that have formed it. To erect a special Mediterranean-type of orogen or orogeny would severely limit the geologist's means of comparison and hypothesis testing by creating *ad hoc* local explanations for the evolution of the circum-Mediterranean orogenic belts, as illustrated by the numerous infertile "Mediterranean" orogenic models since at least the beginning of this century.

4.1. SINUOSITY OF TREND-LINES

The first characteristic of the circum-Mediterranean belts that has led to their being considered special is their extremely sinuous trend-lines. It has not been clear how a steady approach of the two megacratons of Africa and Europe could have created such a highly contorted plan view.

EAF = East Anatolian fault, EI = East Ili basin, GKF = Great Kavir fault, GT = Gerze thrust, H = Hellenides, HF = Herat fault, HRF = Harirud fault, H-RR = Hantaj-Rybninsk rift, IG = Issyk Gol basin, IR = Irkineev rift, KDF = Kopet Dagh fault, KF = Karakorum fault, KKU = Kizil Kum uplift, KTF = Kang Ting fault, MF = Mongolian faults, MR = Main Range of the Greater Caucasus, NAF = North Anatolian fault, NCD = North Caspian Depression, ND = North Dobrudja, PA = Pachelma aulacogen, PNT = Palni-Nilgiri Hills thrust, PT = Polish trough, RG = Upper Rhine graben, RRF = Red River fault, S = Sichuan basin, SF = Sagain fault, SGS = Shanxi graben system, SMÜR + South Mangyshlak - Üst Yurt ridge, SUF = South Ural faults, T = Turkish ranges, TD = Turfan depression, T-LF = Tan-Lu fault, UR = Ura rift, VG = Viking graben, WSB = West Siberian basin, Z = Zagrides.

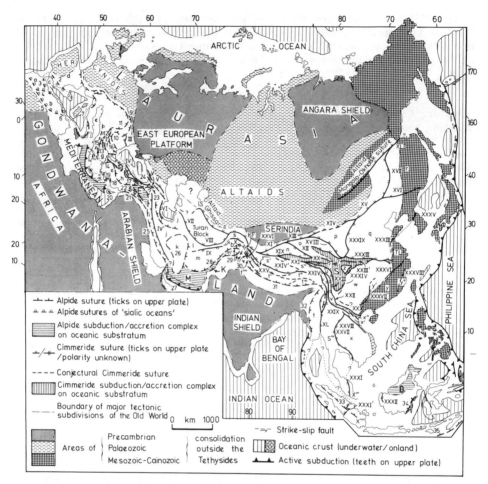

Figure 2. Tectonic map of Eurasia and north Africa showing the major tectonic subdivisions and the distribution of sutures within the Tethysides. Tethysides correspond with white areas on land. Cimmeride sutures: I = Possible(?) Palaeo-Tethyan suture in Greece and Yugoslavia, II = Karakaya, III = Luncavita-Consul, IV = North Turkish, V = Svanetia, V' = Chorchana-Utslevi, VI = Talesh, VII = Kopet Dagh/Mashhad, VIII = Paropamisus/Hindu Kush, IX = Waser, X = Rushan-Pshart, XI, XII, XIII, XIX, XX = lines formerly thought to represent discrete sutures in the Kuen-Lun range (now interpreted simply as ophiolitic fragments within the Kuen Lun Palaeozoic accretionary complex: see Şengör and Okuroğulları, 1991), XIV = Suelun-Hegen, XV = Inner Mongolian, XVI = Suolun-Xilamulun, XVII = Greater (Da) Khingan, XVIII = Tergun Daba Shan/Qinghai Nan Shan, XXI = Hoh Xil Shan/Jinsha Jiang (simply the southwestern limit of the Mesozoic Songpan-Ganze accretionary complex), XXII = Maniganggo, XXIII = Litang, XXIII' = Luochou "arc-trench belt", XXIV = Banggong Lake - Nu River, XXIV' = Chasa, XXV = Shiquan He, XXVI = Southwest Karakorum, XXVII = Nan-Uttaradit/Sra-Kaeo, XXVIII = Tamky-Phueson, XXIX = Song Ma, XXX = Song Da, XXXI = Bentong-Raub, XXXI' = Mid-Sumatra, XXXII = Serabang (West Borneo), XXXIII = Qin Ling/Dabie Shan, XXXIII' = Shandong, XXXIV = Longmen

However, in this respect, the Mediterraean is not unique. Today, at least four other *active* belts share a highly sinuous map view with the Mediterranean chains, namely the Southeast Asian (Indonesian) Alpides (1 in Fig. 3: Hamilton, 1979; also see Hilde and Uyeda, 1983, and Hutchison, 1989), the Melanesian orogen to Tonga islands (2 in Fig. 3: Anonymous, 1977; Doutch et al., 1981; Palfreyman et al., 1988[1]), the Caribbean (3 in Fig. 3: Burke et al., 1983; Pindell et al., 1988; Pindell and Barrett, 1990), and the South Scotia arc system (4 in Fig. 3: Barker and Dalziel, 1983) (Fig. 3). Among fossil orogens, the Altaids of Central Asia (5 in Fig. 3) clearly outdo the Mediterranean chains in terms of their contorted map-views (e.g. Şengör and Okuroğulları, 1991; Şengör, 1992) and the map view of the Hercynian chains of Europe (Matte, 1986) is not markedly less contorted (6 in Fig. 3). I shall discuss the causes of such contortion in the following section, but here suffice it to say that the examples cited above clearly show that there is nothing special about the contortion seen in the trend-lines of the circum-Mediterranean orogens and nor in their being separated by deep oceanic "button-holes" *à la* Argand (Argand, 1924, p. 309: *Boutonnières*), which is also the case in the Southeast Asian (Indonesian) Alpides (e.g. Hamilton, 1979) and in the Altaids (e.g. Aplonov et al., 1992 and the references cited therein; also V. E. Khain, pers. comm., 1992).

Shan/Qionglai Shan (simply the southeastern limit of the Songpan-Ganzi Mesozoic accretionary complex), XXXV = Chugareong, XXXVI = Shaoxing-Pingxiang, XXXVII = Tianyang, XXXVIII = Lishui-Haifeng, XXXIX = Helan Shan (formerly thought to be a Mesozoic suture; it is now interpreted as simply a Mesozoic intracontinental zone of high transpressional strain), XL = Mandalay, XLI = Shilka. Alpide sutures: 1 = Pyrenean, 2 = Betic, 3 = Rif, 4 = High Atlas, 5 = Saharan Atlas, 6 = Kabylia, 7 = Apennine, 8 = Alpine, 9 = Pieniny Klippen Belt, 10 = Circum-Moesian, 11 = Mureş, 12 = Vardar, 13 = Peonias/Intra Pontide, 14 = Almopias/İzmir-Ankara, 15 = Pindos-Budva-Bükk, 16 = Srednogora, 17 = Ilgaz-Erzincan, 18 = Inner Tauride, 19 = Antalya, 20 = Cyprus, 21 = Bitlis (Eocene to Miocene), 22 = Maden, 23 = Zangezur/Karadagh, 24 = Slate/Diabase zone, 25 = Zagros, 26 = Circum-Central Iranian microcontinent (including Nain-Baft, Sabzevar, and Sistan), 27 = Oman, 28 = Waziristan, 29 = Kohistan (North: Chalt, South: Main Mantle Thrust), 30 = Ladakh (North: Shyok, South: Indus), 31 = Indus/Yarlung, 32 = Indoburman, 33 = Woyla, 34 = Meratus, 35 = Timor. Tethyside blocks: a = Moroccan Meseta, b = Oran Meseta, c = Alboran fragment, d = Iberian Meseta, e = African promontory (including the Menderes-Taurus block), e' = Alanya fragment, f = Rhodope-Pontide fragment, g = Sakarya, h = Kırşehir, i = Sanandaj-Sirjan, j = Northwest Iran, k = Central Iranian Microcontinent (consisting of the Lut, Tabas and Yazd blocks), l = Farah, m = Helmand (*sensu* Şengör, 1984), m' = Kohistan, n&o = Kuen-Lun Palaeozoic accretionary complex (formerly thought to be blocks underlain by rigid Precambrian basement), p = Alxa (Ala Shan), q = North China, r = North China Foldbelt (including the Bureya and the Khanka massifs), s = Central Pamir-Qangtang-Sibumasu (s': Central Pamir/West Qangtang, s": East Qangtang, s"': Sibumasu), t = Lhasa-Central Burma (t': Bangthol Tangla, t": Nagqu, t"': Lhasa proper, t"": Ladakh, t""': West Lhasa-South Pamir), u = Shaluli Shan, v = Chola Shan, w = Yangtze, x = Annamia (possibly with two exotic island arcs of early [x'] and late [x"] Palaeozoic age and eastern Malay Peninsula [x"'] and south Sumatra [x""]), y *sensu lato* = Huanan (y *sensu stricto*: Huanan, y': Hunan, y": coastal block), z = Songpan massif. Key to lettering: B = Sarawak accretionary complex, b = Southeast Pamir black slates, E = East Anatolian accretionary complex, K = Katawaz accretionary complex, KF = Karakorum fault, M = Makran accretionary complex, TL = Tan-Lu fault.

---

[1]See also the extremely instructive *Circum-Pacific Council for Energy and Mineral Resources, Earth Science Series* volumes on Melanesia (vols. 3,4,7,8,9)

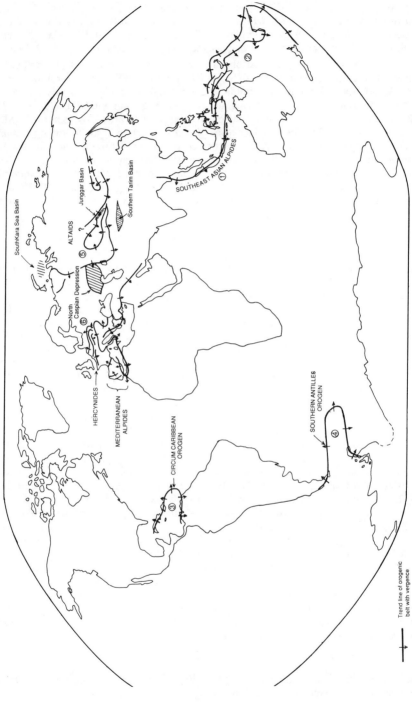

Figure 3. Some examples of extremely sinuous orogenic belts, largely formed through a number of episodes of oroclinal bending, compared with the Mediterranean Alpides. Active examples: 1) Southeast Asian (Indonesian) Alpides, 2) Melanesian orogen, 3) Caribbean orogen, 4) South Scotia orogen. Fossil examples: 5) Altaids, 6) Hercynides. Obliquely ruled areas are regions underlain by oceanic and/or attenuated continental crust now entirey trapped within continents and overlain by large thicknesses of sediment.

4.2. COEVAL AND CODIRECTIONAL SHORTENING AND EXTENSION IN OROGENS

As I mentioned above, recently much has been made about extension in orogenic hinterlands of many of the Mediterranean orogens being simultaneous with shortening in the forelands, in many cases the area of extension invading the area of shortening in time (cf. Burchfiel and Royden, 1991a, and the references cited therein). I think it needs to be emphasized that this observation is an old one, made by Suess already in his epoch-making *Die Entstehung der Alpen* where he wrote: "The...most common mountain form begins with the outline of a major fold striking transeversely to the contraction and inclining in the direction of the contraction. Only then does the fracture follow in the fold along the line of greatest stress. Hereupon, by the continuation of the same force as in the first case, the forward-lying part of the main fold moves farther in the direction of contraction, and piles up sediments in front of it in wide irregular folds; while the part lying behind sinks down and volcanoes appear between its fragments. The Apennine and Carpathian branches of the Alpine System are examples" (Suess, 1875, p. 147). He further elucidated his meaning as follows in 1878: "The best picture that I can give of the origin of the large mountain ranges consists of imagining that when my hand is scraped by accident, folds of skin become piled up in one direction, while behind them the skin is torn and a little blood wells up" (Suess, 1878, p. 5).

This phenomenon, which I here call "Suess' rule", is not peculiar to the Mediterranean, as Suess very early recognised (e.g. Suess, 1885, p. 708). As we have since learned, Suess' rule is a commonplace occurrence in extensional arcs (Dewey, 1980; Jarrard, 1986a, b; Şengör, 1990). In Fig. 4 I compare along cross-sections, the present tectonics of the Calabrian/Tyrrhenian Sea system with that of the Banda arc system/Banda Sea couple and the Marianas systems. In the Banda Sea/Banda Arc couple and the Calabrian/Tyrhhenian systems we are looking at phenomena that seem identical in principle. In both, a pile of nappes is climbing onto a continental margin, while behind extension and volcanism (exactly as Suess had described more than a century ago) is taking place (for the Banda Sea evidence see Anonymous, 1981, esp. p.86 and the references cited there[2]). In these two cases even the scales are very similar.

I put in the Marianas/Parece Vela basin system to emphasize that at a larger scale and in an entirely oceanic setting, the tectonic picture is not at all different. In fact, when one considers the mode of generation and obduction of Miyashiro-type ophiolite nappes, i.e. those generated above subduction zones (Şengör et al., in press), one would be inclined to see in Pearce et al.'s (1984) "pre-arc spreading" the very same marginal basin opening phenomenon as in the Tyrrhenian Sea and in the Marianas, or indeed in the South Scotia Arc (see Şengör, 1990, esp. p. 100), providing further examples of Suess' rule.

What also underlines the likely identity of the back-arc basin opening processes in east and southeast Asia with those in the western Mediterranean is the extremely rapid subsidence of basin floors in both places, that far exceeds expectations according to the empirical Sclater curve. New ODP observations in the Marsili basin of the Tyrrhenian Sea (Leg 107 Shipboard Scientific Party, 1986; Kastens et al., 1988; Kastens and Mascle, 1990) have been intrepreted to imply a subsidence of more than 4 km in less than 2 Ma, i.e. nearly three times as fast as normal oceanic lithosphere! If one assumes that the age vs. depth curve of Parsons and Sclater (1977) applies, one

---

[2] I quote this despite the more recent claim that magnetic anomalies in the Banda Sea are of Mesozoic age by Lee and McCabe (1986), essentially supporting older views about a Mesozoic-early Cainozoic age of the Banda Sea floor. I find this unconvincing, for the depth argument is unreliable and the magnetic anomalies themselves are not clear. If the Mesozoic-early Cainozoic age of the Banda Sea floor is correct, it becomes exceedingly difficult to explain the presence of older continental metamorphic rocks in the Banda Arc islands. I do not believe that the present observations warrant facing such a challenge.

Time	Western Sardinia (Upper margin)	Western Sardinia (Lower margin)	Vavilov (West)	Vavilov (Axial)	Marsili (West)	Southern Apennines	Sicily	Aeolian Archi-pelago	Messina Cone
Holocene	....	....	....	....	....	Left-lateral strike-slip faulting with associated extension and shortening	Strike-slip faulting	Sho-shonitic Calc-alkalic	....
Pleistocene	....	....	....	....	....	End of eastward thrusting and onset of southeast-ward tectonic transport	....	....	....
Late Piacenz.	....	....	....	....	(?) Spreading	Thrusting onto Apulia	Thrusting onto Ionico-Rugosaco platform	....	....
Early Piacenz.	....	....	....	(?) Spreading	....	....	....	....	....
Zanclian	....	....	(?) Spreading	....	....	....	....	....	....
Messinian	....	Rifting	Possible onset of spreading?	....	....	Latest Messinian: Collision of Apennine allochthons with the Gran Sasso-Daunia platform; Ongoing subduction of the Lagonegro-Marsica-Frosolone basin complex.	Thrusting of external Trapanese; subduction of Sicani; thrusting of Saccanese.	....	(?) Onset of the formation of the Messina Cone accretionary complex
Torton	(?) Rifting	....	....	....	....	Continuing thrusting of internal flysch nappes onto the external parts of the Campania-Lucania platform; beginning subduction of Lagonegro basement	Beginning thrusting onto the Panormides	....	....
Serraval	....	....	....	....	....	....	....	....	....
Langhian	....	....	....	....	....	....	....	....	....
Burdigal	....	....	....	....	....	....	....	....	....

Table I : Timing of deformation in the Tyrrhenian Sea, Sicily and the Apennines

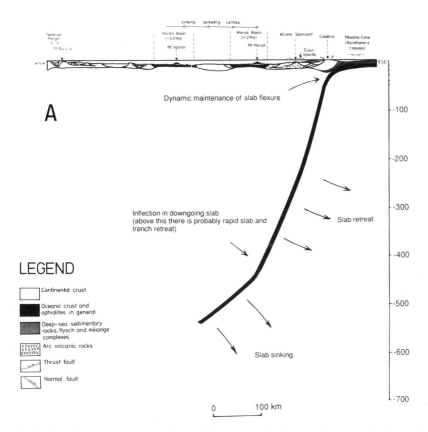

Figure 4. Partly schematic cross-sections, all to true scale (vertical scale = horizontal scale), showing the present structure and likely kinematics of the Tyrrhenian (A) and two analogous arc/back-arc basin systems, namely the Banda Sea (B) and the Marianas (C). In all the cross sections, note the very steep dip of the subducting slab down to an inflection point, beow which it somewhat flattens. Fig. 4D compares all three and underlines that this inflection occurs between the depths of 400 and 600 km. Below the inflection, it is likely that the slab is simply "laid down" onto the mesosphere into which it evidently finds it difficult to penetrate. Above the inflection, the slab is forced to retreat owing to active buckling and thus also forces the trench line to retreat (hinge-line roll-back!). This neatly explains the anomalous observation made by Jarrard (1986a) that the youngest slabs usually show the most rapid retreat. Dip of the Marianas subduction zones from Uyeda and Kanamori (1979) and Taylor and Karner (1983).

In Figure 4D, frames I through IV illustrate how slab buckling through difficulty of penetration into the mesosphere may take place. I shows a young subduction zone, whose slab slides into the asthenosphere with no apparent difficulty. In II, the slab hits the "impenetrable" mesosphere. In III it begins to buckle and the portion above the inflexion begins to "retreat" oceanward. Frame IV shows an advanced stage of this "retreat", during which marginal basins may be rapidly opening within the overlying plate, depending on the motion of the overlying plate with respect to an asthenosphere frame of reference.

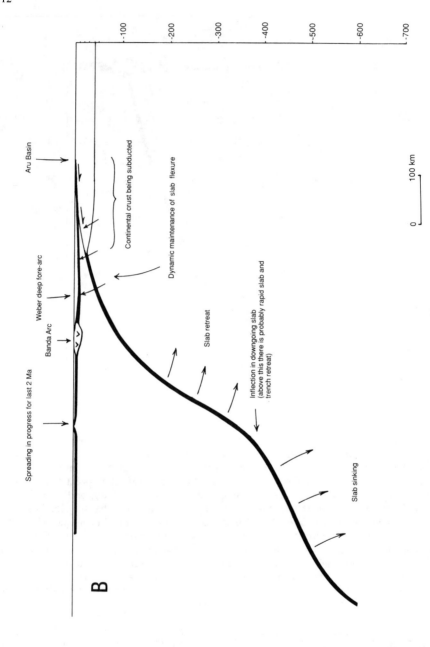

Figure 4. (Cont'd) The Banda Sea/Banda Arc couple.

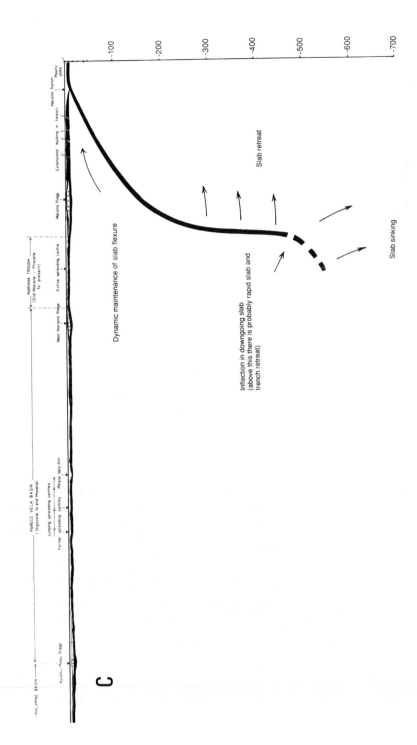

Figure 4. (Cont'd) The Marianas back-arc/arc/trench system.

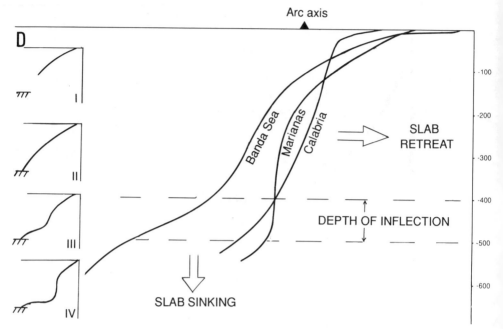

Figure 4. (Cont'd) Depth of inflection and its evolution in downgoing slabs

gets a subsidence of 3.25 km for 2 Ma. However, Hayes (1983) has shown that if one considers the back-arc basins of eastern Asia as a population, one gets a subsidence slightly in excess of 3.5 km. Given the uncertainties of the dating of the east Asian basin floors, this is not a bad discrepancy with the Mediterranean observations.

However, I must underline that not all extension in Italy is a consequence of back-arc extension. Anderson and Jackson (1987) have shown that what underlies the Adriatic Sea moves northeastward both with respect to the Italian peninsula and with respect to the Dinarides, both of which are parts of Eurasia. They have also shown that "The superior data of the last 21 years supports the suggestion from all the 20th century seismicity; that the extension rate in southern peninsular Italy and the shortening rate in coastal Yugoslavia are about equal" (Anderson and Jackson, 1987, p. 976). They believe that this pattern "cannot be explained in terms of the Africa-Eurasia convergence" (Anderson and Jackson, 1987, p. 969).

If find it difficult to imagine how else the circum-Adriatic tectonics can be explained in a way that is not *ad hoc*. Fig. 5 shows the present slip vectors and the rough kinematics of the circum-Adria motions after Anderson and Jackson's (1987) data. Here, it is seen that Adria rotates with respect to Europe around a pole somewhere near Bergamo in northern Italy (45.8°N, 10.2°E). Figs. 5C & D show a simple model illustrating how this rotation may still be directly caused by Europe/Africa convergence. If Adria is assumed to be a broken off piece of Africa embedded in a continuously deforming European mushy zone surrounding it in the east (i.e. in the Dinarides) and in the north (i.e. in the Alps), the shortening of this mushy zone will rotate Adria away from Italy around a pole near its apex in the north. That the lithosphere and the crust of Adria thickens southwards towards a zone of weak seismicity at the Strait of Otranto (see the references given in Anderson and Jackson, 1987, p. 975) may suggest a former and perhaps ongoing crustal

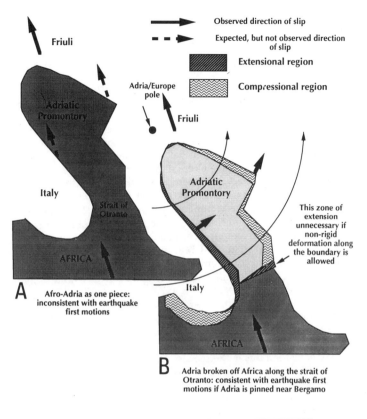
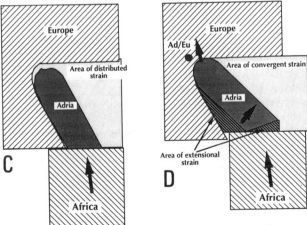

Figure 5. The rotation of Adria. A. If Adria were a rigid part of Africa. B. Actual situation. C&D. Sequential model to explain the present motion of the Adria. Data from Anderson and Jackson (1987)

shortening here, pushing the rest of the "rigid" Adria northeastwards. This would make the *present* extension along much of the Italian peninsula independent of any slab effects associated with the Tyrrhenian subduction.

One might argue that since the Tyrhhenian rifting and rapid opening is entirely *after* the Apennine collision, a mechanism similar to that of Channel and Mareschal (1989) may be needed to accomplish it. However, note that thrusting onto the most external platform in Sicily and in the Apeninnes did not commence until the Pliocene, i.e. until at least the eastern Marsili ocean floor was created (Table I).

The nature of the floor of such "small" basins as the Lagonegro, or Molise in the Apennines, or Imerese, or Sicano in Sicily has long been debated, very especially on analogous basins in the Hellenides (for two recent reviews of ideas see Robertson et al., 1991; Stampfli, 1991), particularly with respect to the petrological nature of the basinal crust. Şengör and Monod (1980) have pointed out however, that beneath a certain thickness, there would be so little difference in the buoyancy of the crust, whether fully oceanic or extremely attenuated continental, that depending on age, both may be negatively buoyant and thus continuously subductable. Now, the basins in Italy had basinal crusts, all created at the latest during the medial Mesozoic, not later that about 140 Ma. Therefore, when their subduction began in the Tortonian and later (see Patacca et al., in press; also Table I), they were all older than 100 Ma, amply sufficient to make them gravitationally unstable if they behaved like oceanic crust (Molnar and Atwater, 1978). So, if they could be treated like ocanic crust and if they had sufficient slab length to create a finite amplitude instability to sustain continuous subduction (cf. McKenzie, 1977), this would be able to sustain the convergence and the opening of the Tyrrhenian Sea with no contribution from the subduction of a major oceanic floor such as that of the Eastern Mediterranean. McKenzie (1977) has shown that a subduction zone becomes self sustaining after about 200 km. or so of subduction, unless the subduction angle is considerably flatter than 45°. I contend that at least the Lagonegro/Molise system may have been this wide (see also Dewey et al., 1989a; Patacca et al., in press) and thus may have contributed to the sustenance of subduction until the late Pliocene when the Tyrrhenian opening was near completion. Lagonegro was probably widest near present Calabria and the greatest width of the Tyrrhenian there may be no coincidence and may have been accomplished without assistance from the subducting Eastern Mediterranean lithosphere. But, if the palaeogeography proposed by Dewey et al. (1989a) and supported here holds, then the Lagonegro floor and the Eastern Mediterranean floor may have been identical near present Calabria, and would provide a very elegant solution not only to the sustenance of subduction in the Apennines and the opening of the Tyrrhenian basin, but also to the problem of how one goes in time smoothly from the subduction system of the northern branch of the Neo-Tethys represented by the Ligurian ocean closure to the subduction system of the southern branch of Neo-Tethys represented by the Eastern Mediterranean (Fig. 6)

I emphasize the identity of the back-arc basins of the western Pacific/Southeast Asian area, as well as in other parts of the world, with the back-arc basins of the Mediterranean, because this would enable us to use the models generated on the example of the former to be employed for the latter. In this regard, it is instructive to look at the age (i.e. density)/rollback correlation in the world's subduction zones to infer the mechanism of back-arc rifting and spreading in the Mediterranean. Since Molnar and Atwater's (1978) suggestion, many authors (e.g. Dewey, 1980; Royden, 1988a; Burchfiel and Royden, 1991a) considered density of the underriding plate in subduction zones to cause roll-back and thus back-arc opening. However, Jarrard (1986a) found an inverse correlation between the age of the subducted lithosphere and the presence of roll-back! Although Şengör (1990) suggested that this may in part be due to the negligence of the rejuvenation age of the oceanic lithosphere by post-generation heating mechanisms (Crough, 1983; Sleep, 1987), its widespread applicability seems to suggest that indeed something else than density may be directly behind rapid roll-back and back-arc rifting. Experiments on mesosphere penetration by subducted slabs by Kincaid and Olson (1987) are in this regard very instructive. If

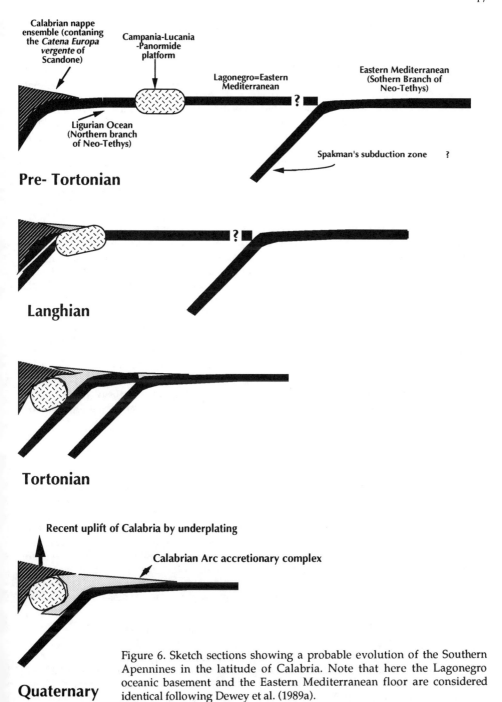

Figure 6. Sketch sections showing a probable evolution of the Southern Apennines in the latitude of Calabria. Note that here the Lagonegro oceanic basement and the Eastern Mediterranean floor are considered identical following Dewey et al. (1989a).

young, and therefore buoyant, lithosphere hits the asthenosphere/mesosphere boundary, it experiences great difficulty in penetration and commonly is laid down onto the boundary, causing the hinge in the subduction zone above to buckle sharply and to retreat in the direction in which the buoyant lithospere is laid down on the boundary, i.e. away from the arc (Fig. 4, esp. D). Indeed, Fig. 4 shows inflexion points in all the subduction zones discussed here suggesting slab buckling between 400 and 600 km. depths (Fig. 4D) likely owing to difficulty of penetration into the mesosphere below the 650-km-discontinuity. Some of the anomalies observed between the plate deflection and the gravity anomalies in Italy (Royden, 1988a) may be explained by a buckling of the subducted lithosphere and dynamic maintenance of small deflection in its upper parts (Fig. 4A).

Spakman's (1987, 1990, 1991) and Spakman et al.'s (1988) work on seismic tomography of subducted slabs in the Mediterranean has indicated that they may extend down to the mesosphere/asthenosphere boundary, in accordance with such geological data as Cretaceous subduction-related granites on Crete (Ercan and Türkecan, 1985), the presence of the Miyashiro-type late Cretaceous Troodos ophiolite (Miyashiro, 1973), and mid-Cretaceous HP/LT metamorphic rocks within the Alanya Massif (Okay and Özgül, 1984), indicating the Mesozoic inception of subduction along the northern boundary of the Eastern Mediterranean, contrary to what is commonly assumed (e.g. Dercourt et al., 1986). These long slabs are found below rapidly extending basins (Tyrrhenian and the Aegean) and lead one to suspect that indeed it is the roll-back associated with slab buckling that may be the cause of the extension in these areas (although in the case of the Aegean, the westward escape of the Anatolian scholle may be an additional factor: see below). It is of prime importance that more work be carried out on the geometry and history of these subducted slabs to be able to test the proposed models on back-arc rifting.

Extension within magmatic arcs related to subduction is not the only cause of extension behind active belts of shortening. Dewey and Şengör (1979) and Şengör (1982) argued that tectonic escape, if checked in the direction of escape, is capable of creating extension within the escaping crustal and/or lithospheric sliver ("scholle"), a model followed more recently by Burchfiel (1980[3]), Royden et al. (1982), Royden, (1988b), Royden and Báldi (1988), and Taymaz et al. (1991). I shall discuss this phenomenon in more detail below, when we consider the significance of tectonic escape in collisional tectonics.

4.3. CONTINENTAL ALLOCHTHONS AS HIGHEST UNITS IN THE MEDITERRANEAN COLLISIONAL OROGENS

Another "peculiarity" of the circum-Mediterranean orogens, and very especially those around the western Mediterranean, is the presence of continental allochthons that override the margins of incoming continents after the collision. The most spectacular, and the earliest-recognised example of such a high overriding sheet is the Austroalpine nappe ensemble in the Alps (Termier, 1903; for an overview and recent data on the highest Alpine allochthonous units, see Flügel and Faupl, 1987). Such continental nappes overriding the margins of other continents for considerable distances (on the order of 100km) are rare in the world's repertoire of orogenic belts. In fact, outside the Mediterranean area, only the Scandinavian Caledonides (the "uppermost allochthon": Roberts and Gee, 1985) and the South Chinese Mesozoic orogens (Hsü et al., 1988;

---

[3]This landmark paper of Burchfiel was published in 1980 and is thus quoted in the literature. But I think it is only fair to point out that it had been ready in its final form by 1975 (this was when I first read it!) and simply had failed to appear where it had been originally intended for (it was found "too technical and too long" for a semi-popular science magazine in the US) and was eventually published in 1980 in the form I had read it in 1975. In fact my reading of it in 1975 was the inspiration and example for the 1981 Şengör and Yılmaz paper that immediately adjoined it in the east.

1990) are known to me to have such high suture-crossing continental nappes among the Phanerozoic orogens. A similar interpretation for the southern Appalachians, where a former Piedmont oceanic arc rode onto the continental margin across a flat suture, of which the Hayesville and the Brevard fault zones are major splays, only applies to a small ocean marginal to early Palaeozoic North America (e.g. Hatcher, 1987).

I have shown elsewhere (Şengör, 1991, 1992) that collisional orogens have three main types (Fig. 7): 1) *Alpine-type collisional orogens* are those in which one continent overrides the other considerably (> 100km) at shallow structural levels (< 15km). They commonly have very poorly developed associated pre-collisional magmatic arcs and small amounts of ophiolitic material in their sutures. All these features point to their generation from the demise of narrow ocean basins (< 1000km). 2) *Himalayan-type collisional orogens* have steep sutures and one continent does not override the other at shallow structural levels. They commonly possess well-developed magmatic arcs and contain abundant ophiolitic material along their sutures. Such orogens develop from the closure of large oceans (> 1000km). 3) *Turkic-type collisional orogens* represent an exaggeration of the Himalayan type by the development of very large subduction accretion complexes into which commonly magmatic arc axes migrate during orogeny. Şengör (1991, 1992) argued that Turkic-type orogens generaly develop from the closure of very large oceanic tracts such as the Pacific, although it is clear that if such large tracts of ocean remain free of large amounts of terrigenous sediment, their disappearance is unlikely to give rise to Turkic-type orogens. Smaller oceans with an overabundant sedimentary fill (such as the Eastern Mediterranean) may, under exceptional circumstances, generate Turkic-type orogens.

Fig. 8 shows the distribution of the three types of collisional orogens within the Tethysides. A significant feature of this map is the exclusively Alpine aspect of the orogens in the western Mediterranean, where the width of the Neo-Tethys was less than 1000km, and the dominance of Himalayan orogens east of the regions, where the widths of both Neo-and Palaeo-Tethys were in excess of 1000km. Palaeo-Tethys was wider than Neo-Tethys, or at least its evolution involved the subduction of longer lengths of oceanic lithosphere than that of Neo-Tethys (see the continental reconstructions in Zonenshain et al., 1991, for example), with which a greater abundance of Turkic orogens in the structure of the Cimmerides correspond.

But Alpine-type orogens are not entirely lacking outside the Mediterranean. In Şengör (1991) I have argued hat the Scandinavian Caledonides may represent a "pseudo-Alpine-type" orogen owing to considerable transpression and the generation of a large flower structure that may have formed the uppermost allochthon. But the Xiangganzhe orogen of South China is a clear Alpine-type orogen (Şengör et al., 1988; Hsü et al., 1988, 1990). The Taconian southern Appalachians also represent an Alpine-type orogen (Hatcher, 1987).

Thus, in terms of their architecture, the circum-Mediterranean orogens do not appear as a unique group either. Their architecture, as well as the poverty of subduction-related magmatics and well-developed ophiolites (especially the Miyashiro-type giant ophiolite nappes) in their western representatives, are a common function of the size of ocean destroyed and of the amount of clastic sediment available.

Figure 7. (following page) Three main types of collisional orogens: A. Alpine type: Highest nappe continetal. B. Himalayan type: highest nappe oceanic. C. Turkic type: Suture is very wide and filled with subduction-accretion material. These three types are commonly a result from the disappearance of narrow (Alpine type: generally < 1000km), wide (Himalayan type: generally > 1000 km) and very wide (Turkic type: generally thousands of kilometres wide) oceanic areas. The Turkic variety also requires a rich clastic supply into the closing ocean.

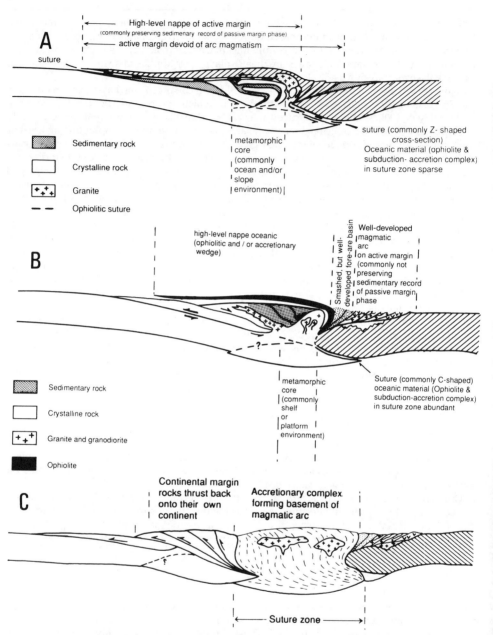

In summary, I emphasize that there is no special "Mediterranean-type orogeny". All the peculiarities of the Mediterranean orogens can be accounted for by comparison with other active and/or fossil orogens in the world.

But this means that comparative tectonics offers us a powerful tool to understand their evolution. It is to that evolution that I turn in the next section.

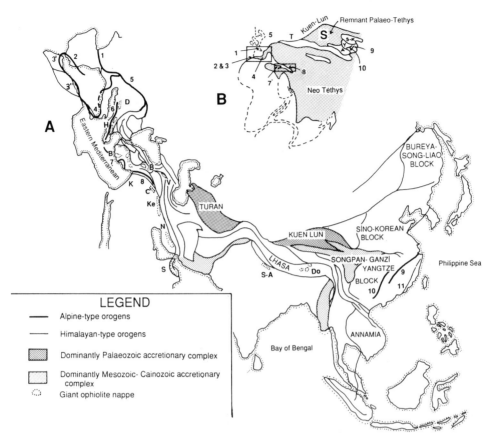

Figure 8. A. The distribution of the three types of collisional orogens within the Tethyside super-orogenic complex. Note that the Alpine-type orogens are almost entirely confined to the circum-Mediterranean mountain ranges and mainly to the west of the Carpathians. B. A highly schematic reconstruction of the early Jurassic Tethyan realm showing the correlation of Alpine-type collisional orogens with narow branches of the Neo-Tethys.

5. How are highly sinuous orogenic belts such as the Mediterranean Alpides created?

In the foregoing section, I pointed out that the extremely contorted plan view of the circum-Mediterranean orogens has been considered one reason to view the Mediterranean chains unique. Now that we have seen that this is not the case, we may inquire how such complex plan views of orogens may be created.

As Argand already realised in the first quarter of this century, the presently contorted appearance of the Mediterranean Alpides is mostly a secondary acquisition. Fig. 9 is a series of four maps illustrating the evolution of the Mediterranean between the Campanian and the Present. In Fig. 9A, we see a united subduction front stretching from southern Spain via Corso-Sardinia and the Alps to the Dinaro-Carpathian realm from whence it enters Asia through

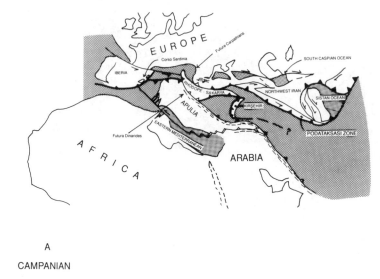

Figure 9. Sequential maps showing the tectonic evolution of the Mediterranean Alpides since the Campanian. A. Campanian palaeotectonics.

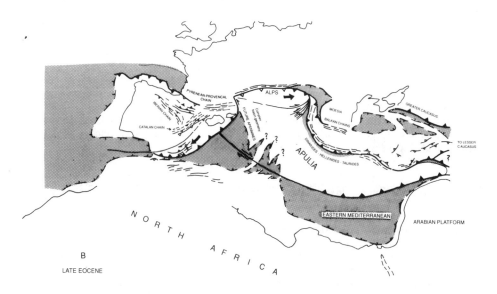

Figure 9. (Cont'd) B. Late Eocene palaeotectonics.

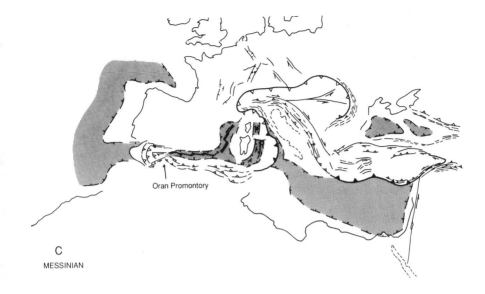

Figure 9. (Cont'd) C. Messinian palaeotectonics.

Figure 9. (Cont'd) D. Present tectonics of the Mediterranean.

Anatolia. It is only gently curvilinear bowing slightly north north of the African (Adriatic) promontory and slightly south while curving down from the Carpathians into the Dinaric/Hellenic/Tauric realm. In this map the united subduction front changes its polarity twice. Once, while crossing from the Corso-Sardinian area to the Alps and the second time while crossing from the Carpathians to the Dinaric system.

Why a certain polarity becomes established in a zone of shortening lies outside the scope of this paper. But subduction zones of opposing polarities must be connected with transform faults or more complicated structures that are their equivalents (Wilson, 1965). Fig. 10 shows the kinematic constraints imposed by such geometries. There are two possible configurations: Two subduction zones that face each other must be connected by a transform fault that becomes shorter with the full spreading velocity (Fig. 10, I). With time, the two subduction zones become colinear for an instant and then evolve into two subduction zones which face away from each other with a transform fault in between that must lengthen with the full spreading velocity (Fig. 10, II). Note that, the first geometry *eliminates* sinuosities in the orogenic front generated by subduction the zone, *but that with time it must evolve into the second geometry*, which *accentuates* the sinuosities along the orogenic front! On the strength of this kinematic argument alone, the subduction zone shown in Fig. 9A is expected to have accentuated the sinuosity of the orogenic front in the Mediterranean Alpides.

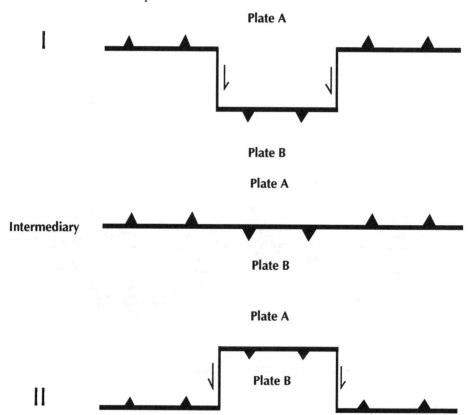

Figure 10. Evolution of a subduction zone (I), where subduction polarity changes twice to create segments whose polarities "face" each other, to one where they face away from each other (II).

While considering the geometry of the main subduction zone in the Mediterranean Alpides, I must touch upon a thorny problem in the Corso-Sardinian segment. In Calabria, Amodio-Morelli et al. (1979) have shown that during the course of the Cainozoic two orogenic edifices of opposing polarity evolved. An earlier *"catena Europa vergente"* involved the obduction of ophiolites and *schistes lustrés* onto the "European" (i.e. Corso-Sardinian + Calabrian fragment) margin along with the collision of a continental fragment identified as "Austroalpine" because of its lithologic and evolutionary similarities to the Sesia Lanzo zone of the western Alps. Then this entire ensemble was thrust southward onto the "African foreland" along with the Kabylian massifs as recognised by Lugeon and Argand (1906) almost a century ago in one of the most prophetic papers ever written in tectonics! These observations and interpretations are in full accord with similar ones made in northeastern Corsica (Cohen et al., 1981; Jolivet et al., 1990, 1991; Fournier et al., 1991), although even the most recent Tethyan reconstructions ignore this part of the story (e.g. Dercourt et al., 1986), except that by Dewey et al. (1989a).

However, if Amodio-Morelli et al.'s (1979) inference - which is, incidentally, identical to that of Argand (1924) - of a complete suturing of the Neo-Tethys in Calabria and Corsica in the early Cainozoic is adhered to, major problems arise in the interpretation of the tectonic evolution of the Apennines, where collision did not occur at least until the Miocene (e.g. Patacca et al., in press). This problem may be solved by assuming that the "Austroalpine" element identified by Amodio-Morelli et al. (1979), i.e. the Centuri gneiss unit (see Coli et al., 1991; Jolivet et al., 1991), was a microcontinental fragment lying between the main Ligurian Neo-Tethys and Corsica (i.e. "Pennine" instead of "Austroalpine" in the Alpine sense), and separated from the latter by a "*Schistes lustrés* marginal ocean" as suggested in Fig. 11. I further suggest that the Centuri unit may have been an analogous object to the higher Penninic Margna Nappe in eastern Switzerland, and to the lower slice of the "lower Austroalpine"Dent Blanche nappe[4] in western Switzerland. Both represent ± lowest Austroalpine elements palaeotectonically, and both sit over serpentinites and *schistes lustrés* (e.g. Trümpy, 1980). Moreover, the gently deformed and slightly metamorphosed oceanic Balagna nappe in northern Corsica (Jolivet et al., 1991) is a still higher unit than the Centuri gneiss and thus makes the latter "sandwiched" between two oceanic units exactly as in the case of the Margna nappe in the Alps (see esp. fig. 3 in Jolivet et al., 1991). I therefore assume that a uniformly south-dipping late Cretaceous subduction zone flipped in Corsica and Calabria after the collision of the loose Centuri unit and thus led to the back-arc opening events of the western Mediterranean.

My proposal to solve Amodio-Morelli et al.'s problem is similar to that by Coli et al. (1991), except that I consider the presece of an initial late Cretaceous eastward (present geographic orientation!) subduction under the Centuri unit.

Back-arc opening events began in the western Mediterranean in the Oligocene and created, succesively, the Algero-Provençal basin (Cohen, 1980) plus the Valencia Trough - copolar with the Algero-Provençal opening in contrast to older interpretations of a trapdoor kinematics (Foucher et al., 1992) - and then the Tyrrhenian Sea (Kastens et al., 1988; Kastens and Mascle, 1990).

Around the Tyrrhenian Sea, a marginal basin opened by disrupting an entirely ensimatic magmatic arc. The continental crust is more than 20 km. thick under Sardinia and nearly 40 km. under Calabria partly forming the forearc of the migratory arc. Under the Tyrrhenian Sea, the crust is not more than 5 km. under the Marsili and the Vavilov basins, that are 2 and 3.5 Ma old respectively (Leg 107 Shipboard Scientific Party; Kastens et al., 1988; Kastens and Mascle, 1990).

---

[4]Remember that in his original scheme of 1911, Argand considered the Dent Blanche as the highest Penninic element, the nappe no. "VI".

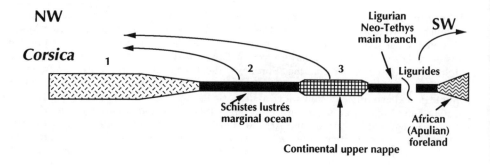

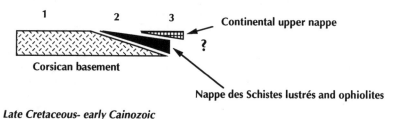

Interpretation of "Catena Europa Vergente" in Corsica and Calabria

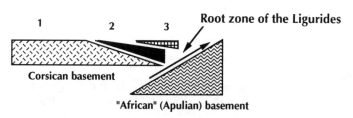

"Catena Europa Vergente" gets thrusted bodily onto the African foreland, building the "Catena Africa Vergente"

Figure 11. Schematic diagrams showing an evolutionary sequence of the Corsican segment of the Alpide orogen to solve the space problem created by Amodio-Morelli et al.'s (1979) interpretation of the catena Europa vergente by considering the "continental upper nappe" in Corsica (the Centuri unit) an equivalent of the Alpine Margne nappe, i.e. "Pennine" and not "Austroalpine". For simplicity, the highest ophiolitic Balagne nappe is not shown, but its provenance according to this scheme would be in the Ligurian Neo-Tethys main branch.

In the Tyrrhenian Sea subsidence progressed from west to east, so that the youngest part of the basin is closest to the active subduction zone, a situation similar to that seen in other active back-arc basins in the world.

Magmatism related to the opening of the Tyrrhenian Sea began in the Oligocene in western Sardinia, associated with a north-west- or north-north-west-dipping subduction zone. Just before the Messinian, the opening of the Tyrrhenian basin began along the eastern margin of Sardinia and migrated eastwards until the Pleistocene. The mafic and ultramafic rocks flooring the magnetic basins of the Tyrrhenian Sea (Heezen et al., 1971) formed at this time. Finally, during the Quaternary, the present island arc system of the Aeolian archipelago developed *on the newly-created marginal basin floor* exactly as in the case of the Mariana arc or the South Scotia Sea!

Fig. 12 illustrates the effect of these marginal basin opening events on the geometry of the Mediterranean orogenic belts. They very strongly accentuated a previously very gently north-concave curve in the trace of the main Neo-Tethyan subduction front and gave it almost a hairpin shape in southern Italy (see also Dewey et al., 1989a; Patacca et al., in press).

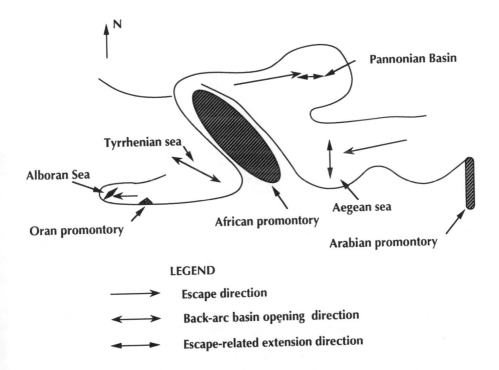

Figure 12. Schematic map of the trend lines of the Mediterranean Alpides illustrating the main mechanisms that likely have created an extremely sinuous mountain chain. They are back-arc basin opening (subduction-related), escape and escape related extension (the latter two processes are collision-related). In some areas (e.g. the Pannonian and the Aegean basins), back-arc basin opening and escape-related extension have created effects that reinforced each other. In the Alboran Sea, "escape" may have been mostly subduction-driven!

The other major basinal areas in the Mediterranean area, Pannonian and the Aegean basins have been described as back-arc basin opening events, but especially in the Aegean, another process clearly later interfered and influenced the extension.

Dewey and Şengör (1979), Şengör (1979), Şengör and Dewey (1985), Şengör et al. (1985) and Şengör (1987b) have ascribed the opening of the Aegean Sea to the obsruction by Greece and Albania of Turkey's westerly escape (Figs. 9C and D, and 16). However, in recent years evidence has accumulated that the onset of the Aegean extension is older than previously assumed (at least Burdigalian as opposed to Tortonian: see Seyitoğlu and Scott, 1991, and Seyitoğlu et al., 1992), which is in accordance with the 25 to 15 Ma K-Ar and Ar-Ar dates that have been obtained on white micas and amphiboles of the Barrovian overprint associated with an extensional crenulation cleavage on Naxos and the 23 to 16 Ma Ar-Ar ages from the Olympos in Greece interpreted to be associated with uplift and normal faulting (Dr. Judith Baker, pes. comm., 1988 and 1992).

This older age makes the interpretation proposed by Şengör and his co-workers questionable, if the data from other parts of the Turkish escape system remain unchanged. Hempton (1985) and later, on the basis of a more comprehensive database, Yılmaz (in press) proposed that the collision between Arabia and Anatolia had taken place much earlier than formerly believed (late Eocene instead of late Miocene!). Although this collision did not immediately give rise to a tight suture system, it juxtaposed continent against continent and led to the deposition of the Çüngüş wildflysch and the Lice molasse.

Inspired by these changes in the timing of collision, the speculation that the westward escape of Anatolia may have commenced earlier than the Tortonian date hitherto given has been under discussion in the last two years by the members of the Department of Geology of the İstanbul Technical University, and to check it Aykut Barka, A. M. Celâl Şengör and Yücel Yılmaz (in preparation) have since recalibrated the Neogene stratigraphy of the basins along the North and the East Anatolian faults according to the Aegean recalibration done by Seyitoğlu and Scott (1991) and Seyitoğlu et al. (1992). The results of this work are discouraging for the model of Şengör and his co-workers, which Taymaz et al. (1991) belatedly adopted. It seems clear that the Aegean extension began earlier than the westerly escape of Turkey. It therefore seems likely that escape related extension in the Aegean only helped along an earlier extension likely related to trench roll-back.

Since Burchfiel's milestone synthesis of the eastern European Alpides (Burchfiel, 1980), he and his co-workers (Royden et al., 1982; Royden, 1988b; Horváth, 1988) have shown that the extension in the basement of the Pannonian basin is also at least partly escape-related, much of northern Hungary being expelled eastwards along the Pustertal Line following the Alpine collision in the Eocene.

The Alboran plate was also expelled westward by the Oran promontory and led to radial thrusting around the Alboran Sea, which itself foundered at the same time (Fig. 13). Fig. 13 shows that the extensional structures here are mostly the products of an east-west extension and the eastward subduction of the lithosphere of the Neo-Tethys creating in the Alboran sea a back-arc environment, much like the Tyrrhenian Sea, the Pannonian Basin, and the Aegean Sea.

These three major escape events have further accentuated the contorted aspect of the trend-lines of the Alpides in the circum Mediterranean area. Fig. 12 illustrates the combined effects of the marginal basin opening events and the escape phenomena on the present geometry of the trend lines in the Mediterranean Alpides.

The most important factors in the generation of the extreme curvature of the trend lines in the Mediterranean area are therefore the marginal basin opening events and the escape events. In the next two sections, I discuss the tectonics of these two processes in the Mediterranean area, because the former has been recently interpreted in terms of extensional orogenic collapse and the latter has been roundy denied, but on the example of the much less-well known Himalayan-Tibetan area.

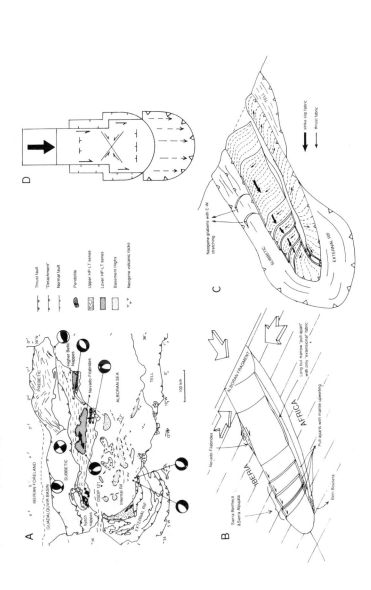

Figure 13. Tectonics of the circum-Alboran fragment. A. Structure and active tectonics of the circum-Alboran fragment orogenic belts. B. The emplacement of the Alboran fragment along transpressional strike-slip systems along which there are a number of pull-apart segments that allowed the rise of mantle peridotites in the Sierra Bermeja, Sierra Alpujata, and the Beni Bousera. Only the "underbelly" of the Alboran fragment is shown. C. Shortening of the Alboran fragment and the conversion of strike-slip boundaries into oblique thrusts. D. Schematic illustration of the tectonics of the Alboran fragment and its boundaries. The motion of the Alboran fragment as shown in D was probably largely maintained by frontal suction by subduction under it and only in a limited way by the push from behind by the Oran promontory.

6. What is the significance of "tectonic escape" in collisional processes?

Tectonic escape (Burke and Şengör, 1986) is the sideways expulsion of strike-slip fault-bounded lithospheric wedges from regions of constriction and has been first recognised by Argand (1920), although Cloos (1928, pp. 303-311, esp. footnote 2, pp. 309-310) seems to have been the first to state it explicitly. McKenzie (1970, 1972) put it in a plate tectonic context by elaborating an old idea of Ketin (1948), when he described the escape westwards of an Anatolian block south of the right-lateral North Anatolian fault. Dewey and Burke (1974) pointed out its importance in explaining the late orogenic structures in many orogenic belts and Molnar and Tapponnier (1975) gave it great popularity when they argued that in Asia it has been occuring on a massive scale since the Himalayan collision.

In the last few years, the importance, even the incidence of tectonic escape associated with the Himalayan collision has been questioned (e.g. England and Houseman, 1988; Dewey et al., 1989b; Burchfiel and Royden, 1991b) for two main reasons: First, field observations in limited areas in northeastern Tibet (e.g. Kidd and Molnar, 1988) have failed to disclose the enormous offsets postulated by Molnar and Tapponier's work. Secondly, theoretical mechanical modelling of continental convergence in the framework of the thin sheet analogy (England and McKenzie, 1982, 1983) has failed to generate the large escape wedges postulated by Molnar and Tapponnier (1975). This has led to much controversy, but the contestants have seldom considered areas where sideways movement of considerable lithospheric wedges away from nodes of constriction is now taking place. The Mediterranean region is an ideal place to form a basis for these discussions, for in it in at least one place a lithospheric wedge *is now moving westward away from a zone of collision*.

England and Houseman (1988) have postulated that the topographic potential may indeed drive such lithospheric chunks away from high areas towards low areas to diminish the high potential energy of elevations. Şengör et al., (1985) have shown, however, that the westward journey of the Anatolian block had commenced, before eastern Turkey even rose above sea-level! It might be argued that the difference of elevation between the bottom of the Eastern Mediterranean and the west Anatolian highland in the Miocene (about 3km above sea-level: Şengör et al., 1985) would have sufficed to "suck" Anatolia westwards, if Anatolia had originally moved onto the Hellenic trench. But Perinçek's work on a recently discovered abandoned branch of the North Anatolian Fault, namely the Thrace Fault System, has made this argument impossible, by showing that the Anatolian block initially moved towards a high continental area, much higher than the submarine east Anatolia at the time (Perinçek, 1991). Only after having found it difficult to cut through the Balkan peninsula at the latitude of south Bulgaria did the North Anatolian fault migrate southwards into a new strand and then cut through Greece, but again with much difficulty. In northern Greece the fault disintegrates into a broad shear zone as shown by Şengör and Dewey (1979) which Şengör (1979) named the Grecian shear zone. McKenzie and Jackson (1983) later have presented a detailed kinematic model of this broad shear zone and re-emphasized the rôle of the old structures in hindering the comfortable development of the western continuation of the North Anatolian fault in northern Greece, although both the behaviour of shear zones as studied along strike-slip zones of various scales and in a range of materials (e.g. Wilson, 1960; Tchalenko, 1970; Wilcox et al., 1973; Gallo et al., 1980) and the later work on pre-fault palaeomagnetism along the North Anatolian Fault zone by Ellen Platzman (personal communication, 1992) have cast substantial doubt on the validity of their approach using a fluid-dynamics based model.

Thus, the elegant arguments brought against tectonic escape or allowing only a limited sideways motion driven by crustal thickness differences by the recent theoretical work on mechanics seem not relevant, since they so manifestly fail to account for the observations in the world's best-studied and the most active escape system.

In Hungary, interdisciplinary studies on the basement of the Hungarian plain have revealed that very considerable escape has indeed happened and inverted the palaobiogeographic realms there. Fig. 14 roughly illustrates the inversion of the early and middle Jurassic brachiopod provinces by tectonic escape in the basement of the Hungarian plain (Vörös, 1988). While this escape was taking place, the elevation of the Alps was not much higher than that today.

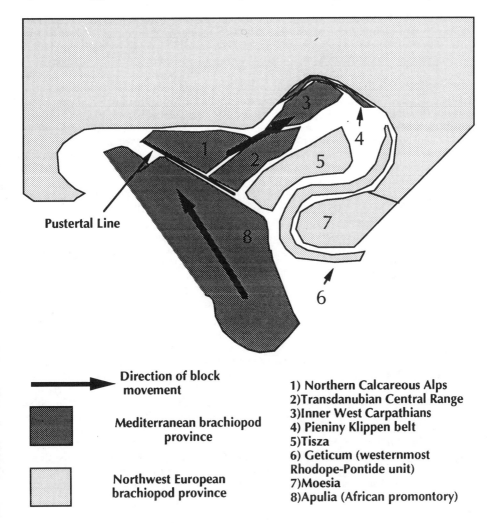

Figure 14. Sketch map showing the distribution of the Middle Jurassic Brachiopod provinces in the Alpine-Carpathian-Balkan realm (after Vörös, 1988), documenting the incidence of the eastward expulsion of a part of Apulia north of the Tisza unit partly along the Pustertal line. Compare this map with Burchfiel (1980).

If continental escape has been and remains so commonplace in the Mediterranean Alpides and, apparently, independent of any topographic augmentation, it is only natural to see whether it also has occurred in a collisional system of a much grander scale, namely the Himalaya, where Molnar and Tapponnier originally proposed a grand scenario of tectonic escape.

Fig. 15 shows three successive reconstructions of the Tibetan/Himalayan area since the late Cretaceous A, B, and C). Although the deatiled justification of this work is going to be published elsewhere, the basis of these reconstructions is very simple and may be briefly summarised as follows: the positions of Eurasia and India were obtained from the magnetic anomaly lineations and fracture zone trends in the Atlantic and Indian oceans as explained in Dewey et al. (1989b). For extra-Tibetan Central Asia, we took the 130-150km. Cainozoic shortening in the Tien Shan as estimated by Prof. Brian F. Windley (pers. comm., 1992) and Dr. Mark Allen (1990). For the Kuen-Lun/Qinghai Nan Shan/Qilian Shan, we used Professor J. F. Dewey's estimate of about 200km. shortening for the Cainozoic (pers. comm., 1992). The northern boundary of the Kuen-Lun/Pamir system has been fixed palaeomagnetically by Bazhenov and Burtman (1986) and the 200 km. stretching to "undo" the Kuen-Lun shortening was applied to the south of that position. Then we allowed a 50% shortening during the Cainozoic for *all* the Tibetan blocks. In other words, we doubled the width of all the Tibetan blocks to reconstruct their pre-collisional width. All these estimates may be wrong by about 30%. Still, as a result of this exercise, we had enough space left in the Tibetan area to fit the entire Indochina block, exactly in the way Tapponnier et al. (1986) had visualised it and similar to the way in which it was reconstructed by Enkin et al. (in press) on the basis of a recent synthesis of palaeomagnetic data.

An important feature of our reconstructions is their stepwise nature: This allows an extremely important point to emerge: The original escape route of Indochina and associated smaller blocks were along suture zones, characterised by soft accretionary complex material. None of the presently active faults in Central Asia and in Indochina (except possibly the Red River) seem to have contributed to the large-scale escape and that is why field work on these active and very spectacular structures has failed to reveal the expected huge offsets: it has been looking at the wrong places![5]

By contrast, looking at the palaeotectonic units of the Tibetan/Himalayan area has proved more fruitful in detecting and tracing the evolution of escape here, simply because of the sharp differences that existed in the eastern Tethysides between the Gondwanian, Cathaysian and Laurasian elements in the late Palaeozoic in terms of lithologic asociations, fauna and flora, and climatic indicators. We have been able to show the tremendous boudinaging of former continental blocks of uniform late Palaeozoic geology, now inserted between blocks with a totally foreign geology (Fig. 15). This sort of positive evidence is difficult to fault, as long as the reported observations are reliable.

It is through this sort of reasoning that we have been able to detect a yet earlier possible episode of escape in the Tibetan/Himalayan/Southeast Asian realm. It looks as if a formerly unified Cathaysian continent was broken up sometime before the late Cretaceous into an east Qangtang, Indochina, and South China blocks. Fig. 15D shows how this may have happened during the collision of the Lhasa block with the rest of Asia during the early Cretaceous, although still earlier disruption during the Triassic collision of the Cathaysian continent with the North China block and the Kuen-Lun is not necessarily ruled out. This hypothesis explains very

---

[5]Given the state of geological knowledge in Central Asia, I doubt whether very large offsets could have been detected by spot field checks, even if every "spot" covered several hundred square kilometres! I thus am very skeptical about the statements made about the large-scale tectonics of South-Central Asia based on spot field checks, however detailed these spot studies may be. We simply need to cover much more ground before anything meaningful can be said about the detailed tectonics of these very poory-known places.

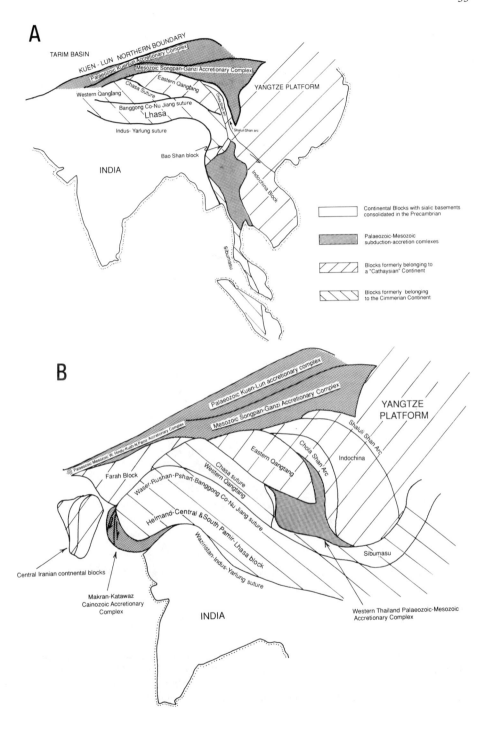

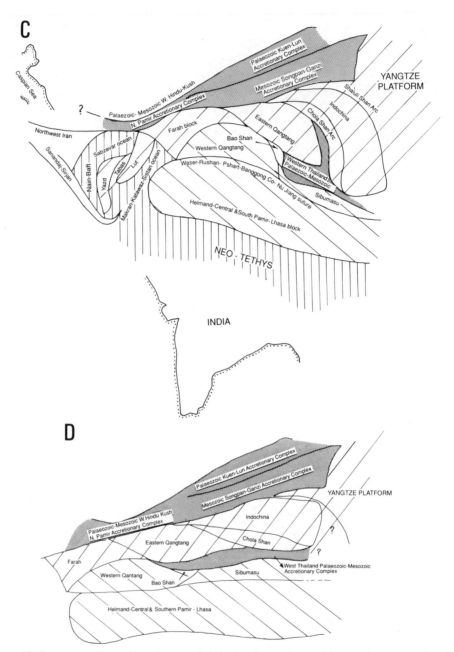

Figure 15. Reconstructions of south-central Asia showing substantial tectonic escape after the India/Asia collision. A. Present, B. Late Eocene, C. Pre-Cainozoic, D. At the time of the collision along the Banggong Lake-Nu River suture, i.e. just prior to the middle Cretaceous. Notice the lubricating role of "suture-filling" subduction-accretion complexes.

elegantly why Indochina and South China share a common late Palaeozoic fauna (cf. Şengör et al., 1988), when "sutures" filled with Triassic accretionary complexes now separate them. All those very linear, accretionary complex material filled "sutures" are here interpreted as very large strike-slip faults, along which suture material had been entrained.

Thus, contrary to much recent argumentation against it, continental escape seems to have occured not only in the Mediterranean Alpides, but also on a gigantic scale in the Tibetan/Himalayan area. In both places sutures, lined with soft schists and slates in addition to ophiolitic rocks of former subduction/accretion complexes served as avenues of initial and large scale escape. Thus all of Dewey et al.'s (1989b) arguments about the sutures being undisrupted by large offset strike-slip faults remain true, because the sutures themselves were the avenues of escape. That is also why it has proved so difficult to detect large strike-slip offsets in Central and South east Asia: the largest offset faults are located between blocks of disparate geology anyway, being themselves former sutures.

In the Mediterranean area, geological evolution becomes unintelligible, if some form of significant continental escape is not taken into account. Escape seems to have been mainly responsible for giving the Mediterranean Alpides their extremely contorted plan view, as already noted by Argand in 1920.

Continental escape has one more important by-product in the Mediterranean area. In the Anatolian Aegean system, the hindernis presented by the Grecian/south Balkan area to the westerly fleeing Anatolian scholle gave rise to roughly north-south extension in the larger region called "Aegea" by Le Pichon and Angelier (1981). Dewey and Şengör (1979) and Şengör (1979) had originally interpreted this as east-west shortening being simply relieved by north-south extension, although McKenzie (1978, p. 247) had criticised this (after a preprint of Dewey and Şengör's 1979 paper): "The westward motion of the Turkish plate is most easily maintained by boundary forces acting on the North and East Anatolian faults....It is however, difficult to understand how such forces can maintain the widespread extension in the Aegean area. Dewey and Şengör (1978) have attempted to do so, but it is hard to believe that blocks in northwestern Greece and Albania have sufficient strength to act in the manner they proposed."

I later pointed out (Şengör, 1982) that the strength of the crust here was not the central issue, as it indeed seemed sufficient to support a model similar to that described by Dewey and Şengör (1979) and Şengör (1979). I pointed out that "the most important difficulty encountered by the Dewey-Şengör model was to accomplish the 50% N-S extension since the Tortonian solely by the effects of E-W shortening." (Şengör, 1982, p. 68). I then asked the question as to what may have been responsible for the *maintenance* of the N-S stretching in the Aegean. Based partly on additional field observations between 1977 (submission date of the Dewey and Şengör, 1979 paper) and 1982, but mainly inspired by the work of Le Pichon and Angelier (1981), I concluded in that paper that "The shear couple indicated by MM in Fig.2 [in the present paper, fig. 16] I think provides the most appropriate answer to this question" (Şengör, 1982, p. 69).

Taymaz, Jackson, and McKenzie (1991) have recently adopted a model very similar to that by Dewey and Şengör and essentially identical to the variant published by Şengör in 1982 (Fig. 16), but without commenting on McKenzie's earlier doubts about the strength of the slats being insufficient to be able to be bent and/or broken. In that paper, the shear model of Şengör (1982) is more elegantly coupled with the E-W shortening model of Dewey and Şengör (1979) than was done in Şengör (1982). Unfortunately, however, the recent revision of the Neogene stratigraphy in Turkey had already made this model suspect before Taymaz et al. (1991) adopted it. As underlined above, it may still be useful as an auxiliary to the trench hinge rollback, but no longer in the main rôle Taymaz et al. so belatedly ascribe to it.

Thus the obstruction of escape apparently is capable of giving rise to basins and related structures of great complexity. The work of Burchfiel (1980) Royden et al. (1982), Royden (1988b), Royden and Báldi (1988) has shown the various geometries of basins in the basement of the Hungarian Plain, which at least in part were generated through mechanisms similar to those that created the Aegean basins.

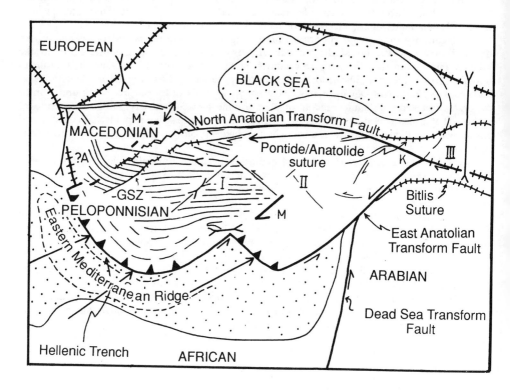

Figure 16. Active tectonics of the Eastern Mediterranean according to Şengör (1982, fig. 2). Here the east-west shortening across the Grecian Shear Zone (GSZ) is postulated to be revealed by north-south extension in the Aegean Sea and surrounding regions following Dewey and Şengör (1979). In addition, the shear couple MM' is believed to sustain the extension by bending the east-west trending crustal boards in the southern and central Aegean in the manner shown. This is substantially the model adopted by Taymaz et al. (1991), but unfortunately the timing of neotectonic events in Anatolia has recently made the mechanism portrayed by this model an unlikely candidate for the main cause of extension in the Aegean since the early Miocene.

7. How does extensional orogenic collapse occur, or what is really the rôle of topography in tectonics?

Consideration of continental escape has brought us into contact with the rôle of topography in orogenic belts. We have seen that topography, or more relevant, crustal thickness differences and associated irregularities in the distribution of different densities in adjacent columns in the lithosphere, do not seem to have a determinant rôle in the initiation and even in the maintenance of tectonic escape. Lithospheric blocks appear to be far stronger than commonly assumed in models of thin sheets of brittle and strong upper crustal material "floating" above a viscous lower crust.

Another significant product of the recent theoretical mechanical models of the continental lithosphere was extensional orogenic collapse scenarios, according to which mountain ranges generally collapse under their own weight like a lump of honey left unsupported on its sides, as originally suggested by Bucher (1956). Fig. 17 shows very schematically and very simplistically how this process is thought to take place: Two major forces act on a piece of lithosphere, namely, buoyancy forces resulting from its own weight and boundary forces, resulting from its interaction with other lithospheric entities. Shortening in a given cross-section may thicken the crust until gravity arrests any further thickening. In other words, boundary forces may go on thickening a piece of crust until its own potential energy tops up the buoyancy forces (F_{bu} in Fig. 17) to a point where they equal the boundary forces (F_{bo} in Fig. 17). In this case further thickening becomes impossible. If, by some means, F_{bu} exceeds F_{bo}, then the region under the influence of this superior F_{bu} begins spreading like honey, in effect overcoming the "lateral support" provided by the inferior F_{bo}.

How F_{bu} may exceed F_{bo} in a given region has been explained by Houseman et al. (1981) by convectively removing a part of the heavy mantle lithosphere beneath a zone of crustal thickening. This "lithospheric detachment" model shows how a piece of lithosphere suddenly may become very light, rise, and augment its potential energy enough to overcome that of its lateral supports.

Although its physics is disarmingly simple and elegant, testing this model is wrought with difficulties. To do so, one needs to find a place where the marginal supports of a collapsing high begin to be pushed away simultaneously with the onset of collapse, and that no other mechanism is interfering to pull the sides away to be aping the effects of extensional orogenic collapse.

Dewey (1988) seggested that most of the major Mediterranaen basins, such as the Alboran Sea, the Tyrrhenian Sea, the Pannonian Basin, and the Aegean Sea formed through extensional orogenic collapse brought about only by *post-collisional* potential energy differences between the orogen and its forelands. I find it difficult to follow him in this interpretation, because the Tyrrhenian clearly extended above a subduction zone, that had been there long before the "collapse" had commenced. In the Marianas, repeated extension occurred in the back-arc area in an oceanic setting, and if anyone is inclined to hold the great depth of the Mariana trench solely responsible for a collapse into it, I may point out to the Banda Sea rifting in a place *lower* than the toe of the forearc of the associated arc in the Aru Trough (Fig. 4: Hamilton, 1979). In the Pannonian basin, the great thicknesses of flysch in the outer Carpathians (e.g. Sandulescu, 1976) probably had kept the trenches full to overflowing (like the present-day Makran) and thus precluded a trench "suction" owing to considerable topographic inferiority, especially in the light of major Pannonian (12-8Ma) extension having taken place after major stretches of the Carpathian suture had formed! Pannonian extension too began in an environment of modest elevation without the towering heights of Tibet or Altiplano, that seem to be capable of inducing only very modest extension (if indeed *any* solely owing to elevation!). Only the Aegean had considerable elevation above sea-level, but it too only collapsed above a pre-existing subduction zone and likely maintained its extension with the aid of the westerly escape of Anatolia.

The Betics, together with the Riff chains of northwestern Africa and the Alboran Sea in between, have long been considered the stronghold of gravity collapse modellers (e.g. van Bemmelen, 1952), and this view has recently regained popularity mainly owing to reported occurrences of two generations of normal faults, the earlier of which had been previously interpreted as thrusts (Platt and Vissers, 1989; Doblas and Oyarzun, 1989a, 1989b). Especially Platt and Vissers (1989) have emphasized the great metamorphic discordance between the high pressure Nevado-Filabrides and the low pressure but structurally higher Alpujárrides. But this interpretation I think is untenable for the following reasons: First, newer observations on the presence of high-pressure minerals such as carpholite indicating pressures up to 7 kb in all the

Conventional wisdom (e.g. England & Houseman, 1988)

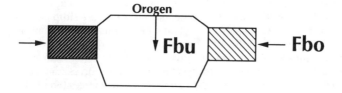

If Fbu > Fbo, extensional orogenic collapse occurs

Other mechanisms capable of extending orogens just after shortening:

a) Across the strike: 1) Back-arc basin opening. Example: Tyrrhenian Sea, 2) Escape-related extension. Example: Aegean Sea.

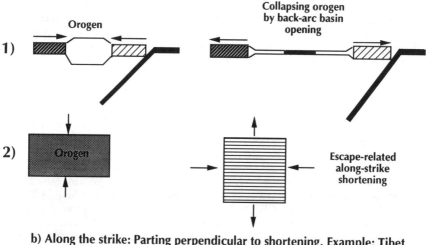

b) Along the strike: Parting perpendicular to shortening. Example: Tibet

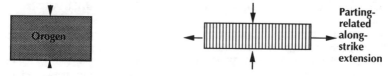

Figure 17. Mechanisms of "extensional orogenic collapse".

Alpujárride structural levels have negated the importance of "lower metamorphics on lower nappes" argument for the presence of extensional faults (Goffé et al., 1989). Secondly, observations on the orientation of mineral stretching lineations in all the important Betic and Riff tectonic windows, with the singular exception of the Sierra Almahilla (Platt et al., 1984), have shown that the direction of tectonic transport is parallel with the trend of the orogen, not across it as would have been expected if tectonic transport had occurred along topographic slopes (Fig. 13A; Darot, 1974; Vissers, 1977; Reuber et al., 1982; Tubía and Cuevas, 1987). Moreover, fault plane solutions of earthquakes in southern Spain and northeastern Morocco show that the present day direction of tectonic transport is the same as that indicated by the stretching lineations (e.g., Buforn and Udías, 1991, esp. fig. 7). Thirdly, most Neogene basins in southern Spain indicate basin opening directions in agreement with those of the stretching lineations and the direction of tectonic transport as shown by fault plane solutions (e.g., Lukowski et al., 1988). Most significantly, Reinoud Vissers has documented that some of the basin-bounding, N-S striking faults turn into the detachments that now mantle the tectonic windows showing the kinematic continuity beween the Neogene basins and the ductile deformation of the basement (Dr. R. L. M. Vissers, verbal comm., 1992 during the Eriche School).

From all these observations it seems clear that tectonic transport since the Miocene in the Betics has been in the direction of the movement of the Alboran fragment as has long been noted (e.g., Andrieux et al., 1971; Wildi, 1983). There is no evidence whatever that the Betic chain collapsed under its own weight to generate its present structure as claimed by Platt and Vissers (1989), but the evolution that the available observations constrain seems more like that shown in Figs. 13 B and C. The Alboran fragment has been moving west since at least the early Miocene, being constricted in the east by the Oran promontory. It has been moving along strike-slip faults into a laterally more spacious area and thus developed a number of pull-aparts along its "escape rails". It is these pull-aparts that localised the mantle injections of the Beni Bousera, Sierra Bermeja, and Sierra Alpujata (Fig. 13B). With continuing Africa-Europe convergence, these escape rails turned into very oblique thrusts and parts of the Alboran fragment became the Alpujárride units overriding the Spanish "mainland" here formed by the Nevado-Filabrides (Fig. 13C). Internal disintegration of the Alboran fragment itself gave rise to the Alboran Sea through a complex, but mainly E-W directed extension (Fig. 13D). As in most other places in the Mediterranean area, all this tectonism was probably driven by "slab pull" under the Alboran fragment that sucked it into its present place.

Nowhere in the circum-Mediterranean chains do we see therefore orogens collapsing solely under their own weight. Post- or late orogenic extension has occurred in many of them, as Suess had recognised more than a century ago (Suess, 1875, 1878, 1885), but this extension has almost always been driven by plate boundary processes and it seems not possible to isolate the effects of gravitational potential in extensional events in the Mediterranean region, although in some cases it may indeed have contributed an auxiliary factor. But topography does indeed interfere and it seems it is in keeping the compressional deformation localised as I discuss in the next section.

8. How is orogenic deformation distributed in a collisional zone at any one time?

One remarkable feature of the maps displayed in Fig. 9 is the extreme narrowness of the compressional deformation from the Betic Cordilleras to the Iranian plateau. It seems that shortening in wide regions by thrust faulting and/or folding at any one time (usually on the scale of several million years) is not common. What these maps show that thrusting concentrates itself in narrow bands essentially at the foot of growing mountain ranges.

That thrusting commonly occurs at the foot of highs in a sense to bring them over lowlands is a generalisation made over a century ago by Suess (1883), who termed it *"Überschiebung der*

Tiefen" (overthrusting of deeps), and has been repeated in countless instances by subsequent workers the world over. Recently Molnar and Lyon-Caën (1983) explained this as an effect of topography, whereby thrusting is deactivated in zones that have risen enough to have sufficient potential energy to resist further thickening and migrates into areas whose potential energy is insufficient to arrest thickening by thrusting (or folding or homogeneous bulk thickening) (Fig. 18).

In the circum-Mediterranean chains we see generally two sub-parallel zones of thrusting developed along the external flank of the Alpine-Carpathian-Balkan- Pontide- Caucasian ranges in the north and the Atlas-Apennine-Dinaride-Hellenide-Tauride-Zagride chains in the south. It gives the impression as if a region of high potential energy is interposed between the two zones from Gibraltar to the Caspian Gates, corresponding essentially to Kober's (1921) "Zwischengebirge". This is, however, not so. In the western Mediterranean, it is the opening marginal basin of the Tyrrhenian Sea and the escape related Alboran sea basin, in the central Mediterranean, the opening of the escape-related Pannonian basin, and in the Aegean first the topographic high and then the escape-related basing opening of the Aegean that separate the two thrust fronts.

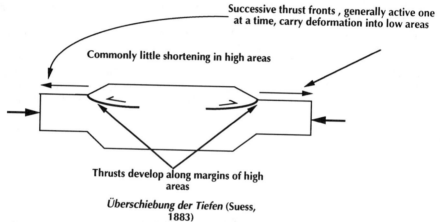

Figure 18. Concentration of thrusting to narrow zones along the foot of highlands in orogens. This phenomenon, recognised by Eduard Suess towards the end of the last century, is called by him the *"Überschiebung der Tiefen"* (thrusting onto lowlands) and is now believed to be caused by the stopping of uplifting under the influence of gravity in uplands.

In contrast to the thrusts and presumably the associated folds, strike-slip faulting is more widespread within and along the Mediterranean orogens, showing no zonal attachment. In the Alps, for example, both the Engadine (Trümpy, 1977) and the Mur-Mürz-Lethaia faults (Arıç, 1981) traverse much of the orogen. In eastern Turkey, a complex set of structures is active, here again thrusts show a strong attachment to certain linear arcuate zones.

The distribution of orogenic deformation in an evolving collisional orogen is therefore not random, but displays the following characteristics:

1) Thrust faulting and folding (i.e. crustal thickening) are at any one time (for several million years) remain confined to narrow and laterally continuous zones. These zones may be relayed along the strike by strike-slip faults possibly functioning as thin-skinned transfer structures within the crust.

2) Complex strike-slip faulting within the orogens is commonly confined to zones located

between active regions of thrusting. This may reflect the propensity of high areas to shorten by strike-slip falting rather than by thickening. But this mode of shortening brings with it along-strike elongation that in places may be very considerable and express itself as lateral expulsion of crustal units (a sort of tectonic escape).

3) Normal faulting also commonly occurs between the two thrust zones. This may have a plethora of reasons, including back-arc rifting, escape, incompatibilities along interacting strike-slip faults, pull-apart along individual strike-slip faults, *but very rarely because of the high gravitational potential of the rising welt.*

9. Discussion and Conclusions

The statements made above concerning the tectonic evolution of the Mediterranean I think inescapably brings us to the conclusion that the continental lithosphere likely has an Argand number (ratio between the stress arising from crustal thickness differences and the stress required to deform the material at the ambient stress rates: England and McKenzie, 1982, 1983) significantly smaller than commonly assumed for the continental lithosphere involved in collision zones, for stresses arising from crustal thickness differences do not seem to have as much affect on the deformation as the stresses communicated across plate and block boundaries. This is a conclusion very different from many recent studies and the difference seems to result from considering the *history* of deformation in the deforming zones in greater detail than hitherto attempted. Particularly, the rapid shifts in the loci and the style of deformation makes it unlikely that a detailed study of the present, or the Quaternary deformation is going to be particularly helpful in illuminating us about the tectonic style that prevailed throughout the Cainozoic or that a simple extrapolation from the Recent or even Quaternary events into the older Cainozoic will provide straightforward answers.

For instance, a high eastern Turkey today sits behind a well-established escape system that feeds a lithospheric wedge into an extending zone that overrides an oceanic trench. When the escape was initiated, however, a low eastern Anatolia propelled a different wedge towards an area at least 2 to 3 km above sea-level by means of faults that attempted to disrupt the high, without reaching a low area. The attempt failed and the earliest escape faults were abandoned. Only careful geological work (including much seismic profiling and subsurface studies) disclosed this earlier history that I think conclusively refutes the view that crustal thickness differences may have initated the escape in the Eastern Mediterranean, provided the observations and the kinematic inferences derived from the geological studies stand.

This helps us understand why the debate in South and Southeast Asia for and against escape continues without the contenders being able to make any concessions to each others' views. I think the reason is that escape routes have changed since the Oligocene and that the initial routes employed largely followed sutures across which offset markers are essentially non-existent. Most of the currently active strike-slip faults in South and Southeast Asia are very young faults, with modest offsets, exactly as Burchfiel and Royden (1991b) have emphasized. But their observations say nothing about earlier escape.

This superior strength of the continental lithosphere must be the reason why orogens do not spread under their own weight like a piece of taffy to any appreciable extent, despite frequent recent claims to the contrary. Burchfiel et al. (1992) have recently documented a very real and spectacular extensional collapse under purely gravitational effects from the Himalaya, but noted that "In our opinion, however, the primary trigger for this extensional collapse of the orogen was a major reduction in crustal strength due to melting of hanging wall rocks within the lower part of the upper plate.....Thus the extensional stresses available to drive extension were generated by the difference in topographic elevation between India and Tibet, but the immediate cause of the extensional collapse was local crustal weakening via melting in the midcrust" (Burchfiel et al., 1992, p. 39). Extensional orogenic collapse seems a lump designation for a diverse set of processes that extensionally disrupt orogens, mostly under the influence of plate boundary processes. As such, it is a misleading and uninformative "all-inclusive" term, similar to the equally uninformative "terrane," and is perhaps best avoided.

My conclusion as a geologist is thus the exact opposite of S. Warren Carey's (1960) presented to the New York Academy of Sciences more than three decades ago. There he complained that geophysicists had been ascribing the crust too much strength. I here maintain the opposite for the current trend. I expect that the too high Argand number now fashionable among geophysicists will soon be disputed widely by geologists working outside the high-grade metamorphic terrains.

If lithospheric, crustal and even intracrustal wedges and slivers are commonly being shoved around by colliding strong promontories, it is hardly surprising that many collisional orogens have a highly contorted map view. The strength of the material making up orogens seems in general enough to keep them from flowing into hopelessly deformed lumps of continental material with simple topography, but bewilderingly complicated internal structure underlying giant plateaux, except in their highly metamorphic, partially molten cores, where even intracrustal convection appears to take place as suggested by very large, polyphase ductile strains, extensive migmatization, and widespread "hair granites" of the Himalayan-type. Comparison of the presently active Mediterranean realm with the dead Altaid edifice, in which a tectonic history as complicated as the Mediterranean can be read with some confidence, strengthens my conviction that continental crustal deformation may be better approximated by dividing it into a finite number of rigid blocks surrounded by commonly relatively narrow mushy zones of deformation[6]. But, the shape and the number of these blocks change fairly rapidly with time (<10 million years), but fortunately not in time intervals smaller than the resolution of graptolite, ammonite or nannoplankton zones! This gives the geologist the power to be able to work out sequentialy these deformations. Combined with comparative tectonics and geophysical methods that enable us to measure accurately sizes, times, and movements, the geologist's traditional methods form an immensely powerful arsenal to attack problems of lithospheric deformation and its history.

In the preceding paragraphs, my frequent reference to Suess and Argand has had more reason than sheer pedantry. Their incredible success in figuring out what had gone on in the Mediterranean by the "old-fashioned" geological tools and reasoning is most instructive. Suess and Argand had no palaeomagnetism, no reflection seismology, no fault-plane solutions to work with. Neither did they have isotopic age dating of any availability. But the kinematics they erected for the Mediterranean deformations, both in general and in detail, stand today, except for timing, which eventualy came out of ocean floors. All the rest was done with a topographic map, a hammer, a compass, a magnifying glass, a notebook, a pencil - and an enormous reservoir of knowledge of the geology of the world (plus their very considerable brains)! If we are to understand the evolution of the world's most civilised sea and its frame of majestic mountains, I

---

[6] It is an interesting observation that "Alpine-type" (sensu Şengör, 1991) orogens commonly have much more sinuous trend lines than either the "Himalayan" or well-developed "Andean" type orogens. This may be because "Alpine-type" orogens have a colder thermal regime than the "Himalayan" or the "Andean" types and thus resist crustal flow more easily. Orogens that are hot flow easily and tend to smooth out the protuberances of colliding promontories by localised lateral flow rather than by tectonic escape of large and coherent lithospheric and/or crustal chunks.

Thus, the degree of contortion in trend lines of an orogen may be one hint to their thermal history!

firmly believe that it is to more geologizing we need to turn.

Acknowledgements

This paper is an outgrowth of my regional tectonic studies of the last two decades. I am indebted to a very large number of organisations and people who assisted me in my work. The thoughts I elaborated above have developed in conversations and discussions especially with Dan P. McKenzie, John F. Dewey, Kevin Burke, Warren B. Hamilton, Philip England, James Jackson, Kenneth J. Hsü, B. Clark Burchfiel, L. H. Royden, Stephen H. Hall, Walter C. Pitman, III, Paul Tapponnier, Peter H. Molnar, Laurent Jolivet, John Platt, İhsan Ketin, and Fuat Şaroğlu. Reinoud Visser's outstanding work on the Betic tectonics, and his kind generosity with his time, data, and ideas combined with a rare open-mindedness, have made it possible for me to understand the Neogene evolution of this complex orogen, although my conclusions are not necessarily parallel with his. This paper would not have originated without Prof. Enzo Mantovani's invitation to speak at the Eriche School and without the extremely generous freedom granted to me by the İstanbul Technical University. In Eriche I enjoyed also the great hospitality of Prof. Enzo Boschi and Dr. Andrea Morelli

References

Allen, M. B. (1990) Tectonics and Magmatism of Western Junggar and the Tien Shan range, Xinjiang province, NW China, unpub. Ph.D. Thesis, Univ. Leicester, England.
Amodio-Morelli, L., Bonardi, G., Colonna, V., Dietrich, D., Giunta, G., Ippolito, F., Liguori, V., Lorenzoni, S., Paglionico, A., Perrone, V., Piccarreta, G., Russo, M., Scandone, P., Zanettin-Lorenzoni, E. and Zuppetta, A. (1979) 'L'Arco Calabro-Peloritano nell'orogene Appenninico-Magrebide', Mem. Soc. Geol. It., 17, 1-60.
Anderson, H. and Jackson, J. (1987) 'Active tectonics of the Adriatic region', Geophys. J. R. astr. Soc., 91, 937-983.
Andrieux, J., Fontbote, J.-M. and Mattauer, M. (1971) 'Sur un modèle explicatif de l'arc de Gibraltar', Earth Planet. Sci. Lett., 12, 191-198.
Anonymous (1977) Geodynamics in South-West Pacific, International Symposium, Noumea - Nouvelle-Calédonie, Editions Technip, Paris.
Anonymous (1981) Studies in East Asian Tectonics and Resources (Seatar), CCOP/TP.7a, Committee for Co-ordination of Joint Prospecting for Mineral Resources in Asian Offshore Areas (CCOP), Intergovernmental Oceanographic Commission, UNESCO, Second Edition, CCOP Project Office, technical support for Regional Offshore Prospecting in East Asia - RAS/80/003, Bangkok, Thailand.
Aplonov, S., Hsü, K. J. and Ustritsky, V. (1992) 'Relict back-arc basins of Eurasia and their hydrocarbon potentials', The Island Arc, 1, 71-77.
Argand, E. (1911) 'Les nappes de recouvrement des Alpes Pennines et leurs prolongements structuraux', Matér. carte géol. Suisse, n.s., XXXIe livr., 1-25 and 3 pl.
Argand, E. (1920) 'Plissements précurseurs et plissements tardifs des chaînes de montagnes', Actes Soc. helv. Scie. Nat., 101(pt.2), 13-39.
Argand, E. (1924) 'La tectonique de l'Asie', Congr. géol. int., Belgique (13e Sess.), 1, 171-372.
Arıç, K. (1981) 'Deutung krustenseismischer und seismologischer Ergebnisse im Zusammenhang mit der Tektonik des Alpenostrandes', Sitzber. Österr. Akad. Wiss., Math.-Naturwiss. Kl., Abt. I, 190, 235-312.
Barker, P. F. and Dalziel, I. W. D. (1983) 'Progress in geodynamics in the Scotia arc region', in S. J. Ramón Cabré (ed.), Geodynamics of the Eastern Pacific Region, Caribbean and Scotia Arcs,

Geodynamics Series, Am. Geophys. Union and Geol. Soc. America, 9, pp. 137-170
Bazhenov, M. L. and Burtman, V. S. (1986) 'Tectonics and paleomagnetism of structural arcs of the Pamir-Punjab Syntaxis', J. Geodyn., 5, 583-596.
Bemmelen, R. W. van (1952) 'Gravity field and orogenesis in the West-Mediterranean region' Geologie en Mijnbouw, 14, 297-306.
Biju-Duval, B., Dercourt, J. and Le Pichon, X. (1977) 'From the Tethys ocean to the Mediterranean seas: A plate tectonic model of the evolution of the western Alpine System', in B. Biju-Duval and L. Montadert (eds.) International Symposium on the Structural History of the Mediterranean Basins, Split (Yugoslavia), 25-29 October 1976, Editions Technip, Paris, pp. 143-164.
Bucher, W. H. (1956) 'Role of gravity in orogenesis', Geol. Soc. America Bull., 67, 1295-1318.
Buforn, E. and Udías, A. (1991) 'Focal mechanism of earthquakes in the Gulf of Cadiz, South Spain and Alboran Sea', in J. Mezcua and A. Udías (eds.), Seismicity, Seismotectonics and Seismic Risk of the Ibero-Maghrebian Region, Monografia Nº 8, Instituto Geográfico Nacional, Madrid, pp. 29-40.
Burchfiel, B. C. (1980) 'East European Alpine System and the Carpathian orocline as an example of collision tectonics', Tectonophysics, 63, 31-61.
Burchfiel, B. C. and Royden, L. H. (1991a) 'Antler orogeny: A Mediterranean-type orogeny', Geology, 19, 66-69.
Burchfiel, B. C. and Royden, L. H. (1991b) 'Tectonics of Asia 50 years after the death of Emile Argand', Eclog. Geol. Helvet., 84, 599-629.
Burchfiel, B. C., Chen, Z. L., Hodges, K. V., Liu, Y. P., Royden, L. H., Deng, C. R. and Xu J. N. (1992) 'The South Tibetan Detachment System, Himalayan Orogen: Extension contemporaneous with and parallel to shortening in a collisional mountain belt', Geol. Soc. America Spec. Pap., 269, 41pp.
Burke, K., Cooper, C., Dewey, J. F., Mann, P. and Pindell, J. L. (1984) 'Caribbean tectonics and relative plate motions', Geol. Soc. America, Mem. 162, 31-63.
Burke, K. and Şengör, A. M. C. (1986) 'Tectonic escape in the evolution of the continental crust', in M. Barazangi, (ed.), Reflection Seismology: The Continental Crust, Geodynamics Series, Am. Geophys. Union and Geol. Soc. America, 14, pp. 41-53.
Carey, S. W. (1960) 'The strength of the Earth's crust', Trans. New York Acad. Sci., ser. II, 22, 303-312.
Channel, J. E. T., D'Argenio, B. and Horváth, F. (1979) 'Adria, the African promontory, in Mesozoic Mediterranean palaeogeography', Earth Sci. Rev., 15, 213-292.
Channel, J. E. T. and Mareschal, J. C. (1989) 'Delamination and asymmetric lithospheric thickening in the development of the Tyrrhenian rift', in M. P. Coward, D. Dietrich and R. G. Park (eds.), Alpine Tectonics, Geol. Soc. London Spec. Pub. No. 45, 285-302.
Cloos, H. (1928) Bau und Bewegung der Gebirge in Nordamerika, Skandinavien und Mitteleuropa, Gebrüder Borntraeger, Berlin.
Coli, M., Peccerillo, A. and Principi, G. (1991) 'Evoluzione geodinamica Recente dell'Appennino settentrionale e attivita' magmatica Tosco-Laziale: Vincoli e problemi', in Studi Preliminari all' Acquisizione Dati del Profilo Punta Ala-Gabicce, Università degli Studi di Camerino, Dipartimento di Scienze della Terra, Studi Geologici Camerti, volume speciale 1991/1, pp. 403-412.
Cohen, C. R. (1980) 'Plate tectonic model for the Oligo-Miocene evolution of the Western Mediterranean', Tectonophysics, 68, 283-311.
Cohen, C. R., Schweickert, R. A. and Odom, A. L. (1981) 'Age of emplacement of the Schistes Lustres Nappe, Alpine Corsica', Tectonophysics, 73, 267-283.
Crough, S. T. (1983) 'Hotspot swells', Ann. Rev. Earth. Planet. Sci., 11, 165-193.
Darot, M. (1974) 'Cinématique de l'extrusion, à partir du manteau, des péridotites de la Sierra Bermeja (Serrania de Ronda, Espagne)', C. R. Acad. Sc. Paris, sér. D, 278, 1673-1676.
Dercourt J., Zonenshain, L. P., Ricou, L.-E., Kazmin, V. G., Le Pichon, X., Knipper, A. L.,

Grandjaquet, C., Sborschikov, I. M., Geyssant, J., Levrier, C., Pechersky, D. K., Boulin, J., Sibuet, J.-C., Savostin, L. A., Westphal, M., Bazhenov, M. L., Lauer, J. P. and Biju-Duval, B. (1986) 'Geologic evolution of the Tethys belt from the Atlantic to the Pamirs since the Lias' Tectonophysics, 123, 241-315.
Dewey, J. F. (1980) 'Episodicity, sequence and style at convergent plate boundaries', Geol. Assoc. Canada Spec. Pap., 20, 553-573.
Dewey, J. F. (1988) 'Extensional collapse of orogens', Tectonics, 7, 1123-1139.
Dewey, J. F., Pitman, W. C., III, Ryan, W. B. F. and Bonnin, J. (1973) 'Plate tectonics and the evolution of the Alpine System', Geol. Soc. America Bull., 84, 3137-3180.
Dewey, J. F. and Burke, K. (1974) 'Hot spots and continental break-up: Implications for collisional orogeny' Geology, 2, 57-60.
Dewey, J. F. and Şengör, A. M. C. (1979) 'Aegean and surrounding regions: complex multiplate and continuum tectonics in a convergent zone', Geol. Soc. America Bull., pt I, 90, 84-92.
Dewey, J. F., Helman, M. L., Turco, E., Hutton, D. H. W. and Knott, S. D. (1989a) 'Kinematics of the western Mediterranean' in M. P. Coward, D. Dietrich and R. G. Park (eds.), Alpine Tectonics, Geol. Soc. London Spec. Pub. No. 45, 265-283.
Dewey, J. F., Cande, S. and Pitman, W. C., III, (1989b) 'Tectonic evolution of the India/Eurasia Collision Zone', Eclog. Geol. Helvet., 82, 717-734.
Doblas, M. and Oyarzun, R. (1989a) 'Neogene extensional collapse in the western Mediterranean (Betic-Rif Alpine orogenic belt): Implications for the genesis of the Gibraltar Arc and magmatic activity', Geology, 17, 430-433.
Doblas, M. and Oyarzun, R. (1989b) '"Mantle core complexes" and Neogene extensional detachment tectonics in the western Betic Cordilleras, Spain: an alternative model for the emplacement of the Ronda peridotite', Earth Planet. Sci. Lett., 93, 76-84.
Doutch, H. F., Packham, G. H., Rinehart, W. A., Simkin, T., Siebert, L., Moore, G. W., Golovchenko, X., Larson, R. L. and Pitman, W. C., III (1981) Plate tectonic map of the Circum Pacific Region, Southwest Quadrant, scale 1:10,000,000, American Association of Petroleum Geologists, Tulsa, Oklahoma
England, P. C. and Houseman, G. A. (1988) 'The mechanics of the Tibetan Plateau' Phil. Trans. R. Soc. London, A326, 301-320.
England, P. C. and McKenzie, D. P. (1982) 'A thin viscous sheet model for continental deformation', Geophys. J. R. astr. Soc., 70, 295-321.
England, P. C. and McKenzie, D. P. (1983) 'Correction to: a thin viscous sheet model for continental deformation', Geophys. J. R. astr. Soc., 73, 523-532.
Enkin, R., Yang, Z. Y., Chen, Y. and Courtillot, V. (in press) 'Paleomagnetic costraints on the geodynamic history of the major blocks of China from Permian to Present' Jour. Geophys. Res.
Ercan, T. and Türkecan, A. (1985) 'Batı Anadolu - Ege Adaları - Yunanistan ve Bulgaristan'daki plütonların gözden geçirilişi', in T. Ercan and M. A. Çağlayan (eds.), Ketin Simpozyumu, Türkiye Jeoloji Kurumu, Ankara, pp. 189-208.
Finetti, I. (1982) 'Structure, stratigraphy and evolution of Central Mediterranean', Boll. Geof. Teor. App., 24, 247-315.
Flügel, H. W. and Faupl, P. (eds.) (1987) Geodynamics of the Eastern Alps, Franz Deuticke, Vienna.
Foucher, J. P., Mauffret, A., Steckler, M., Brunet, M. F., Maillard, A., Rehault, J. P., Alonso, B., Desegaulx, P., Murillas, J. and Ouillon, G. (1992) 'Heat flow in the Valencia Trough: geodynamic implications', Earth Planet. Sci. Lett., 203, 77-97.
Fournier, M., Jolivet, L. and Goffé, B. (1991) 'Alpine Corsica metamorphic core complex', Tectonics, 10, 1173-1186.
Gallo, D. G., Kidd, W. S. F., Sloan, H. S. and Şengör, A. M. C. (1980) 'Large angular rotations of blocks along strike-slip zones as shallow décollement features' EOS, 61, 1120.
Goffé, B., Michard, A., Garcia-Dueñas, V., Gonzalez-Lodeiro, F., Monié, P., Campos, J., Galindo-

Zaldivar, J., Jabaloy, A., Martinez-Martinez, J. M. and Simancas, J. F. (1989) 'First evidence of high-pressure, low-temperature metamorphism in the Alpujárride nappes, Betic Cordilleras (S.E. Spain)' Eur. J. Mineral., 1, 139-142.
Hamilton, W. B. (1979) 'Tectonics of the Indonesian Region', U. S. Geol. Surv. Prof. Pap., 1078, 1-345.
Hatcher, R. D., Jr. (1987) 'Tectonics of the Southern and Central Appalachian internides', Ann. Rev. Earth Planet. Sci. , 15, 337-362.
Hayes, D. E. (1983) 'Global studies of age-depth relationships', abs., EOS, 64, 760.
Heezen, B., Gray, C., Segre, A. G. and Zarudski, E. F. K. (1971) 'Evidence of foundered continental crust beneath the central Tyrrhenian Sea', Nature, 229, 327-329.
Hempton, M. R. (1985) 'Structure and deformation history of the Bitlis suture near Lake Hazar, southeastern Turkey', Geol. Soc. America Bull., 96, 233-243.
Hilde, T.W. C. and Uyeda S. (eds.) (1983) Geodynamics of the Western Pacific-Indonesian Region, Geodynamics Series, Am. Geophys. Union and Geol. Soc. America, 11, Washington, D. C.
Horváth, F. (1988) 'Neotectonic behaviour of the Alpine-Mediterranean region', in L. H. Royden and F. Horváth (eds.), The Pannonian Basin, A Study in basin Evolution, Am. Assoc. Petrol. Geol. Mem. 45, 49-55.
Houseman, G. A., McKenzie, D. P. and Molnar, P. (1981) 'Convective instability of a thickened boundary layer and its relevance for the thermal evolution of continental convergence belts', Jour. Geophys. Res., 86, 6115-6132.
Hsü, K. J. (1977) 'Tectonic evolution of the Mediterranean basins', in A. E. M. Nairn, W. H. Kanes and F. G. Stehli (eds.), The Ocean basins and Margins, 4A, The Eastern Mediterranean, Plenum Press, New York, pp. 29-75.
Hsü, K. J. (1989) 'Time and place in Alpine orogenesis - the Fermor Lecture', in M. P. Coward, D. Dietrich and R. G. Park (eds.), Alpine Tectonics, Geol. Soc. London Spec. Pub. No. 45, 421-443.
Hsü, K. J., Sun, S., Chen, H.H., Pen, H.P. and Şengör, A. M. C. (1988) 'Mesozoic overthrust tectonics in south China,' Geology, 16, 418-421.
Hsü, K.J. Li J.L., Chen H. H., Wang Q. C, Sun S. and Şengör, A. M. C. (1990) 'Tectonics of South China: Key to understanding of west Pacific geology', Tectonophysics, 183, 9-39.
Hutchison, C. S. (1989) Geological Evolution of South-East Asia, Clarendon, Oxford.
Jackson, J. and McKenzie, D. P. (1984) 'Active tectonics of the Alpine-Himalayan belt between western Turkey and Pakistan', Geophys. J. R. astr. Soc., 77, 185-264.
Jarrard, R. D. (1986a) 'Relations among subduction parameters', Rev. Geophys., 24, 217-284.
Jarrard, R. D. (1986b) 'Causes of compression and extension behind trenches', Tectonophysics, 132, 89-102.
Jolivet, L., Daniel, J.-M. and Fournier, M. (1991) 'Geometry and kinematics of extension in Alpine Corsica', Earth Planet. Sci. Lett., 104, 278-291.
Jolivet, L., Dubois, R., Fournier, M., Goffé, B., Michard, A. and Jourdan, C. (1990) 'Ductile extension in alpine Corsica', Geology, 18, 1007-1010.
Kastens, K., Mascle, J., Auroux, C., Bonatti, E., Broglia, C., Channel, J., Curzi, P., Emesisi, K.-C., Glaçon, G., Hasegawa, S., Hieke, W., Mascle, G., McCoy, F., McKenzie, J., Mendelson, J., Müller, C., Réhault, J.-P., Robertson, A., Sartori, R., Sprovieri, R. and Torii, M., (1988) 'ODP Leg 107 in the Tyrrhenian Sea: Insights into passive margin and back-arc evolution', Geol. Soc. America Bull., 100, 1140-1156.
Kastens, K. and Mascle, J. (1990) 'The geological evolution of he Tyrrhenian Sea: An introduction to the scientific results of ODP Leg 107', in K. A. Kastens and J. Mascle et. al., Proceedings of the Ocean Drilling Program, Scientific Results, 107, pp. 3-26.
Ketin, İ. (1948) 'Über die tektonisch-mechanischen Folgerungen aus den grossen anatolischen Erdbeben des letzten Dezenniums', Geol. Rundsch., 36, 77-83.
Kidd, W. S. F. and Molnar, P. (1988) 'Quaternary and active faulting observed on the 1985 Academia Sinica-Royal Society geotraverse of Tibet', Phil. Trans. R. Soc. London, A327, 337-363.

Kincaid, C. and Olson, P. (1987) 'An experimental study of subduction and slab migration', Jour. Geophys. Res., 92, 13,832-13,840.
Kober, L. (1921) Der Bau der Erde, Gebrüder Borntraeger, Berlin
Kozur, H. (1991a) 'The evolution of the Meliata-Hallstatt ocean and its significance for the early evolution of the Eastern Alps and Western Carpathians', Palaeo 3, 87, 109-135.
Kozur, H. (1991b) 'The geological evolution at the western end of the Cimmerian ocean in the Western Carpathians and the Eastern Alps', Zbl. Geol. Paläont. Teil I, H.1, 99-121.
Leg 107 Shipboard Scientific Party (1986) 'A microcosm of ocean basin evolution in the Mediterranean', Nature, 321, 383-384.
Le Pichon, X. and Angelier, J. (1981) 'The Aegean Sea', Phil. Trans. R. Soc. London, 300A, 357-372.
Lugeon, M. and Argand, E. (1906) 'La racine de la nappe sicilienne et l'arc de charriage de la Calabre', C. R. Acad. Sc. Paris, 142, 1107-1109.
Lee, C.-S. and McCabe, R. (1987) 'The Banda-Celebes-Sulu basin: a trapped piece of Cretaceous-Eocene oceanic crust?' Nature, 322, 51-54.
Lukowski, P., Wernli, R. and Poisson, A. (1988) 'Mise en évidence de l'importance des dépôts messiniens dans le bassin Miocène de Fortuna (Prov. de Murcia, Espagne)', C. R. Acad. Sc. Paris, sér. II, 307, 941-947.
Matte, P. (1986) 'Tectonics and plate tectonics model for the Variscan belt of Europe', Tectonophysics, 126, 329-374.
McKenzie, D. P. (1970) 'Plate tectonics of the Mediterranean region', Nature, 226, 239-243.
McKenzie, D. P. (1972) 'Active tectonics of the Mediterranean region', Geophys. J. R. astr. Soc., 30, 109-185.
McKenzie, D. P. (1977) 'The initiation of trenches: a finite amplitude instability' in M. Talwani and W. C. Pitman, III, (eds.), Island Arcs, Deep Sea Trenches and Back-Arc Basins, Maurice Ewing Series, 1, Am. Geophys. Un., pp. 57-61.
McKenzie, D. P. (1978) 'Active tectonics of the Alpine-Himalayan belt: the Aegean Sea and surrounding regions' Geophys. J. R. astr. Soc., 55, 217-257.
McKenzie, D. P. and Jackson, J. (1983) 'The relationship between strain rates, crustal thickening, palaeomagnetism, finite strain and fault movements within a deforming zone', Earth Planet. Sci. Lett., 65, 182-202.
Miyashiro, A. (1973) 'The Troodos ophiolitic complex was probably formed in an island arc', Earth Planet. Sci. Lett., 19, 218-224.
Molnar, P. and Atwater, T. (1978) 'Interarc spreading and Cordilleran tectonics as alternates related to the age of subducted oceanic lithosphere', Earth Planet. Sci. Lett., 41, 330-340.
Molnar, P. and Lyon-Caën, H. (1989) 'Some simple physical aspects of the support, structure and evolution of mountain belts', in S. P. Clark, Jr., B. C. Burchfiel and J. Suppe (eds.), Processes in Continental Lithospheric Deformation (Rodgers Symposium Volume), Geol. Soc. America Spec. Pap. 218, 178-207.
Molnar, P. and Tapponnier, P. (1975) 'Cenozoic tectonics of Asia: effects of a continental collision', Science, 189, 419-426.
Morelli, C. (1975) 'Geophysics of the Mediterranean', Bulletin n°7 de l'Étude en commun de la Méditerranée, Monaco, pp. 27-111.
Morelli, C., Pisani, M. and Gantar, G. (1975) 'Geophysical studies in the Aegean Sea and in the Eastern Mediterranean', Boll. Geophys. Teor. Appl., 17, 127-168.
Mountrakis, D. (1985) 'The Pelagonian zone in Greece: A polyphase deformed fragment of the Cimmerian continent and its role in the geotectonic evolution of the Eastern Mediterranean' Jour. Geol., 94, 335-347.
Mountrakis, D., Patras, D., Kilias, A., Pavlides, S. and Spyropoulos, N. (1987) 'Structural geology of the internal Hellenides and their role to the geotectonic evolution of the Eastern Mediterraneean' Acta-Nat. "L'Ateneo Parmense", 23, 147-161.

Okay, A. İ. and Özgül, N. (1984) 'HP/LT metamorphism and the structure of the Alanya Masisf, southern Turkey: an allochthonous composite tectonic sheet' in J. E. Dixon and A. H. F. Robertson (eds.), The Geological Evolution of the Eastern Mediterranean, Geol. Soc. London, Spec. Pub. 17, pp. 429-439.
Palfreyman, W. D., Doutch, H. F., Nozawa, T., Craddock, C., McCoy, F. W., Swint-Iki, T.R., Richards, P. W. and Moore, G. W. (1988) Geologic Map of the Circum-Pacific Region, Southwest Quadrant, scale 1:10,000,000, American Association of Petroleum Geologists, Tulsa, Oklahoma.
Parsons, B. and Sclater, J. G. (1977) 'An analysis of the variation of ocean floor bathymetry and heat flow with age', Jour. Geophys. Res., 82, 803-827.
Patacca, E., Sartori, R. and Scandone, P. (in press) 'Tyrrhenian Basin and Apenninic arcs: Kinematic relations since the Tortonian times', Mem. Soc. Geol. It., 45
Pearce, J. A., Lippard, S. J. and Roberts, S. (1984) 'Characteristics and tectonic significance of supra-subduction zone ophiolites', in B. P. Kokelaar and M. F. Howells (eds.) Marginal Basin Geology, Geol. Soc. london, Spec. Pub. 16, pp. 77-94.
Perinçek, D. (1991) 'Possible strand of the North Anatolian fault in the Thrace Basin, Turkey - An interpretation', Am. Ass. Petrol. Geol. Bull., 75, 241-257.
Pindell, J. L., Cande, S. C., Pitman, W. C. III, Rowley, D. B., Dewey, J. F., LaBrecque, J. and Haxby, W. (1988) 'A plate kinematic framework for models of Caribbean evolution', Tectonophysics, 155, 121-138.
Pindell, J.L. and Barrett, S. F. (1990) 'Geological evolution of the Caribbean region; A plate-tectonic perspective' in The Geology of North America, Vol. H, The Caribbean Region, Geol. Soc. America, Boulder, pp. 405-432.
Platt, J. P., Behrmann, J. H., Martínez, J.-M. M. and Vissers, R. L. M. (1984) 'A zone of mylonite and related ductile deformation beneath the Alpujarride nappe complex, Betic Cordilleras, S. Spain', Geol. Rundsch., 73, 773-785.
Platt, J. P. and Vissers, R. L. M. (1989) 'Extensional collapse of a thickened continental lithosphere: a working hypothesis for the Alboran sea and Gibraltar Arc' Geology, 17, 540-543.
Rakús, M., Dercourt, J. and Nairn, A. E. M. (eds.) (1990) Evolution of the Northern Margin of Tethys, vol. III, Mém. Soc. Géol. France, Paris, Nouvelle Série No. 154 (III, pt. 1-2).
Reuber, I., Michard, A., Chalouan, A., Juteau, T., Jermoumi, B. (1982) 'Structure and emplacement of the Alpine-type peridotites from Beni Bousera, Rif, Morocco: A polyphase tectonic interpretation', Tectonophysics, 82, 231-251.
Roberts, D. and Gee, D. G. (1985) 'An introduction to the structure of the Scandinavian Caledonides' in D. G. Gee and B. A. Sturt (eds.), The Caledonide Orogen - Scandinavia and Related Areas, John Wiley & Sons, Chichester, pp. 55-68.
Robertson, A. H. F., Clift, P. D., Degnan, P. J. and Jones, G. (1991) 'Palaeogeographic and palaeotectonic evolution of the Eastern Mediterranean Neotethys', Palaeo3, 87, 289-343.
Rögl, F. and Steininger, F. F. (1983) 'Vom Zerfall der Tethys zu Mediterran und Paratethys', Ann. Naturhist. Mus. Wien, 85/A, 135-163.
Royden, L. H. (1988a) 'Flexural behavior of the continental lithosphere in Italy: Constraints imposed by gravity and deflection data', Jour. Geophys. Res., 93, 7747-7766.
Royden, L. H. (1988b) 'Late Cenozoic tectonics of the Pannonian basin system', in L. H. Royden and F. Horváth (eds.), The Pannonian Basin, A Study in basin Evolution, Am. Assoc. Petrol. Geol. Mem. 45, 27-48.
Royden, L. H. and Báldi, T. (1988) 'Early Cenozoic tectonics and paaleogeography of the Pannonian and surrounding regions', in L. H. Royden and F. Horváth (eds.), The Pannonian Basin, A Study in basin Evolution, Am. Assoc. Petrol. Geol. Mem. 45, 1-16.
Royden, L. H., Horváth, F. and Burchfiel, B. C. (1982) 'Transform faulting, extension and subduction in the Carpathian-Pannonian region', Geol. Soc. America Bull., 73, 717-725.
Sandulescu, M. (1975) 'Essai de synthèse structurale des Carpates', Bull. soc. géol. France, sér. 7, 17, 1266-1274.

Şengör, A. M. C. (1979) 'The North Anatolian Transform Fault: its age, offset and tectonic significance', Jour. Geol. Soc. London, 136, 269-282.

Şengör, A. M. C. (1982) 'Ege'nin neotektonik evrimini yöneten etkenler', in O. Erol and V. Oygür (eds.), Batı Anadolu'nun Genç Tektoniği ve Volkanizması Paneli, Türk. Jeol. Kur., Ankara,pp. 59-71.

Şengör, A. M. C. (1984) 'The Cimmeride orogenic system and the tectonics of Eurasia', Geol. Soc. America Spec. Paper 195, xi+82.

Şengör, A. M. C. (1985a) 'The Story of Tethys: How many wives did Okeanos have?' Episodes, 8, p.3-12.

Şengör, A. M. C. (1985b) 'Die Alpiden und die Kimmeriden: Die verdoppelte Geschichte der Tethys', 74, 181-213.

Şengör, A. M. C. (1986) 'The dual nature of the Alpine-Himalayan System: Progress, problems, and prospects', Tectonophysics, 127, 177-195.

Şengör, A. M. C. (1987a) 'Tectonics of the Tethysides: Orogenic collage development in a collisional setting:', Ann. Rev. Earth Planet Sci., 15, 213-244.

Şengör, A. M. C. (1987b) 'Cross-faults and differential stretching of hangingwalls in regions of low-angle normal faulting: examples form western Turkey', in M. P. Coward, J. F. Dewey and P. L. Hancock (eds.), Continental Extensional Tectonics, Geol. Soc. London Spec. Pub. 28, pp. 575-589.

Şengör, A. M. C. (1990) 'Plate tectonics and orogenic research after 25 years: A Tethyan perspective', Earth Sci. Rev., 27, 1-201.

Şengör, A. M. C. (1991) 'Orogenic architecture as a guide to size of ocean lost in collisional mountain belts', Bull. Tech. Univ. Istanbul (Ketin Festschrift), 44, 43-74.

Şengör, A. M. C. (1992) 'The Palaeo-Tethyan suture: A line of demarcation between two fundamentally different architectural styles in the structure of Asia', The Island Arc, 1, 78-91.

Şengör, A. M. C., Altıner, D., Cin, A., Ustaömer, T. and Hsü, K. J. (1988) 'Origin and assembly of the Tethyside orogenic collage at the expense of Gondwana-Land', in M. G. Audley-Charles and A. Hallam (eds.), Gondwana and Tethys, Geol. Soc. London Spec. Pub. 37, 119-181.

Şengör, A. M. C., Cin, A., Rowley, D. B. and Nie S. Y. (in press) 'Space-time patterns of magmatism along the Tethysides: a preliminary study' Jour. Geol.

Şengör, A. M. C. and Dewey, J. F. (1985) 'Post-Oligocene tectonic evolution of the Aegean and neighbouring regions: relations to the North Anatolian Transform Fault' in E. İzdar and E. Nakoman (eds.), Sixth Colloquium on Geology of the Aegean Region, PİRİ REİS Int. Cont. Ser. Pub. 2, pp. 639-646.

Şengör, A. M. C., Görür, N. and Şaroğlu, F. (1985) 'Strike-slip faulting and related basin formation in zones of tectonic escape: Turkey as a case study', in K. T. Biddle and N. Christie-Blick (eds.), Strike-slip Deformation, Basin Formation, and Sedimentation, Soc. Econ. Paleont. Min. Spec. Pub. 37 (in honor of J.C. Crowell), pp. 227-264.

Şengör, A. M. C. and Monod, O. (1980) 'Océans sialiques et collisions continentales' C. R. Acad. Sc. Paris, sér. D., 290, 375-386.

Şengör, A. M. C. and Okuroğulları, A. H. (1992) 'The rôle of accretionary wedges in the growth of continents: Asiatic examples from Argand to plate tectonics', Eclog. Geol. Helvet., 84, 535-597.

Şengör, A. M. C., Yılmaz, Y. and Sungurlu, O. (1984) 'Tectonics of the Mediterranean Cimmerides: nature and evolution of the western termination of Palaeo-Tethys', in J.E. Dixon and A.H.F. Robertson (eds.), Geological Evolution of the Eastern Mediterranean, Geol. Soc. London Spec. Pub., 17, pp. 77-112.

Seyitoğlu, G. and Scott, B. C. (1991) 'Late Cenozoic crustal extension and basin formation in west Turkey', Geol. Mag., 128, 155-166.

Seyitoğlu, G., Scott, B. C. and Rundle, C. C. (1992) 'Timing of Cenozoic extensional tectonics in west Turkey', Jour. Geol. Soc. London, 149, 533-538.

Sideris, C. (1989) 'Late Paleozoic in Greece', Geol. práce, 88, 191-202.

Sleep, N. H. (1987) 'Lithospheric heating by mantle plumes', Geophys. J. R. astr. soc., 91, 1-11.
Smith, A. G. and Woodcock, N. H. (1982) 'Tectonic syntheses of the Alpine-Mediterranean region: A review', in H. Berckhemer and K. J. Hsü, (eds.), Alpine-Mediterranean Geodynamics, Geodynamics Series, Am. Geophys. Union and Geol. Soc. America, 7, 15-38.
Spakman, W. (1986) 'Subduction beneath Eurasia in connection with the Mesozoic Tethys', Geologie en Mijnbouw, 65, 145-153.
Spakman, W. (1990) 'Tomographic images of the upper mantle below central Europe and the Mediterranean', Terra Nova, 2, 542-553.
Spakman, W. (1991) 'Delay-time tomography of the upper mantle below Europe, the Mediterranean, and Asia Minor', Geophys. J. Int., 107, 309-332.
Spakman, W., Wortel, M. J. R. and Vlaar, N. J. (1988) 'The Hellenic subduction zone: A tomographic image and its geodynamic implications' Geophys. Res. Lett., 15, 60-63.
Stampfli, G., Marcoux, J. and Baud, A. (1991) 'Tethyan margins in space and time', Palaeo3, 87, 373-409.
Steininger, F. F., Senes, J., Kleemann, K. and Rögl, F. (eds.) (1985) Neogene of the Mediterranean tethys and Paratethys, Stratigraphic Correlation Tables and Sediment Distribution Maps, vols. 1 & 2 (IGCP 25), Institut für Paläontologie, Univ. Wien, Wien.
Suess, E. (1875) Die Entstehung der Alpen, W. Braumüller, Wien.
Suess, E. (1878) Die Heilquellen Böhmens, A. Hölder, Wien.
Suess, E. (1883) Das Antlitz der Erde, 1a, Freytag, Leipzig.
Suess, E., (1885) Das Antlitz der Erde, 1b, Freytag, Leipzig.
Tapponnier, P., Peltzer, G. and Armijo, R. (1986) 'On the mechanics of the collision between India and Asia', in M. P. Coward and Ries, A. C. (eds.), Collision Tectonics, Geol. Soc. London, Spec. Pub. 19, pp. 115-157.
Taylor, B and Karner, G. D. (1983) 'On the evolution of marginal basins', Rev. Geophys. Space Phys., 21, 1727-1741.
Taymaz, T., Jackson, J. and McKenzie, D. (1991) 'Active tectonics of the north and central Aegean Sea', Geophys. J. Int., 106, 433-490.
Tchalenko, J. S. (1970) 'Similarities between shear zones of different magnitudes' Geol. Soc. America Bull., 81, 1625-1640.
Termier, P. (1903) 'Les nappes des Alpes orientales et la synthèse des Alpes', Bull. Soc. géol. France, sér. 4, 3, 711-765.
Trümpy, R. (1977) 'The Engadine Line: a sinistral wrench fault in the Central Alps' Mem. Geol. Soc. China, 2 (Bing volume), 1-12.
Trümpy, R. (1980) Geology of Switzerland, A Guide-Book, Part A. An Outline of the Geology of Switzerland, Wepf & Co., Basel.
Tubía, J. M. and Cuevas, J. (1987) 'Structures et cinématique liées à la mise en place des péridotites de Ronda (Cordillères Bétiques, Espagne)', Geodinamica Acta (Paris), 1, 59-69.
Uyeda, S and Kanamori, H. (1979) 'Back-arc opening and the mode of subduction', Jour. Geophys. Res., 84, 1049-1061.
Vissers, R. L. M. (1977) 'Deformation of pre-Alpine age in the Nevado-Filabride complex of the central Sierra de los Filabres, SE Spain: Macroscopic and microtextural evidence', Proc. Kon. Nederl. Akad. Wetensch., Amsterdam, Ser. B, 80, 302-311.
Vörös, A. (1988) 'Conclusions on Brachiopoda', in M. Rakús, J. Dercourt and A. E. M. Nairn (eds.), Evolution of the Northern Margin of Tethys, vol. I, Mém. Soc. Géol. France, Nouvelle Série No. 154, pp. 79-83.
Wildi, W. (1983) 'La chaîne tello-rifaine (Algérie, Maroc, Tunisie): Structure, stratigraphie et évolution du Trias au Miocène', Rev. Géol. Dyn. Géogr. Phys., 24, 201-297.
Wilcox, R. E., Harding, T. P. and Seely, D. R. (1973) 'Basic Wrench Tectonics', Bull. Am. Asso. Petrol. Geol., 57, 74-96.
Wilson, G. (1960) 'The Tectonics of the 'Great Ice Chasm', Filchner Ice Shelf, Antarctica' Proc.

Geol. Ass., 71, pt 2, 130-138.
Wilson, J. T. (1965) 'A new class of faults and their bearing on continental drift', Nature, 207, 343-347.
Wortel, M. J. R. and Spakman, W. (1992) 'Structure and dynamics of subducted lithosphere in the Mediterranean region', Proc. Kon. Akad. v. Wetensch., 95, 325-347.
Yılmaz, Y. (in press) 'New evidence and model on the evolution of the southeast Anatolian orogen', Geol. Soc. America Bull.
Zonenshain, L. P., Kuzmin, M. I. and Natapov. L. M. (1991) Geology of the USSR: A Plate Tectonic Approach, Geodynamics Series, 21, American Geophysical Union and The Geological Society of America, Washington, D. C.

RATES OF ACTIVE DEFORMATION IN THE EASTERN MEDITERRANEAN

JAMES JACKSON
Bullard Laboratories
Madingley Road
Cambridge CB3 0EZ
United Kingdom

ABSTRACT. The active tectonics of the eastern Mediterranean is dominated by the relative motions between the major African, Arabian and Eurasian plates and the smaller, relatively aseismic, blocks of central Turkey and the southern Aegean. In the next decade the relative motions of these aseismic regions will be measured directly by space-based geodetic techniques. We have some estimates already of what we expect these motions to be. This article summarizes those estimates, and the arguments on which they are based.

1. Introduction

Since the early days of Plate Tectonics it has been clear that the active deformation of the eastern Mediterranean is too complex to be described by simple narrow boundaries between the African, Arabian and Eurasian plates (McKenzie, 1972). The deformation is distributed and varied, but encloses some relatively flat and aseismic regions such as central Iran, central Turkey and the southern Aegean Sea, which are probably *relatively* rigid today. A first step towards understanding the tectonics of the eastern Mediterranean is to know the relative motions between these aseismic regions and the major plates; although the relative motions of the aseismic blocks do not predict in detail the deformation within the wide seismic belts that separate them (see e.g., England and Jackson, 1989).

Until very recently, our estimates of the rates and styles of deformation in the eastern Mediterranean have come from arguments based on seismicity, geology and geomorphology. We are now entering an era when we can measure the motions directly with space-based geodetic techniques. This paper summarizes briefly the motions we expect these new techniques to be able to see, as this will help plan the campaigns of geodetic observation and measurement that will eventually reveal how good the non-geodetic estimates were. The arguments have changed little from the early pioneering work of McKenzie (1972), though some of the seismological observations are much improved.

The newly revised plate global plate model NUVEL-1 (DeMets et al., 1990) predicts significantly different overall motions between Eurasia and Africa-Arabia from the earlier models PO71 (Chase, 1978) or RM2 (Minster and Jordan, 1978). A comparison between NUVEL-1 and PO71 is shown in Figure 1. Although the rates are not much changed in the eastern Mediterranean, the NUVEL-1 convergence directions are up to 20° anticlockwise of the earlier models in the Caucasus and Iran, and 15° anticlockwise in Italy and Greece. Particularly in the east, where the strike of the deforming zone in the Caucasus and eastern Turkey is roughly WNW-ESE, the convergence direction across the whole orogenic belt is more oblique, with a greater right-lateral strike-slip component, than was previously thought.

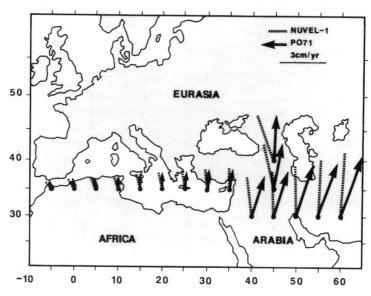

Figure 1. NUVEL-1 (DeMets et al., 1990) and PO71 (Chase, 1978) velocities between Africa and Eurasia and between Arabia and Eurasia in the Mediterranean region.

2. The Caucasus and Turkey-Iran border region

The NUVEL-1 convergence direction between Arabia and Eurasia in the Caucasus region is about 338°: significantly oblique to the topographic trend of the region, which is about 300°. Earthquake focal mechanisms show a dominance of right-lateral strike-slip faulting in the Turkey-Iran border region with an average slip vector of about 307°, and a dominance of thrust faulting with an average slip vector of about 017° in the Greater Caucasus to the north (Figure 2, and Jackson, 1992). The hypothesis that the oblique convergence is largely partitioned into right-lateral strike-slip in the south and almost perpendicular shortening in the north is consistent with most of the earthquake focal mechanisms (but not all: the faulting in the lesser Caucasus is more varied and complicated) and the observation that the elevation, and hence probable crustal thickness, is much greater in the high Caucasus than in the Turkey-Iran border region (Jackson, 1992). The expected rates of strike-slip and shortening are shown in Table 1, together with the rates accounted for by earthquakes in the period 1911-1991. It appears that much of the shortening in the Caucasus occurs by aseismic creep or folding, while faulting in earthquakes probably accounts for most of the strike-slip motion. The configuration of partitioning between strike-slip and shortening in this region is not stable, and must evolve with time (see Jackson, 1992).

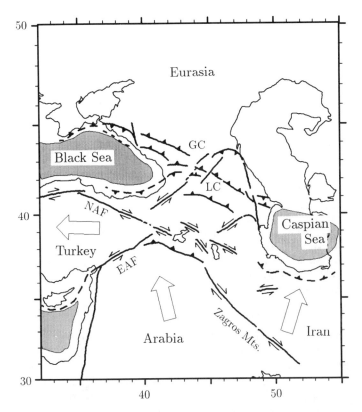

Figure 2. Summary tectonic map of the eastern Turkey and Caucasus regions. Large open arrows are the approximate directions of motion of Turkey, Arabia and central Iran relative to Eurasia. GC: Greater Caucasus; LC: Lesser Caucasus; NAF: North Anatolian Fault; EAF: East Anatolian Fault.

	Eastern Turkey and NW Iran Right-Lateral Strike-Slip Parallel to Direction 120°, mm/yr		
	minimum	probable	maximum
Estimated from earthquakes	8.7	20.4	48.2
Predicted from NUVEL-1		21.9	

	Caucasus Shortening in direction 030°, mm/yr		
	minimum	probable	maximum
Estimated from earthquakes	1.6	3.4	7.8
Predicted from NUVEL-1		17.1	

Table 1. Velocities in the Caucasus and Turkey-Iran border region estimated from the hypothesis that the predicted NUVEL-1 oblique convergence between Arabia and Eurasia is partitioned and from the moment tensors of earthquakes in the period 1911-1991 using the global moment-magnitude relation of Ekström and Dziewonski (1988); see Jackson (1992).

3. The North and East Anatolian fault zones

The central Anatolian plateau is a relatively flat and aseismic region, bounded by the North Anatolian fault zone on its northern side and by the East Anatolian fault zone to the SE. The right-lateral character of the North Anatolian fault zone is beyond doubt, since it ruptured virtually its entire length in a series of large earthquakes this century (Ketin, 1948; Ambraseys, 1970). The East Anatolian fault zone is more diffuse, containing a variety of active faults and other structures (Lyberis et al., 1992). The historical record reveals several large earthquakes in this fault zone, even though the seismicity of this century has been low (Ambraseys, 1989; Figure 3). The few moderate-sized earthquakes in the period 1964–86 show a variety of focal mechanisms, but all of them have slip vectors directed roughly 063°, parallel to the strike of the zone (Taymaz et al., 1991a). There seems little doubt that the zone accommodates principally left-lateral strike-slip motion, as envisaged by McKenzie (1976), though it may also involve some shortening.

McKenzie (1972) estimated the motions on the North and East Anatolian fault zones by constructing a velocity triangle for the motions between Arabia, Eurasia and Turkey at 41°E. This is redrawn in Figure 4(a) using the revised NUVEL-1 motion for Eurasia-Arabia. It assumes that the shortening between Eurasia and Arabia west of 41°E is accommodated purely by strike-slip motion on the North and East Anatolian fault zones, and hence by the westward motion of Turkey relative to Eurasia (Figure 5). This argument yields slip rates across the North and East Anatolian fault zones of ∼38 mm/yr and ∼29 mm/yr (Table 2), which are likely to be upper bounds, as some shortening may occur between the two strike-slip faults. This analysis is simplistic but is the basis all estimates of deformation rates farther west in the Aegean. As yet, there is little substantial evidence to test these rates. The seismicity between 1909–81 accounts for ~39 mm/yr on the North Anatolian fault zone (Jackson and McKenzie, 1988a), but may not be representative of longer periods. Preliminary results from GPS surveys are consistent with the predicted senses of motion on these fault zones, but the rates are not yet well resolved (Oral et al., 1991, 1992a).

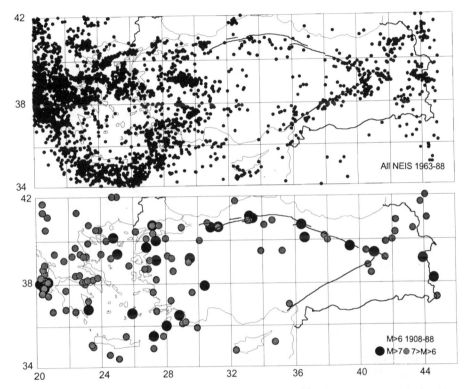

Figure 3. Seismicity of Turkey and the Aegean Sea. The top map shows all events from the National Earthquake Information Service (NEIS) in the period 1963–1988. The bottom map shows the earthquakes of $M_s \geq 6$ in the period 1908–1988.

North Anatolian Fault (mm/yr)				
	minimum	probable	maximum	
Kinematic (Figure 4a)	31	38	48	(upper bound)
Seismicity 1909–1981	25	39	80	(representative?)
GPS 1988–90 (western part)	18	28	38	(Oral et al. 1991)
GPS 1989–91 (eastern part)	12	19	26	(Oral et al. 1992a)

East Anatolian Fault (mm/yr)				
	minimum	probable	maximum	
Kinematic (Figure 4a)	25	29	39	(upper bound)
Seismicity 1909–1981		?		(no earthquakes!)
GPS 1989-91	5	10	15	(Oral et al. 1992a)

Table 2. Estimated velocities on the North and East Anatolian fault zones, from the velocity triangle in Figure 4(a), from the seismicity between 1909–91, and from preliminary GPS results.

4. The Aegean Sea

McKenzie (1972) realized that the westward motion of Turkey relative to Eurasia was ultimately accommodated in the Hellenic trench system, though the North Anatolian fault system does not cross the northern Aegean Sea in the simple manner indicated by Figure 5, as the seismicity in Figure 3 shows. The deformation in the north Aegean region is distributed and involves east-west shortening, north-south extension and crustal thinning.

The seismicity of the north-central Aegean Sea is dominated by right-lateral strike-slip motion on a series of sub-parallel faults that strike NE to ENE and are distributed over a N-S distance of about 200–300 km. These faults end abruptly near the eastern coast of mainland Greece, where the deformation is taken up by a series of NW- to WNW-striking normal faults that dominate the active tectonics of central Greece. These normal faults are thought to rotate clockwise about a vertical axis as they move. Summaries of these observations are given by Kissel and Laj (1988), Roberts and Jackson (1991), and Taymaz et al. (1991b).

The first estimates of the present day rates of motion in the Aegean region were made by McKenzie (1972). He observed that the southern Aegean Sea was relatively aseismic (Figure 3), and thought that this block was moving SW relative to Eurasia, because of the distributed right-lateral strike-slip faults with a NE-SW strike in the northern Aegean. He also observed that the distributed deformation in SW Turkey, which separates the Anatolian plateau from the southern Aegean block, contained mostly normal faults striking E-W. He therefore reasoned that the motion of the southern Aegean relative to Turkey can not have an easterly component, or some E-W shortening would occur in SW Turkey. These arguments led to the construction of a velocity triangle, which is produced in Figure 4(b), updated with the revised Turkey-Eurasia motion from Taymaz et al. (1991a), and the revised average slip vector in the northern Aegean (047°) from Taymaz et al. (1991b). The predicted velocity of the southern Aegean block relative to Eurasia is 52 mm/yr in the direction 227°. Le Pichon and Angelier (1979) also estimated an extension rate in the Aegean using different arguments based on finite motions. Their rate of ∼35 mm/yr is averaged over ∼13 Ma, and is probably a lower bound (see Jackson and McKenzie, 1988b).

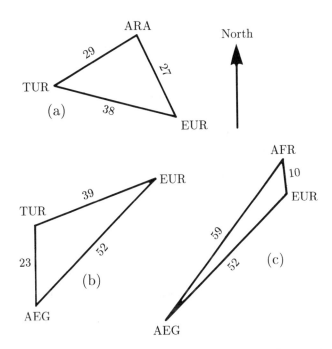

Figure 4. Velocity triangles used to estimate the velocities of aseismic blocks in (a) eastern Turkey, (b) the southern Aegean, (c) the Hellenic trench. Thick lines use NUVEL-1 and updated seismological data. ARA: Arabia; EUR: Eurasia; TUR: Turkey; AEG: southern Aegean; AFR: Africa. Velocities are in mm/yr.

Much of the motion between the southern Aegean and Eurasia appears to be accommodated seismically. Jackson and McKenzie (1988a, 1988b) and Ekström and England (1989) examined how much motion could be accommodated by earthquakes of $M_s \geq 6.0$ in the period 1909–1981. They used the method of Kostrov (1974) to obtain the average extension rate across the region. Their results depend on an empirical relation between seismic moment (M_0) and surface wave magnitude (M_s), the precise nature of which is the main source of uncertainty. The $M_s - M_0$ relation in the Aegean appears to be different from the global average relation. The seismic extension rate in Table 3 is obtained from the $M_s - M_0$ relation that gives the smallest seismic moment, and hence that rate is a lower bound. Jackson et al. (1992) used earthquakes of $M_s \geq 6.0$ in the period 1909–1983 to estimate the spatial variations in seismic strain rates, from which they calculated a horizontal velocity field for the deforming Aegean region using the technique of Holt et al. (1991). Their velocity field shows the southern Aegean moving SW relative to Europe at ~30 mm/yr (Figure 6), which is also a lower bound.

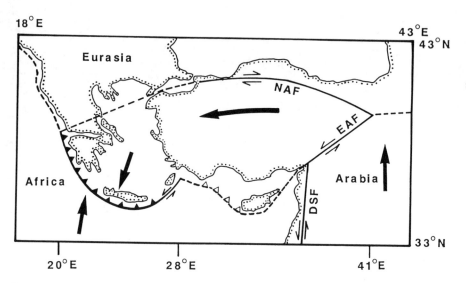

Figure 5. Sketch of the motions in eastern Turkey and the Aegean, showing how the westward motion of Turkey relative to Eurasia is accommodated by strike-slip faulting on the North and East Anatolian fault zones and by shortening in the Hellenic trench. Black arrows are the approximate directions of motion of Arabia, Turkey, the southern Aegean and Africa relative to Eurasia.

Taymaz et al. (1991b) proposed a simple block model consisting of two sets of rotating fault blocks (slats) in an attempt to describe the kinematics of faulting in the north Aegean region. The inputs to their model are the slip rate on the North Anatolian fault zone, the rotation rate of the western margin of Greece, the observed fault strikes, and the dimensions of the zone. Their model does not include the observed moment release in earthquakes, but they show that most of the motion required by their model can be accommodated by the earthquakes in the period 1909–1981. Their model shows a motion of the central Aegean towards the SW at ~38 mm/yr.

All the analyses above are consistent with extension rates of around 40–50 mm/yr, with the southern Aegean moving SW relative to Europe. It is too early yet for geodetic measurements to unequivocally refine or correct these estimates. A re-measurement in 1988 of a triangulation network installed around 1890–1900 in central Greece found an extension rate of 10–20 mm/yr, but this covers only a portion of the deforming region (Billiris et al., 1991). A modern GPS survey found western Turkey moving at 50 ± 20 mm/yr relative to Eurasia, but covers only a short time interval from 1988–1990 (Oral et al., 1992b). Results

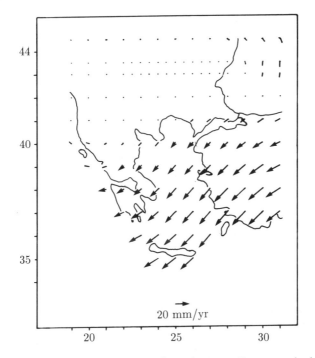

Figure 6. The velocity field in the Aegean Sea relative to Europe, calculated by Jackson et al. (1992). The scale is given at the bottom.

Extension in the Aegean Sea (mm/yr)				
	minimum	*probable*	*maximum*	
Kinematic (Figure 4b)		52		(upper bound)
Seismicity 1909–1981	18	29	40	(lower bound)
Velocity field from earthquakes 1909–83		30		(lower bound)
Broken slat model		38		(Taymaz *et al.* 1991b)
Geodetic (GPS, SLR)		30–50?		(but not yet clear)

Table 3. Estimated extension rates across the Aegean Sea.

from satellite laser ranging (SLR) are starting to look encouraging, but involve only three epochs of measurement in Greece and two in Turkey. Geodesists are understandably reluctant to over-interpret initial results. However, it is clear even from preliminary analyses in

Greece that extension rates across the Aegean Sea are at least 30–50 mm/yr, and that the extension parallel and adjacent to the Hellenic trench, emphasized by Armijo et al. (1992), is also significant (Sellers et al., 1991; Ambrosius et al., 1991).

5. The Hellenic Trench

The extension in the Aegean Sea must be accommodated by shortening in the Hellenic trench system, since there is no doubt that Africa and Eurasia are converging at this longitude. The shortening rate in the trench system is easily estimated (Figure 4(c)), but depends almost entirely on the estimated extension rate in the Aegean, since this is so much greater than the 10 mm/yr expected convergence rate between Africa and Eurasia. It is difficult to escape a present-day shortening rate in the trench of less than 50–60 mm/yr, though the length of time over which this rate has applied is not clear. The present extension rates in the Aegean are probably applicable to about the last 5 Ma (Kissel and Laj, 1988; Mercier et al., 1992). The present shortening rate of ∼60 mm/yr needs only 5 Ma to produce a subducted slab 300 km long, which is about the length of the seismically active slab today. Yet the seismic tips of subducted slabs elsewhere generally take about ∼10 Ma in the mantle to heat up sufficiently to become aseismic (Wortel, 1986). This poses one of the fundamental questions of eastern Mediterranean tectonics: is the presently subducting slab in the Hellenic Trench no older than 5 Ma, or is it much older, as some tomographic studies suggest (Spakman et al., 1988)? It is clear that the seismicity in the Hellenic Trench is able to account for only a small fraction of the present-day shortening rate (Table 4), and that much of the deformation must be aseismic (Jackson and McKenzie, 1988a, 1988b). There are not yet any direct geodetic measurements of shortening between the Hellenic arc and Africa.

	Shortening in the Hellenic Trench (mm/yr)			
	minimum	*probable*	*maximum*	
Kinematic (Figure 4c)		59		(upper bound)
Seismicity 1909–1981	2	6	14	(mostly aseismic)

Table 4. Shortening velocities in the Hellenic Trench.

Acknowledgements.

I thank Dan McKenzie for producing Figure 3. This is Cambridge Earth Sciences contribution number 3050.

Ambraseys, N. 1970. Some characteristic features of the North Anatolian fault zone. *Tectonophys.*, **9**, 143-165.

Ambraseys, N., 1989. Temporary seismic quiescence: SE Turkey. *Geophys. J. Int.*, **96**, 311-331.

Ambrosius, B.A.C., Noomen, R. and Wakker, K.F. 1991. European station coordinates and motions from LAGEOS SLR observations, presented at the Crustal Dynamics Project Principal Investigators Meeting, Greenbelt, MD, Oct. 22-24, 1991.

Armijo, R., Lyon-Caen, H. and Papanastassiou, D. 1992. East-west extension and Holocene normal-fault scarps in the Hellenic arc. *Geology*, **20**, 491-494.

Billiris, H., Paradissis, D., Veis, G., England, P., Featherstone, W., Parsons, B., Cross, P., Rands, P., Rayson, M., Sellers, P., Ashkenazi, V., Davison, M., Jackson, J. and Ambraseys, N., 1991. Geodetic determination of tectonic deformation in central Greece from 1900 to 1988. *Nature*, **350**, 124-129.

Chase, C.G., 1978. Plate kinematics: the Americas, east Africa, and the rest of the world. *Earth Planet. Sci. Lett.*, **37**, 355-368.

DeMets, C., Gordon, R.G., Argus, D.F. and Stein, S. 1990. Current plate motions. *Geophys. J. Int.*, **101**, 425-478.

Ekström, G. and Dziewonski, A.M., 1988. Evidence of bias in the estimation of earthquake size. *Nature*, **332**, 319- 323.

Ekström , G. and England, P.C., 1989. Seismic strain rates in regions of distributed continental deformation. *J. geophys. Res.*, **94**, 10, 231-10,257.

England, P.C. and Jackson, J.A., 1989. Active deformation of the continents, *Ann. Rev. Earth Planet. Sci.*, **17**, 197-226.

Holt, W.E., Ni, J.F, Wallace, T.C. and Haines, A.J., 1991a. The active tectonics of the eastern Himalayan syntaxis and surrounding regions, *J. geophys. Res.*, **96**, 14,595-14,632.

Jackson, J.A., 1992. Partitioning of strike-slip and convergent motion between Eurasia and Arabia in eastern Turkey and the Caucasus. *J. geophys. Res.*, **97**, 12,471-12,479.

Jackson, J.A. and McKenzie, D., 1988a. The relationship between plate motions and seismic moment tensors, and the rates of active deformation in the Mediterranean and Middle East, *Geophys. J.*, **93**, 45-73.

Jackson, J.A. and McKenzie, D., 1988b. Rates of active deformation in the Aegean Sea and surrounding regions. *Basin Res.*, **1**, 121-128.

Jackson, J.A., Haines A.J. and Holt, W.E., 1992. Determination of the horizontal velocity field in the deforming Aegean Sea region from the moment tensors of earthquakes, *J. geophys. Res.*, **97**, 17,657-17,684.

Ketin, I., 1948. Uber die tektonischmechanischen Folgerungen aus den grossen anatolischen Erdbeben des letzten Dezenniums. *Geol. Rundsch.*, **36**, 77-83.

Kissel, C. and Laj, C., 1988. The Tertiary geodynamical evolution of the Aegean arc: a paleomagnetic reconstruction, *Tectonophys.*, **146**, 183-201.

Kostrov, V.V., 1974. Seismic moment and energy of earthquakes, and seismic flow of rock. *Izv. Acad. Sci. USSR Phys. Solid Earth*, **1**, 23–44.

Le Pichon, X. and Angelier, J., 1979. The Hellenic arc and trench system: a key to the evolution of the eastern Mediterranean area. *Tectonophys.*, **60**, 1–42.

Lybéris, N., Yürür, T., Chorowicz, J., Kasapoğlu, E., and Gündoğdu, N., 1992. The East Anatolian Fault: an oblique collisional belt. *Tectonophys.*, **204**, 1–15.

McKenzie, D., 1972. Active tectonics of the Mediterranean region. *Geophys. J. R. astr. Soc.*, **30**, 109–185.

McKenzie, D.P., 1976. The East Anatolian Fault: a major structure in eastern Turkey. *Earth Planet. Sci. Lett.*, **29**, 189–193.

Mercier, J.L., Vergeley, P., Simeakis, C., Kissel, C. and Laj, C., 1992. The continuation of the North Anatolian dextral strike-slip fault into the oblique fault zone of the North Aegean Trough (W. Turkey and N. Greece): timing, tectonic regimes, fault kinematics and rotations, *Tectonics*, (in review).

Minster, J.B. and Jordan, T.H., 1978. Present-day plate motions. *J. geophys. Res.*, **83**, 5331–5354.

Oral., M.B., Reilinger, R.E., Toksöz, M.N., Barka, A.A., and Kınık, I., 1991. Preliminary results of 1988 and 1990 GPS measurements in western Turkey (abstract). *EOS, Fall AGU meeting supplement*, **72**, 115.

Oral., M.B., Reilinger, R.E., and Toksöz, M.N., 1992a. Deformation of the Anatolian block as deduced from GPS measurements (abstract). *EOS, Fall AGU meeting supplement*, **73**, 120.

Oral., M.B., Reilinger, R.E., and Toksöz, M.N., 1992b. Preliminary results of 1988 and 1990 GPS measurements in western Turkey and their tectonic implications. In: *Crustal Dynamics Project*, AGU monograph, (in press).

Roberts, S. and Jackson, J.A., 1991. Active normal faulting in central Greece: an overview. In: *The geometry of normal faults*, edited by Roberts, A.M., Yielding, G. and Freeman, B., *Spec. Publ. Geol. Soc. Lond.*, **56**, 125–142.

Sellers, P.C, Rands, P.N. and Cross, P., 1991. Crustal deformation in Europe and the Mediterranean determined by the analysis of Satellite Laser Ranging Data, presented at the Crustal Dynamics Project Principal Investigators Meeting, Greenbelt, MD, Oct. 22-24, 1991.

Spakman, W., Wortel, M.J.R., and Vlaar, N.S., 1988. The Hellenic subduction zone: a tomographic image and its geodynamical implications. *Geophys. Res. Letts.*, **15**, 60–63.

Taymaz, T., Eyidoğan, H. and Jackson, J.A., 1991a. Source parameters of large earthquakes in the East Anatolian Fault zone (Turkey), *Geophys. J. Int.*, **106**, 537–550.

Taymaz, T., Jackson, J. and McKenzie, D., 1991b. Active tectonics of the north and central Aegean Sea, *Geophys. J. Int.*, **106**, 433–490.

Wortel, R., 1986. Deep earthquakes and the thermal assimilation of subducted lithosphere. *Geophys. Res. Letts.*, **13**, 34–37.

POST-TORTONIAN DEFORMATION PATTERN IN THE CENTRAL MEDITERRANEAN: A RESULT OF EXTRUSION TECTONICS DRIVEN BY THE AFRICA-EURASIA CONVERGENCE.

E. MANTOVANI, D. ALBARELLO, D. BABBUCCI, C. TAMBURELLI
University of Siena
Dept. of Earth Sciences
Via Banchi di Sotto, 55
53100 Siena - Italy

ABSTRACT. The tectonic activity which has occurred in the Central Mediterranean since the late Tortonian is explained as a result of the Africa-Eurasia convergence roughly along a SSW-NNE direction. This convergence has been first accommodated by a considerable reduction of the Adriatic foreland, through the consumption of its eastern and western margins, and then by the lateral escapes of crustal wedges, accompanied by crustal thickening, in the zone comprised between the Adriatic and African forelands. The lateral escapes of the Calabria and Sicily blocks, towards SE and NW respectively, have been allowed by the presence, at the sides of the most strongly compressed zone, of poorly constrained boundaries, corresponding to the thinned Ionian foreland and, to the zone of crustal stretching in the Tyrrhenian basin. This interpretative scheme allows physically plausible explanations of a considerable amount of geological, geophysical and volcanological evidence in the framework of relatively simple and coherent tectonic mechanisms.

1. Premise

A deterministic approach of some problems which may have important social implications, such as the estimate of seismic hazard or middle-long term prediction of earthquakes, can be attempted only if a reasonably good knowledge of the tectonic process which are taking place in the zone under consideration is available.

This knowledge can be hardly derived from present-day deformation estimates, such as those which can be obtained by seismological investigations or geodetic measurements. Strain rates are too slow with respect to our observation periods and, furthermore, the deformations which can be directly observed are only the surface expression of a more complex process which involves a much thicker zone of the Earth. The chances to recognize the ongoing tectonic processes increases considerably if the set of geological, volcanological and geophysical information allows a reliable reconstruction of the recent geodynamic evolution. This work describes our attempt in the above direction.

2. Introduction

The post-Tortonian evolution of the central Mediterranean region (Fig.1) has been characterized by several major deformation events:
- Around the late Tortonian, the Giudicarie fault system of Southern Alps was activated as a left lateral transpressional fault and since then intense compressional deformations have only occurred in the eastern part of the Southern Alps (see, e.g., Semenza, 1974; Castellarin and Vai, 1986).
- Around the late Tortonian, the stress field in the zone lying between the Corsica-Sardinia block and the Adriatic foreland, which was mainly occupied by Alpine and Apenninic thrust belts, underwent a drastic change, passing from a compressional to an extensional regime, which caused crustal stretching with E-W extensional trend. This led to the formation of the Tyrrhenian basin lying North of the Selli line (see Fig.1).
- A continental collision took place on the Adriatic-Balkan border (Outer Hellenides) around the late Miocene (Mercier et al., 1979, 1989).
- From roughly the Messinian to the middle-upper Pliocene, the Apenninic belt experienced an overall migration of 100-200 km towards the Adriatic foreland (see, e.g., Di Nocera et al., 1976; Ortolani, 1979; Sartori, 1989; Patacca and Scandone, 1989). This migration was associated with compressional deformations and accretion along the

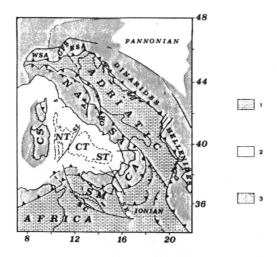

Figure 1. Main structural-tectonic elements in the central Mediterranean region 1) African-Adriatic domain 2) European domain 3) Main deformation belts. The dashed lines in the Tyrrhenian contour the zones of crustal stretching. WSA=Western part of Southern Alps, ESA=Eastern part of Southern Alps, GFS=Giudicarie fault system, NA=Northern Apennines, SA=Southern Apennines, AA=Ancona-Anzio line; OR=Ortona-Roccamonfina line, SL=Selli line, CA=Calabrian Arc, SM=Sicily microplate, SE=Siracusa escarpment, SC=Sicily Channel, CS=Corsica-Sardinia microplate, NT,CT,ST=Northwestern, central and southern parts of the Tyrrhenian basin.

external front of the chain and by extensional tectonics in the internal, Tyrrhenian, margin. In the Southern Apennines, this phenomenon was much more intense than in the northern part of the belt.
- Simultaneously, a large part of the Adriatic-Ionian foreland was consumed, through a downward flexure, beneath the outward extruding Apenninic belt (Casnedi et al., 1982; Patacca and Scandone, 1989; Patacca et al., 1990).
- From the Messinian to the middle-upper Pliocene, intense crustal stretching, with E-W extensional trend, took place in the central part of the Tyrrhenian zone, the present Magnaghi-Vavilov basin (see, e.g., Rehault et al., 1987; Sartori, 1989).
- Important deformations started in the zones surrounding Sicily around the late Messinian-early Pliocene. Extensional tectonics occurred in the Siracusa escarpment (Carbone et al., 1982; Grasso and Lentini, 1982; Sartori et al., 1992). A number of grabens began to develop in the Sicily Channel, in the framework of a transcurrent stress regime (Finetti, 1984; Boccaletti et al., 1987; Cello, 1987; Jongsma et al., 1987; Reuther, 1987, 1990).
- Around the middle-upper Pliocene, the outward migration of the Southern Apennines and the crustal stretching in the central Tyrrhenian bathyal plain underwent a considerable slowdown (see, e.g., Ortolani and Aprile, 1977; Ciaranfi et al., 1983; Sartori, 1989; Patacca et al., 1990). Since then the Calabrian block experienced a fast SEward drifting, at the expense of the thinned Ionian lithosphere and crustal stretching took place in the southernmost Tyrrhenian, i.e. the present Marsili basin, with exposure of oceanic crust (Finetti and Del Ben, 1986; Kastens et al., 1988; Sartori, 1989).
- Since the Pleistocene, a fast uplifting, with maximum rates of about 1,5 mm/y occurred in the Eastern Sicily, Calabria, Southern Apennines and the adjacent foredeep (Ghisetti and Vezzani, 1982; Ciaranfi et al., 1983).
- In the Quaternary a renewal of shortenings took place in the Outer Hellenides (Mercier et al., 1979).
- The Corsica-Sardinia block has not been affected by any appreciable compressional deformation or lateral migration after the Tortonian.

The above mentioned tectonic events have never been explained all together in the framework of a coherent geodynamic evolutionary model. A detailed discussion about the major outstanding problems of earlier attempts is given by Mantovani et al. (1992) and is partly resumed in the last section of this work.

Here we suggest that all the major features listed above can be coherently interpreted as consequences of a succession of shortening mechanisms and block readjustments, driven by the convergence between Africa and Eurasia.

3. Proposed evolutionary model

It is assumed that the overall driving force of post-Tortonian deformations in the Central Mediterranean has been the convergence between Africa and Eurasia along a SSW-NNE direction. The arguments supporting this hypothesis are described by Mantovani et al. (1992) and Albarello et al. (this volume). The block movements mentioned in the text refer to a fixed Eurasia plate, unless different indications are given.

Paleomagnetic observations in the Mediterranean area have not been taken into account to reconstruct the positions of intermediate plates, as we are afraid that this kind of information may be affected by significant uncertainty (see, e.g., Marton, 1987 and this volume). The proposed evolution of the central Mediterranean region is schematically illustrated by five paleogeographic maps (Fig.2 A,B,C,D,E) which cover the time span from the upper Tortonian, prior to the Tyrrhenian extension, up to the present. The evolutionary model is divided in three main phases: late Tortonian-early Messinian, middle Messinian-middle Pliocene, upper Pliocene-present. The beginning of each phase is determined by a key tectonic event which produces a significant change of the shortening mechanism and related deformation pattern.

Magmatic and geological evidence suggests that large portions of the Adriatic-Ionian lithosphere have been consumed since the upper Miocene (see, e.g., Patacca et al., 1990). This implies that extended sunk lithospheric edifices have been present in the last 10 My beneath the zones here considered and that they might have significantly influenced the evolution and kinematics of shallow structures. As a consequence, we think that an attempt at reconstructing the evolution of the Central Mediterranean cannot neglect the above problem and should at least provide some plausible hypotheses on how the presence of deep lithospheric roots may be reconciled with the proposed shallow deformation patterns.

In order to illustrate the speculative considerations given in the text about this connection, a perspective sketch of the sunk lithosphere is reported at the bottom of each paleogeographic map.

3.1 TORTONIAN (PRE-TYRRHENIAN)

Fig.2A shows the presumed structural-tectonic setting in the Central Mediterranean zone after the opening of the Balearic basin and before the extensional phase in the Tyrrhenian region.

The African/Adriatic promontory was much larger than at present. The present shape is shown for reference in each phase by the dark brown area. The width of the Africa-Adriatic foreland which has been successively consumed has been tentatively chosen on the basis of shortening estimates across the Alps, the Dinarides, the Hellenides, the Apennines and the Maghrebides (see, e.g., Mercier et al., 1979, 1989; Burchfiel, 1980; Laubscher, 1983; Ghisetti and Vezzani, 1984; Horvath, 1984; Castellarin and Vai, 1986; Philip, 1987; Catalano et al., 1989; Patacca and Scandone, 1989; Sartori, 1989; Schmid et al., 1989; Patacca et al., 1990).

During this phase, the relative approach between the Adriatic promontory (driven by Africa) and Eurasia was mostly absorbed by the consumption of the eastern Adriatic margin beneath the Dinarides and Hellenides (Mercier et al., 1979, 1989; Burchfiel, 1980) and by shortening processes in the Alps (Laubscher, 1983; Castellarin, 1984).

The northwestern protuberance of the Adriatic promontory was deeply indented into the Eurasian domain (see Fig.3) after the main continental collision in the Western Alps (see, e.g., Semenza, 1974; Channell and Horvath, 1976; Laubscher, 1983). Due to this embedding, the northwestern Adriatic edge was most probably characterized by a very low mobility and, consequently, it could have represented a sort of hinge zone for a counterclockwise rotation of the Adriatic block.

Figure 2 A,B,C,D,E. Proposed evolutionary scheme of the central Mediterranean area since the Tortonian time. 1,2)African-Adriatic foreland, the light tone indicates the zones which will be consumed during the successive evolution, bricks identify the continental parts of the present foreland area 3)Paleozoic crystalline basement in the Southern Alps 4)Deformed European margin 5)Corsica-Sardinia microplate 6)Deformation belts 7)Calabria 8,9)Zones affected by extension, the dark tone indicates intense stretching with exposure of oceanic crust 10)Subduction related magmatism 11)Main trends of compressional deformations 12)External front of deformation belts 13)Main tensional features 14)Main trascurrent or transpressional fault systems. The red arrows tentatively indicate the major motion trends with respect to Eurasia. The movements of Africa are compatible with an Africa-Eurasia rotation pole located at 41.6° N 11.8° W, offshore the Northern Portugal (see Albarello et al., this volume). Present geographic contours and grid (thin black lines) are reported for reference. Dotted geographic contours indicate the presumed positions of some significant foreland zones during each evolutionary phase. The speculative considerations which are reported in the text about deep tectonics are tentatively illustrated by perspective views of the subducted lithosphere. The accretionary belts which built up around the Adriatic plate during the period considered are represented by imbrication nappes.

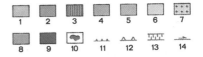

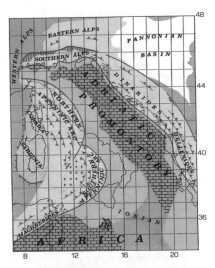

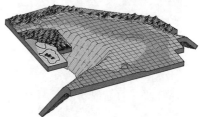

2A. Upper Tortonian

AA=Ancona-Anzio line, AE=Apulian escarpment, AL=Abruzzi-Lazio units, AM=Aeolian magmatic arc, AP=Adventure plateau, B=Basilicata region, C=Campania region, CA=Catanzaro fault system, EG=Egadi fault, GA=Gargano-Apulia zone, GF=Giudicarie fault system, GN=Gela nappes, IB=Iblean zone, KE=Kefallinia line, M=Molise region, MB=Marsili basin, MD = Medina seamounts, MV=Magnaghi-Vavilov basin, OR=Ortona-Roccamonfina line, P=Padanian region, PE=Peloritani block, SC=Sicily Channel, SE=Siracusa escarpment, SL=Selli line, SM=Sicily microplate, SV=Schio-Vicenza line, TG=Taranto gulf, TL=Taormina line, TW=Tauern window, VU=Vulcano fault.

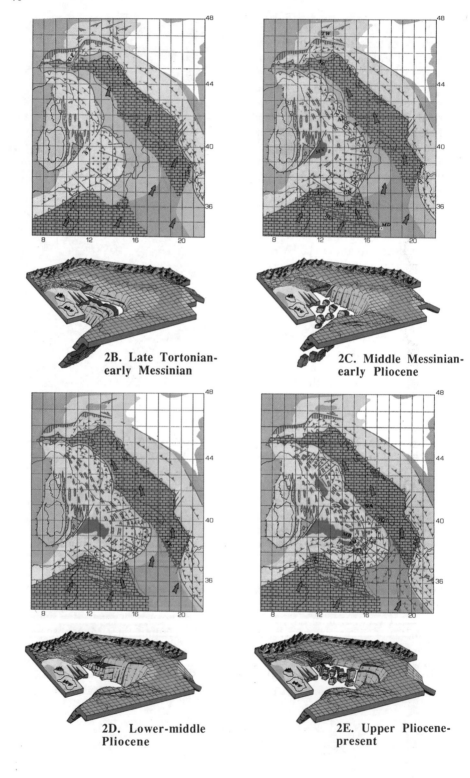

2B. Late Tortonian-early Messinian

2C. Middle Messinian-early Pliocene

2D. Lower-middle Pliocene

2E. Upper Pliocene-present

It is assumed that the Corsica-Sardinia microplate was already in its present position and that it was separated from the African-Adriatic foreland by an orogenic belt, constituted by Alpine and pre-Messinian Apenninic units (see, e.g., Alvarez et al., 1974; Biju-Duval et al., 1977; Dewey and Sengor, 1979; Scandone, 1979; Cohen, 1980; Montigny et al., 1981; Rehault et al., 1985, 1987; Dercourt et al., 1986; Patacca and Scandone, 1989, Sartori, 1989)

A feature of the Tortonian structural setting which significantly influenced the successive evolution of the Central Mediterranean was the presence of thinned lithosphere in a relatively narrow zone, the Ionian, between the African and Adriatic continental areas (see, e.g., Rossi and Sartori, 1981; Scandone et al., 1981; Dercourt et al., 1986; Malinverno and Ryan, 1986).

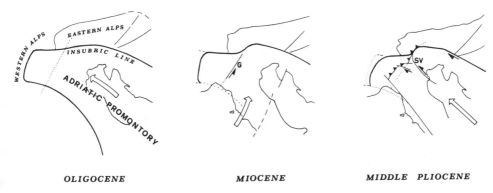

Figure 3. Schematic tectonic evolution of the Adriatic-Eurasia interaction zone which has been proposed by Semenza (1974) to explain the geological evidence in the Alps and surrounding regions. G: Giudicarie fault system, SV: Schio-Vicenza line. The big arrows indicate the dominant motion of the Adriatic foreland with respect to Eurasia.

The Iblean foreland, i.e. the present southern part of Sicily, was still closely connected with Africa. The region corresponding to the present Sicily Channel and surrounding zones was affected by SW-NE compressional stresses (see, e.g., Reuther, 1987; Boccaletti et al., 1990; Barrier, 1992). Illies (1981) suggested that "an uparching of about 200 m has preceded the physiographic rifting which occurred in the post-middle Miocene times".

The distribution of Oligo-Miocenic calc-alkaline magmatic activity in Sardinia and North Africa, which ended about 10-15 My ago (see, e.g., Savelli et al., 1979; Bellon, 1981), suggests that after the opening of the Balearic basin an extended arcuated edifice of subducted lithosphere was present beneath the Apenninic and Maghrebian belts, as tentatively shown in the perspective view (Fig.2A).

3.2 LATE TORTONIAN - EARLY MESSINIAN

The key event which, around the late Miocene, caused a profound change in the deformation pattern of the Central Mediterranean, was the activation, in the Southern

Alps, of a transpressional sinistral fault system, the Giudicarie belt (see Fig.2B), which allowed the decoupling of the Adriatic promontory from the Eurasian block. This hypothesis, advanced by Semenza in the 1974 (see Fig.3), can satisfactorily account for the pattern of deformations which have occurred in the Alps since the late Miocene and, in particular, for the fact that only weak deformations have affected the sector of the chain lying West of the Giudicarie belt, whereas the eastern Southern Alps have been interested by significant shortening activity (Semenza, 1974; Laubscher, 1983; Castellarin and Vai, 1986; Castellarin and Sartori, 1986). The time-space distribution of compressional deformations in the Southern Alps (Fig.4) points out another significant feature which supports the supposed left lateral transcurrency along the Giudicarie fault system, i.e. the relative shifting of crystalline Paleozoic basement units lying East and West of this line. Laubscher (1988) argued that the offset along the Giudicarie fault system was probably accommodated by strong shortening, partly in the area of the Tauern Window and partly in the Southern Alps. Shortening in the Tauern Window area is indirectly confirmed by white mica and biotite ages fission track ages, which suggest that the western portion of the Tauern Window underwent acceleration of uplift from 10 My onwards (Grundmann and Morteani, 1985).

The age of the displacement along the Giudicarie fault system is in good agreement with that required to accommodate continued thrusting in the eastern block, after the termination of thrusting in the western one (Massari, 1990). A similar role was played by the sinistral NE-SW fault system bounding the Vienna Basin (Royden, 1985) accommodating part of the displacement between the active and inactive parts of the thrust belt in the western Carpathians.

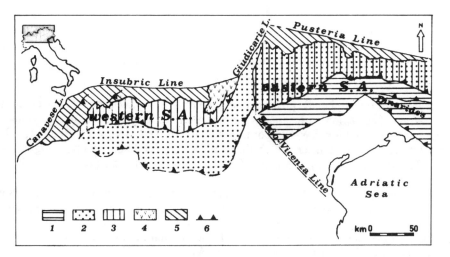

Figure 4. General, chronological classification of compressional deformations inside the Southern Alps (S.A.), after Castellarin and Vai, 1986, modified. 1) post-Tortonian deformations 2)Middle-Upper Miocene deformations 3)Cretaceous-Paleogene deformations 4) Main tertiary plutons 5) Crystalline basement and Paleozoic 6) Thrust fronts.

After the wrenching along the Giudicarie fault system and the consequent decoupling from its hinge zone in the Western Alps, the Adriatic block moved in a closer connection with Africa. As a consequence, the Adriatic-Eurasia rotation pole ceased to be located around the Western Alps and shifted towards a position nearer to the Africa-Eurasia pole. This new pole implied a roughly S-N relative motion of the northern Adriatic block (see Fig.2B), which caused the squeezing of the Eastern Alps against the Eurasian foreland. In response to this compression, the Alpine nappes underwent a lateral escape towards the weakly constrained eastern border, in correspondence to the northern edge of the Pannonian basin (see Fig.2B). Crustal thickening and lateral flow of orogenic units, driven by sets of conjugate strike slip faults, caused tectonic denudation of the zone which had been affected by most intense squeezing and uplifting, i.e. the present Tauern Window (see Fig.2C). A very interesting modeling of this extrusion pattern has been described by Ratschbacher et al. (1991a,b).

Further South, the collision of the Adriatic plate with the Balkan zones was mainly accommodated by the consumption of the remaining thinned Adriatic margin. The most intense shortenings occurred along the Outer Hellenides (Mercier et al., 1979, 1989), while minor deformations took place in the Dinaric sector of the belt (Channell and Horvath, 1976; Burchfiel, 1980). This difference might be explained by the fact that along the Dinarides the compression induced by the Adriatic drifting interplayed with the extensional regime which was affecting the Pannonian basin (see, e.g., Royden et al., 1983; Horvath, 1984).

In the Outer Hellenides the displacement of the Adriatic block was mainly absorbed by the underthrusting of its eastern margin beneath the North Aegean block, whereas in the Dinaric sector it might also have been accommodated by a rearrangement of crustal wedges within the Pannonian region. The differentiated shortening patterns along the above two sectors of the Adriatic margin, were accommodated by a dextral transcurrent fault system, which now corresponds to the evident lateral shift of deformation belts in the Albania zone (see Fig.5).

The drifting of the Adriatic block also had significant consequences along its western margin. The divergence between the Adriatic foreland and the Corsica-Sardinia block produced a tensional regime in the region corresponding to the present northwestern Tyrrhenian basin, with the formation of several S-N trending troughs (see, e.g., Zitellini et al., 1986; Finetti and Del Ben, 1986; Kastens, Mascle et al., 1987; Sartori, 1989; Mascle and Rehault, 1990). This extensional tectonics was most probably responsible for the magmatic activity which occurred in the Northwestern Tyrrhenian from roughly 10 to 6 My (CNR-PFG, 1989).

During this phase, the Southern Apenninic Arc was lying between the African and Adriatic forelands which were both moving NNEward (Fig. 2B) and thus no significant extension occurred in that part of the belt. In the Messinian, the Selli line was separating a deep basin, to the NW, from a more or less emerged land to the SE (Fabbri and Curzi, 1979; Sartori, 1989).

The hypothesis that the Adriatic block underwent a displacement during this phase raises a major problem concerning the behavior of its western subducted margin. Has this slab moved in close connection with the Adriatic platform? Or has it remained trapped in the mantle and then detached from the shallow Adriatic lithosphere? The first hypothesis seems to be unlikely, since the displacement through the mantle of such an extended lithospheric body would have required a lateral flow of a huge volume of asthenospheric

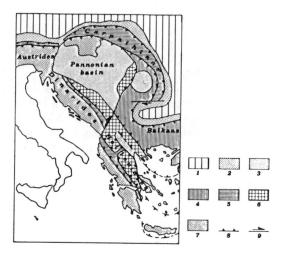

Figure 5. Organization of the Alpine chains in the central Mediterranean area (after Aubouin, 1984, modified). 1) European foreland 2,3) Internal and external molasse foredeep 4) Tectonic margin of the European domain 5) Flysch nappes of Tethyan ocean 6) Ophiolite nappes and relative units 7) Tectonic margin of the African domain 8) Main overthrusts 9) Main transcurrent fault systems.

material. The second hypothesis, instead, appears to be more plausible from the physical point of view and, furthermore, it might explain the time of occurrence (roughly between 6 and 2.5 My ago) and the space distribution (see Fig.6) of calc-alkaline magmatic activity in the Tyrrhenian area (Sartori et al., 1989; CNR-PFG, 1989).

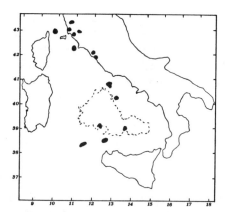

Figure 6. Messinian-Pliocene "subduction related" magmatism in and around the Tyrrhenian Sea (black spots), after Sartori (1986) and CNR-PFG (1989). The dashed line contours the Tyrrhenian bathyal plain.

This magmatic episode might have been generated by a relatively fast lowering of pressure which should have accompanied the hypothesized stearing and stretching of the slab beneath the Tyrrhenian area (Fig.2B). Assuming a reasonable velocity of some cm/y (White and McKenzie, 1989) for the magma uprising, from the source in depth (presumably 100-200 km) to the surface, the rupture of the slab should have started about 1-2 My before the oldest calc-alkaline episode, i.e. about 7-8 My ago.

3.3 MIDDLE MESSINIAN - MIDDLE PLIOCENE

This phase involved considerable deformations in the shallow and deep structures so, in order to make their progressive development over time clearer, we thought it advisable to illustrate them by two paleogepraphic maps and relative deep structural sketches (Figs.2C and 2D).

The key event which determined the beginning of this phase was the occurrence of a continental collision along the border between the Adriatic plate and the Balkan regions, in the Outer Hellenides (Mercier et al., 1976; 1979; 1989). This collision strongly reduced the eastward drifting of the Adriatic plate, which only maintained a minor movement roughly towards North.

A major effect of this kinematic change was the end of the divergence between the Adriatic foreland and the Corsica-Sardinia block, with the consequent cessation of crustal stretching in the northwestern Tyrrhenian basin (Finetti and Del Ben, 1986; Sartori, 1989).

No longer absorbed by the lithosphere consuming process in the Hellenides, the NNEward displacement of Africa induced an intensification of compressional stresses and strains in the zone lying between the Adriatic and African continental forelands, i.e. the Southern Apenninic Arc and the Ionian zone.

This dynamic context caused the outward extrusion of Apenninic units, at the expense of the downward flexure of the Adriatic-Ionian lithosphere (see Figs.2C,D). The occurrence of these deformations, from about the Messinian to the middle Pliocene, is well documented by geological data (Di Nocera et al., 1976; Casnedi et al., 1982; Mostardini and Merlini, 1986; Patacca and Scandone, 1989; Patacca et al., 1990). Fig.7 shows the time patterns of the lithospheric flexure and eastward migration of the thrust belt-foredeep system, which has been reconstructed on the basis of sedimentary data along three cross-sections in the Southern Apennines.

The above compressional mechanism was responsible for the most intense tectogenetic pulses that occurred in the Apennines after the period of minor activity which followed the last orogenic phase in the Tortonian (see Di Nocera et al., 1976; Vai, 1987; Torre et al., 1988).

The violence of this tectonic phase is testified by the great amount of shortening, estimated in about 150-300 km (Patacca and Scandone, 1989), by the counterclockwise rotation of the Southern Apennines with respect to the Northern Apennines and by the fact that the Burdigalian-Tortonian units were thrusted over the Messinian-Pliocene units through duplex mechanisms (see, e.g., Ortolani, 1979; Casnedi et al., 1982; Mostardini and Merlini, 1986).

The most evident eastward allochtonous character of the Southern Apenninic units and the development of the adjacent foredeep seem to be mostly confined to the region lying South of the Ortona-Roccamonfina (OR) line (Casnedi et al., 1982). Given that the outward migration of the belt had a close relationship with the downward flexure of the Adriatic foreland, one can reasonably suppose that the OR line was reflected deep down by a lithospheric tear fault, that allowed the different flexural patterns of the parts of the Adriatic slab which were lying South and North of this fault (Royden et al., 1987; Patacca et al., 1990).

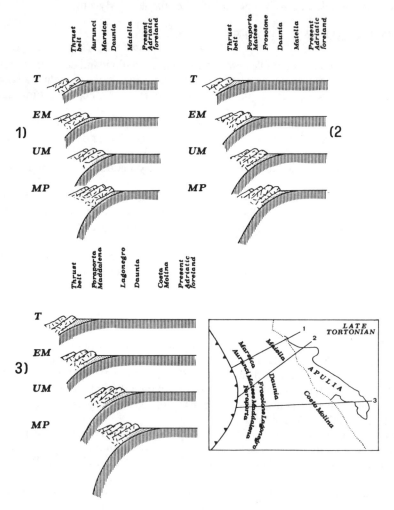

Figure 7. Tentative reconstruction of the outward migration of the thrust belt-foredeep system in the Southern Apennines and related flexural pattern of the adjacent Adriatic foreland, from the Tortonian (T) to the middle Pliocene (MP), through the early Messinian (EM) and the upper Messinian (UM), based on geological analysis (Patacca et al., 1990). The inset shows the location of the three cross-sections in the Late Tortonian paleogeographic setting (after Patacca et al., 1990).

The SEward shift over time of the tectogenetic pulses and foredeep development in the Southern Apennines (Casnedi et al., 1982) seem to delineate a sort of migrating compressional phase which progressively affected more and more southern zones of the belt, from the OR line to the Gulf of Taranto (Casnedi et al., 1982; Van Dijk and Okkes, 1991). The above pattern might indicate that inside the Southern Apenninic Arc there was a body, most probably corresponding to the Calabrian crystalline massif, which was characterized by a greater rigidity with respect to the surrounding orogenic units. The position over time of this rigid body can be reconstructed through the analysis of the deformation pattern in the Southern Apennines. Around the early Pliocene, this body may be placed just South of the Selli line, to account for the fact that the most intense compressional deformations in the Apennines were affecting the Molise sector (see Fig.2C). Then this body, in response to the Africa-Adriatic compressional regime, underwent a progressive extrusion towards SE, guided by transpressional fault systems. Coherently, the NEward compressional deformation in the Southern Apennines shifted from the Molise to the Campania sector, finally to reach the Basilicata zone and the Gulf of Taranto in the upper Pliocene.

In the wake of the outward migrating Apenninic units, crustal stretching occurred in the central Tyrrhenian basin. The morphology of the troughs which opened up during this phase and the analysis of ODP/DSDP drilling data indicate that the dominant trend of extension was about E-W (Kastens, Mascle et al., 1987). This implies that the opening of the Magnaghi and Vavilov basins was mainly connected with the eastward displacement of Southern Apennines units (Moussat et al., 1986; Sartori, 1989).

Significant constraints on the driving mechanism responsible for the evolution of Apennines can be inferred from the time-space evolution of deformation trends in the whole chain (Fig.8). This pattern clearly indicates that the direction of the orogenic forces considerably changed from the Aquitanian-Tortonian phases to the Messinian-middle Pliocene phases.

During the Aquitanian-Tortonian, the orogenic activity in the Apennines was dominated by a W-E compressional stress field, which was most probably connected with the convergent motion between the Corsica-Sardinia block and the Adriatic foreland.

This phase determined the shape of the Northern Apenninic Arc (see, e.g., Parotto and Praturlon, 1975; Boccaletti et al., 1980; Catalano and D'Argenio, 1982).

Around the Late Tortonian, the dynamic context and stress pattern in the Apennines changed drastically (Figs.2A,B). In the internal part of the Northern Arc, E-W compressional deformations ceased and were substituted by a tensional stress regime in connection with the stopping of the Corsica-Sardinia microplate and the beginning of the Adriatic's NEward drifting. This phase determined the formation of the northwestern Tyrrhenian basin.

Around the Messinian, the Abruzzi-Latium carbonate platform started being affected by a roughly NEward compressional regime, which caused the progressive closure of all preexistent SE-NW troughs and the formation of thrust fronts (Castellarin et al., 1978, 1982; Ghisetti and Vezzani, 1986). The Laga Flysch underwent strong bending and translation (Casnedi et al., 1982). Meanwhile, the zones lying just North of the Ancona-Anzio line (AA) were affected by a roughly S-N compressional regime which caused torsion of previous orogenic fronts. During this phase the AA line behaved as a lateral ramp for the further eastward migration of the Northern Apenninic units (Calamita and Deiana, 1988; Lavecchia et al., 1988).

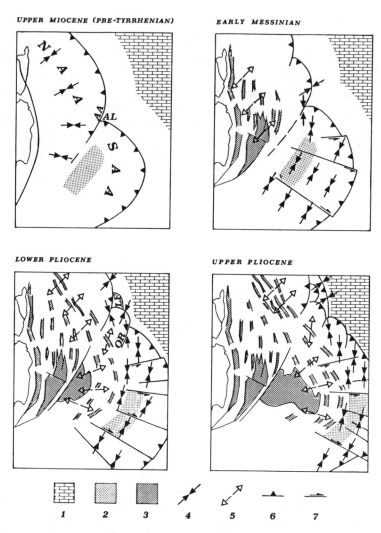

Figure 8. Proposed evolutionary pattern of stress regimes in the Apennines. 1) Present Adriatic foreland 2) Calabrian Massif 3) Extensional zones 4,5) Main trends of compressional and extensional stresses 6) External fronts of the Apenninic belt 7) Main transcurrent or transpressional faults. NAA = Northern Apenninic Arc, SAA = Southern Apenninic Arc, AA = Ancona-Anzio Line, AL = Abruzzi-Latium platform, LF = Laga Flysch, OR = Ortona-Roccamonfina Line.

Since the Messinian, the Northern Apenninic Arc (NAA) has been affected by compressional tectonics along the external fronts and tensional deformations in the internal area. Both types of phenomena presented a progressive migration towards E/NE (Elter et al., 1975; Marinelli, 1975; Bartolini et al., 1983; Lavecchia, 1988). This deformation pattern in the Central-Northern Apennines was caused, in our opinion, by the S-N to SE-NW compression exerted by the Southern Apenninic Arc (SAA) and the Adriatic platform, which produced a progressive outward extrusion of Northern Apenninic units over the adjacent Adriatic foreland. This extrusion was guided by a number of transpressional faults transversal to the chain.

The hypothesis that the Plio-Quaternary deformations in the Northern Apennines (Fig.9), with particular reference to the series of arcs which formed along the external Padanian fronts (Pieri and Groppi, 1981), were caused by roughly S-N shortenings has also been advanced by Castellarin and Vai (1986).

The fact that the most intense post-Tortonian orogenic pulses in the Southern and Northern Apennines have occurred during the same period (Messinian-Pliocene), supports the hypothesis that these deformations were closely connected with a unique driving mechanism: i.e. the SSW-NNE convergence between the African and Adriatic forelands.

The occurrence of significant Plio-Quaternary shortenings along the northern Adriatic border, principally in the eastern Southern Alps (Castellarin, 1979; Castellarin and Vai, 1986), and the minor activity in the Outer Hellenides, from about early Pliocene to early Pleistocene (Mercier et al., 1989), suggest that after the continental collision in the Outer Hellenides, the kinematics of the Adriatic plate was mainly characterized by a N/NWward motion.

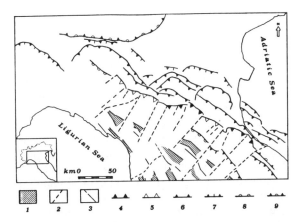

Figure 9. Tentative chronological classification of compressional fronts in the outer and buried Northern Apenninic units (after Castellarin and Vai, 1986). 1) Major tectonic depressions 2) Transversal faults 3) Normal faults 4) pre-Terminal Quaternary 5) pre-Upper Quaternary 6) pre-Quaternary 7) pre-Middle Upper Pliocene 8) pre-Lower Pliocene 9) pre or intra-Messinian.

We advance the hypothesis that the strong compressional phase which affected the Africa-Adriatic interaction zone since the Messinian was also absorbed by another major tectonic event, i.e. the NW ward extrusion of an African crustal wedge, here after called Sicily microplate (Fig.2C).

The transpressional lateral guides of this extrusion were most probably represented, on one side, by a fault system located between the Iblean zone and Tunisia (the present Sicily Channel) and, on the other side, by the Taormina fault system, an old discontinuity which formed in the Miocene, in connection with the overriding of the Calabrian Alpine units over the Maghrebian-Apenninic belt and was then reactivated as a transpressional fault (Amodio-Morelli et al., 1976; Scandone, 1982).

As discussed by Ratschbacher et al. (1991 a,b), the tectonic conditions which can produce lateral extrusion of crustal wedges are:

1) an overall compression regime 2) a strong foreland in front of it 3) the presence of a weak constraint along a lateral boundary and 4) an "extruding" body constituted by a previously thickened, gravitationally unstable, thermally weakened crust.
Such conditions can be recognized in the lateral extrusion of Sicily: 1) the compressional stress field was kinematically induced by the SSW-NNE Africa-Adriatic convergence 2) the strong foreland corresponded to the Adriatic platform and the Balkan massifs lying behind 3) the weak lateral constraint was represented by the stretched crust in the Northwestern Tyrrhenian basin 4) the extruded crustal wedge was mainly constituted by units of the Maghrebian-Apenninic belt, remnants of the Alpine chain and a tectonized fragment of Africa, for which it seems reasonable to suppose thermal conditions and gravitational instability of the type mentioned above.

The hypothesized NWward extrusion of the Sicily microplate can account for the deformations which occurred, starting in the late Messinian - early Pliocene, in the zones surrounding the Iblean foreland:

- The structural border between the Iblean foreland and the thinned Ionian area, i.e., the present Siracusa escarpment, was activated as normal fault, with a throw of several hundred metres (Carbone et al., 1982; Grasso and Lentini, 1982; Sartori et al., 1992). Extensional activity was accompanied by the occurrence of basaltic magmatism along the Siracusa escarpment (see, e.g., Barberi and Innocenti, 1980). The part of the Ionian zone which was interested by subsidence during this phase was bounded to the South by the seamounts of the Medina Rise, which are interpreted as "horsts of sedimentary rocks standing up after the general lowering of the Ionian bathyal plain" (Rossi and Zarudski, 1978; Jongsma et al., 1987).This extensional event may be interpreted as an effect of the diverging motion between the Sicily microplate, which was drifting roughly NWward, and the Ionian region, which was moving in close connection with Africa (Fig.2C).

- Compressional deformations trending SW-NE, i.e. perpendicular to the drifting direction of the Sicily microplate, developed in the Adventure plateau. In addition, the Gela nappes underwent further bowing and migration towards the Iblean foreland (Lentini, 1982; Ghisetti and Vezzani, 1984; Argnani et al., 1986; Grasso et al., 1990; Argnani, this volume).

- The border between the Iblean foreland and the Calabrian massif, i.e. the Taormina fault, was characterized by transpressional movements (see, e.g., Patacca and Scandone, 1989). This allowed the decoupling between the Sicily and Calabria blocks, which were moving towards NW and SE respectively.

- The Sicily Channel was affected by strike-slip tectonics in the framework of a compressional stress regime. This type of deformation was accompanied by the formation of pull-apart troughs (Finetti, 1984; Boccaletti et al., 1987; Cello, 1987; Jongsma et al., 1987; Reuther, 1987,1990). The complex fracturation pattern in the Sicily Channel has led to different interpretations of the stress orientation and the regional shear mechanism. Argnani (this volume) suggests that the Plio-Quaternary extensional tectonics which affected the Sicily Channel and the simultaneous compressional deformations in the adjacent Gela belt-foredeep system can be explained as consequences of a roll-back mechanism or of a mantle delamination process. These types of interpretations, however, encounter great difficulties in explaining some major features in the zones surrounding the Sicily Channel, as argued in the last section of this work. In our opinion, this deformation belt represented the transpressional decoupling zone between the Sicily microplate and the African foreland. The formation of grabens was connected with the irregular shape of the transcurrent fault system.

As regards deep structures, this phase was characterized by the building of a new slab beneath the Tyrrhenian basin, through the consumption, by a downward flexure, of the Adriatic and Ionian forelands (see Figs.2C, D). As argued earlier, the two sectors of the slab lying South and North of the Ortona-Roccamonfina line underwent different flexure patterns, with a greater consumption of the southern part with respect to the northern one.

3.4 UPPER PLIOCENE-PRESENT

Around the middle-upper Pliocene the southern Adriatic foreland reached a critical stage of deformation, beyond which any further downward bending began to encounter a noticeable resistance. This condition developed after that the southern part of the Adriatic plate, i.e. the one most directly stressed by the African displacement, had experienced a bilateral consumption (see Fig.10), first beneath the Hellenides, in the late Tortonian-Messinian phase, and then beneath the Apennines, in the Messinian-middle Pliocene phase (see Moretti and Royden, 1988). Around the upper Pliocene, the southern Adriatic plate was reduced to its bulge zone, i.e. the present structural high in the Apulia region (the bricked area in Fig.10). Consequently, the downward flexure of lithosphere beneath the Southern Apennines underwent a progressive slowdown finally to cease almost completely.

The above hypothesis is suggested by the fact that in the middle-upper Pliocene the outward migration of the rift basin-thrust belt-foredeep system in the southern Apennines and in the adjacent Tyrrhenian basin has gradually come to an end (Patacca et al., 1990; Sartori, 1989). After this event, the Africa-Adriatic convergence, no longer absorbed by the downward bending of the Adriatic margin beneath the southern Apennines, was mainly accommodated by the lateral escape of crustal wedges, with particular reference to the SEward migration of Calabria, at the expense of the downward bending Ionian foreland (see e.g., Rossi and Sartori, 1981; Sartori, 1989; Patacca et al., 1990; Van Dijk and Okkes, 1991).

In the wake of Calabria, crustal stretching, with NW-SE extensional trend, occurred in the southernmost Tyrrhenian, i.e. the Marsili basin, with exposure of oceanic crust (Finetti and Del Ben, 1986; Kastens et al., 1988; CNR-PFG, 1989; Sartori, 1989).

Since the Pleistocene, a fast uplifting, with maximum rates of about 1,5 mm/y, took

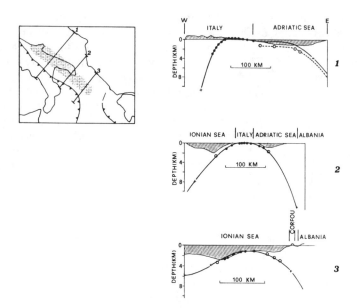

Figure 10. Observed (dots) and calculated (lines) deflection patterns of the basal Pliocene surface along the three profiles, shown in the inset, through the southernmost Adriatic plate. Hatching shows topography and bathymetry. The calculated deflection patterns refer to a thin elastic sheet subject to end loads at both plate ends (after Moretti and Royden, 1988, modified). The bricked zone in the inset identifies the structural high in the Adriatic foreland.

place in the eastern Sicily, Calabria, Southern Apennines and the adjacent foredeep (Ciaranfi et al., 1983; Ghisetti and Vezzani, 1982). This phenomenon might imply that the lateral extrusion of crustal wedges during this phase was accompanied by crustal thickening.

Around the late Pliocene-early Quaternary, the northern part of Calabria collided with the Apulian margin, in the Gulf of Taranto. After this event, Southern Calabria only has been affected by significant mobility, in terms of SEward drifting (Barone et al., 1982). The transpressional lateral guides which were decoupling this last sector from the adjacent ones, i.e. Northern Calabria and the Peloritani block, were respectively corresponding to the Catanzaro fault system and the Messina sphenocasm (Ghisetti and Vezzani, 1982; Finetti and Del Ben, 1986).

Since the middle-upper Pliocene, normal movements have occurred along the Apulian escarpment (Auroux et al., 1985; Finetti and Del Ben, 1986). This feature could mark the vertical decoupling between the continental Adriatic foreland, which was not affected any longer by downward bending, and the Ionian foreland, which continued to sink under the SEward migrating Calabrian block.

The intensity of the SW-NE Quaternary compressional phase (Barbano et al., 1978; Bousquet and Philip, 1986; Van Dijk and Okkes, 1991) which affected the Calabrian Arc and the southern Adriatic area is also testified by the reactivation of compressional deformations along the Adriatic-Balkan border, in the Outer Hellenides (Mercier et al., 1989).

We advance the hypothesis that since the middle-upper Pliocene the drifting trend of the Sicily microplate has undergone a significant variation passing from about NWward (see Fig.2D) to Northward (Fig.2E). This hypothesis can account for the tectonic events which have occurred around this block :

- Lateral motions ceased along the Taormina line, while transpressional tectonics began at the near Vulcano fault, roughly oriented SSE-NNW (Finetti and Del Ben, 1986; Patacca and Scandone, 1989). Something similar might have occurred on the other lateral guide of Sicily, in the northern part of the Sicily Channel, where an almost S-N trending transcurrent discontinuity, the Egadi fault, was activated after the middle Pliocene (Finetti and Del Ben, 1986; Reuther, 1987).

- Geological observations indicate that post-Late Pliocene thrust fronts in the Maghrebian belt lying North of Sicily have developed at high angles with respect to the previous phases and that SW-NE folding ceased in the Adventure plateau and Western Sicily (Catalano et al., 1989; CNR-PFG, 1989).

- Along the Siracusa escarpment extensional activity slowed down considerably (Reuther, 1987). This effect might be a consequence of the fact that the new motion of Sicily was more parallel to that of the Ionian region, and thus it has implied a lower diverging rate (see Fig.2E) at the border between these two blocks, i.e. the Siracusa escarpment.

- In the Sicily Channel extensional activity has slowed down since the middle Pliocene (Finetti, 1984; Cello, 1987; Jongsma et al., 1987; Calanchi, et al., 1989). In particular, Argnani (1990) suggested that only minor extension has occurred in this zone after the upper Pliocene. This phenomenon might be another consequence of the post-middle Pliocene motion of Sicily, which, being more parallel to the drifting direction of Africa, involved a lower transcurrent motion along the SE-NW fault system in the Sicily Channel and, consequently, a reduced extensional activity in the pull-apart troughs associated to these faults.

- Around the upper Pliocene, extensional activity began in the Messina Straits, leading to a progressive decoupling between the Peloritani block and Southern Calabria (Barbano et al, 1978; Ghisetti and Vezzani, 1982). Before the upper Pliocene, the Peloritani block was moving in close connection with Calabria and was separated from the sicilian Maghrebian units by the transpressional Taormina fault. Since the upper Pliocene, when Sicily began to move northward (Fig.2E), the transcurrency along the Taormina fault has been replaced by a prevalently compressional regime, which led to the detachment of the Peloritani block from Calabria and to the opening of an angular trough, i.e. the Messina sphenocasm.

- The middle Pliocene change in the drifting trend of Sicily might also have had some effects on the deformation pattern of deep lithospheric structures. This speculative hypothesis could explain the occurrence and the space-time evolution of the Aeolian calc-alkaline magmatic episode (see e.g., Beccaluva et al., 1982; 1985a). To this regard, it is interesting to note that the Aeolian magmatic activity started and developed along an arc roughly comprised between the Taormina and Vulcano faults, i.e. just in the zone where one could expect the maximum deformations, in terms of wrenching and steepening, of the Ionian subducted lithosphere stressed by the lithospheric roots of the Sicily microplate. This deep collisional pattern only occurred after the change in the drifting direction of the Sicily microplate. Before this event, the motion of Sicily was parallel to retreating direction of the Ionian foreland, towards SE, and, thus, no compressional interaction was taking place between the Ionian slab and the Sicilian lithospheric block. The hypothesized deformation pattern of the Ionian sunk lithosphere is consistent with the chronological zonation, the ring-like distribution and the general tendency of both calc-alkaline and shoshonitic volcanism to become younger as one moves counterclockwise in the Aeolian Arc. These features, in fact, have been interpreted as the results of a considerable torsion, segmentation and lateral stretching, accompanied by progressive steepening of the deep plate (Beccaluva et al., 1982; 1985a).

We suggested earlier that the extrusion of the Sicily microplate in the Messinian, and the direction of this lateral escape, was allowed, or at least strongly favoured, by the presence of a weak constraint in the Northwestern Tyrrhenian. Coherently with this way of thinking, one could suppose that the change in the escape direction of the Sicily microplate, was determined by the fact that around the middle-upper Pliocene the zone of crustal stretching in the Tyrrhenian (and thus the weak lateral constraint) was reaching more to the East, after the formation of the Magnaghi-Vavilov basin (Fig.2D).

During this phase, the stress regimes and the consequent deformation patterns were significantly different in the Southern and Northern Apennines. The southern part of the chain was no longer directly involved in the Adriatic-Africa interaction zone and started experiencing a transtensional tectonics, driven by the diverging motion between the Adriatic plate and the Tyrrhenian region (Fig.2E). The Northern Apennines, instead, being stressed by the N/NW displacement of the Adriatic platform, have continued to undergo a NEward extrusion, at the expense of the downward flexure of the adjacent foreland.

Fig.2E shows a tentative reconstruction of the remnant subducted lithosphere beneath the region under consideration, which is also based on structural seismological investigations and on the distribution of deep earthquakes in the southern Tyrrhenian area.
Over the past few years, seismic tomographic analyses have been carried out in the Central Mediterranean (Spakman, 1990; Amato and Alessadrini, 1991; Amato et al., this volume). The results reported by Spakman (1990) indicate the presence of relatively rigid lithosphere down to depths of some hundreds of km, beneath the Southernmost Tyrrhenian in correspondence to the region where deep seismicity occurs, and of a low velocity zone in the depth range 0-200 km beneath the Tyrrhenian-Apennines system. A low rigidity of the slab beneath the Southern Apennines has also been suggested by Giardini and Velonà (1991) on the basis of travel time residuals from deep Tyrrhenian events.
The results of Amato and Alessandrini (1991) suggest higher velocities, with respect to Spakman (1990), beneath the northernmost Apennines, which could account for the occurrence of some subcrustal earthquakes (h = 50-100 km) in that zone.

To explain the uprising of a considerable amount of "subduction related" magmas in the Roman and Neapolitan provinces (Fig.2E) during the Quaternary (see, e.g., Di Girolamo, 1978; Peccerillo and Manetti, 1985; Beccaluva et al., 1985b; 1989; Conticelli et al., 1986; Di Girolamo et al.,1988; Civetta et al., 1989; CNR-PFG, 1989; Serri et al., 1991) we advance the speculative hypothesis that since the middle-Upper Pliocene the slab has undergone a considerable deformation and fracturing beneath Central Italy.

This new deformation pattern of the Adriatic sunk lithosphere was determined by two tectonic events. One is the cessation of the downward flexure beneath the Southern Apennines, and the other is the change of the Sicily drifting trend from NWward to roughly Northward. The first event interrupted a mechanism which was allowing the Adriatic-Ionian subducted foreland lying South of the O.R. line to absorb the motion of Africa and the second event has strengthened the SSW-NNE compression of the Sicily-Africa blocks on the above slab. As discussed earlier, this new dynamic context caused deep compressional interactions at the contact of the Adriatic Ionian subducted lithosphere with the Sicilian lithosphere. We suppose that a similar collisional pattern occurred beneath central Italy, i.e. at the contact between the two decoupled parts of the Adriatic sunk margin which were lying beneath the Southern and Northern Apenines respectively. The sketch shown in Fig.2D illustrates a possible geometry of the slab's fracturation compatible with the hypothesized dynamic setting. The stretching and wrenching of the deep lithosphere which is implied by the mechanism described above might be responsible for the generation of the magmas which reached the surface in the Quaternary.

The uprising of magmas through and around the sunk lithosphere might have caused a significant acceleration in the heating and disruption of the slab (Nur et al., 1991). This effect could explain the presence, beneath the Western margin of Italy of an extended low velocity zone, evidenced by tomographic investigations, and the lack of intermediate earthquakes beneath the Southern Apennines..

The presence of rigid lithosphere beneath the southern Tyrrhenian, as implied by the occurrence of intermediate deep earthquakes (Fig.11), is consistent with the fact that along this sector of the belt the building of the slab has continued through the Quaternary, by the downward flexure of the Ionian lithosphere.

Recent analyses of deep shocks (Anderson and Jackson, 1987; Giardini and Velonà, 1991) have pointed out two major features of the "Tyrrhenian Benioff zone", i.e. the concentration of most energy release in the depth range 250-300 km and a significant change in the dip of the slab, from about 70°, in the shallowest part, to 50°, in the lower one, roughly located where the maximum energy release occurs (see Fig.11B).

These features could be interpreted as consequences of an impending fracturation of the sunk lithosphere, as schematically indicated in Fig.2E. This fracturation could be due to the fact that the shallowest part of the Ionian slab, being squeezed between the African and Adriatic forelands, tends to undergo steepening and downward bending. This deformation, however, cannot easily be absorbed by the deeper part of the slab, which is probably trapped in the mantle and thus some decoupling has to occur between the shallow and deep parts of the subducted lithosphere.

The most intense seismicity in the depth range 250-300 km might be a consequence of this decoupling mechanism, which seems to be dominated by almost horizontal shear stresses, as suggested by focal mechanisms of deep shocks (Anderson and Jackson, 1987; Giardini and Velonà, 1991). A similar fracturing in the slab beneath the Calabrian

Arc has also been suggested by Van Dijk and Okkes (1991) on the basis of independent considerations.

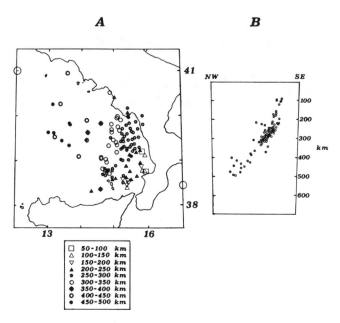

Figure 11. Deep seismicity in the Tyrrhenian Sea. (A) Map view, different symbols indicate different depth ranges in 50 km intervals; large circles indicate the limits of the NW-SE cross-section (B) Lateral NW-SE cross-section of the seismicity in (A) (after Giardini and Velonà, 1991).

4. Conclusions and discussion

- *Proposed evolution*

It is hypothesized that a great number of major and minor deformation events which have occurred in the Central Mediterranean region since the Late Miocene can be coherently interpreted in the framework of a SSW-NNE convergence between Africa and Eurasia. The shortening required by the Africa-Eurasia convergence has been accommodated by a complex distribution in space and time of lithosphere consuming processes and lateral extrusion of crustal wedges towards weakly constrained lateral borders (Fig.12).

Before the Late Tortonian the northwestern protuberance of the Adriatic promontory, was deeply indented into the Western Alps and thus little freedom of movement was allowed to this block. The event which determined the starting of the post-Tortonian evolutionary pattern in the Central Mediterranean was the activation of the Giudicarie transpressional fault system, that caused the decoupling of the Adriatic plate from its hinge

Figure 12. Sketch of the shortening mechanisms which accomodate the SSW-NNE Africa-Eurasia convergence.in the Central Mediterranean during the three main post-Tortonian evolutionary phases.
For each phase, two sketches are reported to illustrate the structural tectonic settings corresponding to the early and late parts of the period considered. The large grid identifies the present size of the Adriatic and African continental domains. The darker mountains inside the Southern Apenninic Arc tentatively indicates the position of the Calabrian massif. The arrows indicate the presumed motion of blocks with respect to Eurasia. CS: Corsica-Sardinia microplate.

FIRST PHASE (late Tortonian to early Messinian)
SHORTENING PROCESSES ALONG THE NORTHERN-EASTERN ADRIATIC BORDER

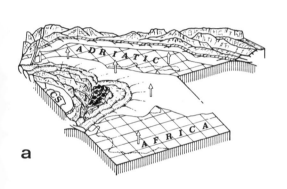

a) The activation of the Giudicarie transpressional fault system in the Southern Alps allows the decoupling of the Adriatic block from Eurasia. Driven by the SSW-NNE motion of Africa, the Adriatic plate drifts towards the Balkan zones causing shortening processes in the Alps, Dinarides and Hellenides. The remaining thinned margin of the Adriatic plate along its eastern border is almost completely consumed during this phase.

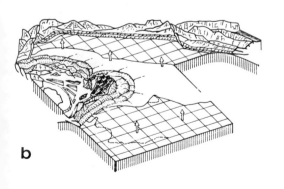

b) The divergence between the Adriatic block and the stable Corsica Sardinia microplate produces extensional tectonics in the interposed belts, causing the formation of the Northwestern Tyrrhenian basin. The progressive slowdown/cessation of this tectonic phase is determined by the development of a continental collision in the Hellenides, which considerably decreases the eastward drifting rate of the Adriatic block.

SECOND PHASE (middle Messinian to middle Pliocene)
OUTWARD EXTRUSION OF THE APENNINIC BELT, CONSUMPTION OF THE ADRIATIC-IONIAN FORELAND AND EXTRUSION OF THE SICILY BLOCK.

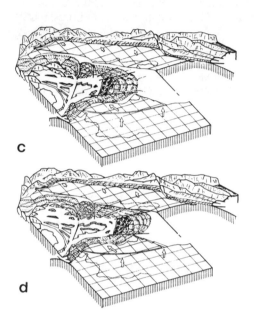

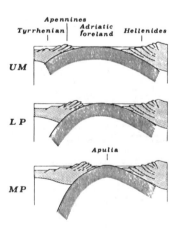

c) After the continental collision in the Hellenides, most of the shortening activity required by the Africa-Eurasia convergence passes in the zone comprised between the Adriatic and African blocks. This shortening is mainly accomplished by the lateral escape of an African crustal wedge (the Sicily block) toward NW and by the extrusion of the Apenninic belt toward East and SE. The escape of Sicily is allowed by the presence of crustal stretching in the Northwestern Tyrrhenian basin, while the extrusion of the Apenninic wedges occurs at the expense of the Adriatic-Ionian foreland which undergoes a downward bending.

d) This phase progressively come to an end as the southern part of the Adriatic platform approaches the final stage of its bilateral flexure, as shown in the cross sections (UM=upper Messinian, LP=lower Pliocene, MP=middle Pliocene). This gradually produces the slowdown/cessation of the eastward migration of the rift-thrust belt-foredeep system in the Southern Apennines.

THIRD PHASE (upper Pliocene to present)
LATERAL ESCAPES OF CALABRIA AND SICILY AND CRUSTAL THICKENING

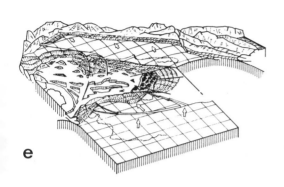

e) No longer absorbed by the consumption of the southern Adriatic foreland, the Africa-Adriatic convergence can only be accomodated by the opposite lateral escapes of the Calabrian block, at the expense of the downward flexuring Ionian foreland, and of the Sicily block towards the stretched Tyrrhenian zone. During this phase, the extension in the Tyrrhenian area can only occur towards SE, in the wake of the Calabrian block.

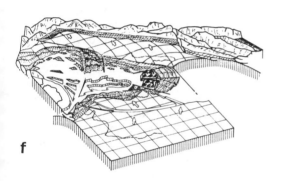

f) The rapid uplift which has affected Sicily, Calabria, Southern Apennines and the adjacent foredeep since roughly the Pleistocene seems to suggest that during this last phase the Africa Adriatic convergence has been also accomodate by crustal thickening. This tectonic phase is still going on, even though the major mobility of the Calabrian block is now confined to its southern part.

zone in the Western Alps. The above decoupling allowed a greater mobility of the Adriatic block which, driven by Africa, shifted roughly towards NE/NNE. This displacement induced intense compressional stresses along the northern and eastern Adriatic borders. The main consequences of this stress regime along the northern Adriatic were the compressional deformations in the Southern Alps and the eastward extrusion processes in the Eastern Alps. Along the eastern border, the displacement of the Adriatic was accommodated by the consumption of its remaining thinned margin beneath the Hellenides-Dinarides thrust front.

These shortening processes along the northern and eastern Adriatic boundaries absorbed almost completely the Africa-Eurasia convergence in the Central Mediterranean over about 3 My, from about the late Tortonian to the Messinian. Along the western Adriatic margin, an important extensional event took place, i.e. the opening of the Northwestern

Tyrrhenian basin, driven by the diverging motion between the Adriatic foreland and the steady Corsica-Sardinia block.

The end of the first evolutionary phase, in the Messinian, was determined by the occurrence of a continental collision along the Adriatic-Balkan border in the Hellenides. The displacement rate of the Adriatic plate towards the East decreased considerably and, consequently, compressional stresses increased in the zone lying between the African and Adriatic forelands. This SSW-NNE compressional regime produced the extrusion of crustal wedges in the Southern Apenninic Arc towards the adjacent Adriatic and Ionian forelands. The forced outward migration of the Apenninic belt caused the scratching of the adjacent Adriatic foreland, with the accumulation of light crustal material along the external front of the Apenninic belt. Once decoupled from its shallow buoyant part, the Adriatic lithosphere presented a much less resistance to subduct and, stressed by horizontal forces connected with the Africa-Adriatic convergence and possibly by a negative buoyancy with respect to the surrounding mantle, underwent a relatively fast downward bending.

One could wonder on whether the estimated migrations of the Southern Apennines and the Calabrian Arc towards the Adriatic-Ionian foreland (see Fig.2) are compatible with the SSW-NNE shortening implied by the Africa-Eurasia convergence (about 150-200 km) we have assumed in our reconstruction. A rough estimate of the extrusion rates which can be produced by a SSE-NNW shortening of 200 km in a more simplified structural context, suggests that the hypothesized SEward migration of the Calabrian Arc for about 400 km cannot be straightforwardly accounted by such a driving mechanism. However, one must consider that the extrusion of the Southern Apenninic Arc has developed in a geometrical situation which involved a progressive narrowing of the lateral confinement, in between the Adriatic and African forelands. In Fig.2B it can be noted, for example, that in the Messinian the extruding belt was lying in between lateral continental walls (the Adriatic and the Africa-Sicily forelands) about 500 km apart, whereas in the Quaternary the same belt was being extruded through a "corridor" about 300 km wide. As known from Fluid-Dynamics, a funnel shape of the confining lateral guides emphasizes the rate of the extrusion flow. The plausibility of this effect is not clearly demonstrated for the extrusion of rocks, but it seems reasonable to suppose that this mechanism might have played a significant role in the evolution of the Calabrian Arc.

Another important change in the deformation pattern of the zone under study was determined, around the upper Pliocene, by the fact that the southern part of the Adriatic platform, after a bilateral downward bending, first beneath the Hellenides and then beneath the Apennines, reached a crucial stage of deformation, for which any further flexure became extremely difficult. In this new dynamic context, the Adriatic-Africa convergence had to look for other shortening mechanisms. Considering that around the middle Pliocene the only two "weak zones" left in the central Mediterranean were the Tyrrhenian stretched area and the thinned Ionian lithosphere, it can be easily understood why the upper Pliocene -present deformation pattern was dominated by the opposite lateral escapes of the Sicily and Calabria crustal wedges towards the above mentioned "weak zones".

The above extrusion processes, however, were probably not sufficient to absorb the Africa-Adriatic convergence. This hypothesis is suggested by the occurrence in the Africa-Adriatic interaction zone of a fast uplifting since approximately one million years ago, which could be an effect of crustal thickening in the zone under compression.

Some deformations have also been produced by the minor N/NWward motion that the Adriatic plate as maintained after the late Messinian. The imprints of this displacement are mainly recorded along the northern Adriatic border, in the eastern Southern Alps, and in the Northern Apennines.

In our reconstruction, we have assumed, for simplicity, that the Balkan zones have behaved as a coherent part of the Eurasian block. However, this assumption is most probably unreliable since it does not take into account the deformations which have occurred in the Carpatho-Balkan-Pannonian system. The kinematic indications reported in Fig.2 could therefore be corrected for this effect. We suspect, in particular, that the eastward component of motion of the Adriatic plate with respect to stable Eurasia may be slightly greater relative to that shown in Fig.2.

In our Africa-Eurasia kinematic model, we have assumed that the convergence between these two plates has progressively slowdown in the last 10 My. This hypothesis is suggested by the fact that the late Miocene-middle Pliocene deformation rates can be satisfactorily accounted by an Africa-Eurasia convergent motion of about 2 cm/y, whereas the post-middle Pliocene deformations suggest a convergence rate lower than 1 cm/y, i.e. comparable with the one indicated by most reconstructions of present day global plate motion. A slowdown of the Africa-Eurasia convergence might represent a plausible consequence of the fact that, from the late Miocene to middle-upper Pliocene, the "consumable" margins of the Adriatic promontory have progressively disappeared and thus, shortening processes have become ever more difficult and slow. A structural feature, evidenced by tomographic analyses, which certainly behaves as a strong obstacle to the occurrence of further shortening on the Adriatic-Balkan border is the concentration of high velocity lithosphere beneath the Northwestern Aegean region (Spakman, 1990).

- Deep tectonics and magmatic activity

As regards deep tectonic processes, it is advanced the speculative hypothesis that the episodes of subduction-related magmatism in the Tyrrhenian and surrounding regions have been connected with two main kinds of circumstances: the occurrence of profound fractures or deformations in the underlying slab, caused by tensional and wrenching mechanisms, and the contemporaneous presence of extensional tectonics in the overlying crustal structure.

Different geodynamic interpretations of magmatic evidence in the Tyrrhenian and periTyrrhenian regions have been proposed in the recent literature (see, e.g., Serri, 1990; Serri et al., 1991). These hypotheses suggest that the petrogenesis and the time-space distribution of the upper Tortonian to present magmatism is consistent with a model of roll-back subduction and back-arc extension, driven by gravitational sinking of the subducted lithosphere. Following this hypothesis, the formation and eruption of calc-alkaline and shoshonitic magmas from deep mantle sources previously modified by subduction would have been produced by intense rifting, developed along the internal zones of the Apenninic belt during a post-collisional extensional phase (Di Girolamo et al., 1988; Beccaluva et al., 1991). The different composition of magmas between the magmatic provinces lying North and South of the Ortona-Roccamonfina line is explained by supposing that they relate to the subduction of an oceanic lithosphere in the South and to subduction/deformation of the continental lithosphere in the North (Serri, 1990; Serri et al., 1991).

The major problem of the above interpretative scheme is that it assumes that the generation of magma was produced by tectonic processes which have also occurred in other regions where no or scarce magmatic activity is observed. For example, geological data clearly indicate that the Southern Apennines have been affected by a roll-back subduction of the Adriatic lithosphere accompanied by an eastward migration of the rift basin-thrust belt foredeep system (Patacca and Scandone, 1989). It is not easy to understand why this type of tectonic mechanism, which has been described as responsible for "subduction related" volcanism in the Central-Northern Apennines, has produced almost no magmatic activity in the Southern Apennines, south of the Campanian province.

Moreover, the hypothesis that the subduction process South of the Ortona-Roccamonfina line has involved oceanic lithosphere contrasts with the conclusions of Casero et al. (1988) and Patacca et al. (1990) which, on the basis of geological data, have supposed that a continental lithosphere was consumed beneath the Southern Apennines.

More in general, we think that if the generation of calc-alkaline and shoshonitic magmas was simply connected with roll-back subduction and extensional tectonics in the overlying crust, one should expect to observe a more regular and continuous distribution in space and time of volcanic activity along the western side of Italy and in the Tyrrhenian basin. On the contrary, the subduction-related activity has mainly occurred in relatively small zones (central Italy and Aeolian Arc) and during relatively short time intervals: a Pliocene phase in the Tyrrhenian (Sartori, 1989) and a Quaternary phase, around 1 My, in central Italy and in the southernmost Tyrrhenian.

We advance the hypothesis that the discontinuous distribution of volcanic activity in space and time can be better explained as an effect of particular thermo-baric conditions which developed in the subducted lithosphere, during its tectonic evolution. In our scheme we have supposed that these "particular conditions" have been connected with stretching or wrenching of the slab, since such kind of mechanisms seem to be the most probable in the geodynamic framework here proposed. The Pliocene calc-alkaline magmatic activity in the Tyrrhenian just followed the period of greater mobility of the Adriatic plate (Fig.2C). It appears plausible, from the mechanical point of view, to suppose that the displacement of the Adriatic plate induced E-W extensional stresses and stretching in its subducted margin, favouring melting processes and consequent uprising of magmas through the Tyrrhenian crust, which by that time was experiencing an extensional regime.

A new phase of subduction-related magmatism extensively occurred in the Quaternary, along the Tyrrhenian side of Italy and in the southernmost Tyrrhenian (see Fig.13). We suppose that the causal mechanism of this new magmatic episode was analogous to that of the Pliocene episode, i.e. connected with stretching and wrenching of subducted lithosphere. In this last case the fracturing slab was the one built up during the Messinian-middle Pliocene tectonic phase. The stress regime which led to this fracturation of the slab beneath central Italy started approximately in the middle Pliocene, when the downward bending of the southern Adriatic foreland ended up (see Fig.2D). After this event, the displacement of Africa, no longer accommodated by the consumption of the Adriatic lithosphere beneath the Southern Apennines, caused an increase of stresses on the whole subducted Adriatic margin, which could have produced the deep deformations tentatively reconstructed in Fig.2D,E with the consequent melting processes and generation of magmas.

Since the middle Pliocene, arc-trench migration and downward flexure of the adjacent foreland has only continued in the Calabrian Arc, at the expense of the thinned Ionian lithosphere. The presence of a relatively young and still rigid slab beneath this zone is indicated by tomographic data and by the occurrence of deep earthquakes.

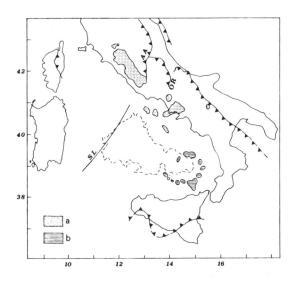

Figure 13. Pleistocene subduction related volcanism in and around the Tyrrhenian Sea (CNR-PFG, 1989). a) Roman and Neapolitan Province; b) Southern Tyrrhenian Province. AA= Ancona-Anzio line; OR= Ortona-Roccamonfina line; SL= Selli line. The dashed line contours the Tyrrhenian bathial plain.

- *Previous evolutionary models*

Whoever presents a new evolutionary hypothesis, as we do in this work, has to provide some plausible explanations of why the model under consideration should be preferred to the ones previously proposed. Otherwise any new interpretation risks of increasing the ambiguity around the Mediterranean geodynamic evolution rather than mitigating it.

To this regard, we make reference to the discussion of alternative models which is reported by Mantovani et al. (1992). Here we only devote some new remarks to the rollback mechanism that seems to be the most popular in the recent literature.(see e.g., Malinverno and Ryan, 1986; Royden et al., 1987; Patacca and Scandone, 1989)

This type of interpretation supposes that the post-Tortonian deformations in the Tyrrhenian-Apennines system have been mainly caused by the gravitational sinking of the subducted Adriatic-Ionian lithosphere beneath this region. This hypothesis can satisfactorily account for the downward flexure of the Adriatic-Ionian foreland beneath the Apenninic Arc, but cannot easily explain some other major Plio-Quaternary deformation events in the Calabrian Arc, Sicily, Ionian area and southern Adriatic plate.

As the opening of the Tyrrhenian basin has developed through three main phases, characterized by rather different tectonic conditions, it seems opportune to analyse the plausibility of the "gravitational sinking" model for each single phase.

The first extensional phase, which approximately occurred from the late Tortonian to the Messinian in the Northwestern Tyrrhenian, can hardly be interpreted as an effect of gravitational sinking of the subducted lithosphere. In fact, for the above time interval there is no clear evidence of intense outward migration of the chain or of orogenic activity in the Northern Apennines and downward bending of the adjacent Adriatic foreland (see, e.g., Vai, 1987; Castellarin et al., 1992). Moreover, from the geometry of the Oligo-Miocene calc-alkaline arc and other tectonic considerations about the opening of the Balearic basin, it seems reasonable to believe that at the Tortonian time, prior to the Tyrrhenian opening, the subducted lithosphere was mainly developed beneath the most arcuate sector of the belt, i.e. the one lying South of the Selli line (see Fig.2A,B). Given this structural premise, it is difficult to understand why the presumed roll-back mechanism started in the Northwestern Tyrrhenian and occurred only there till around the Messinian.

The second extensional phase developed in the central Tyrrhenian region from the Messinian to the middle-upper Pliocene. If this phenomenon is explained as an effect of gravitational sinking of the subducted Adriatic foreland, it is hard to reconcile this driving mechanism with the presence of a SW-NE compressional stress regime in the zone comprising Sicily, the Calabrian Arc, the Southern Apennines and the Southern Adriatic platform. Furthermore, it seems improbable, from the physical point of view, that a roll-back mechanism can have produced a radial distribution of the belt migration and a downward flexure of the adjacent Adriatic-Ionian foreland, such as the one which have developed around the Southern Apenninic Arc during the Plio-Quaternary (see Fig.2). Another major evidence which does not seem compatible with the "gravitational sinking" mechanism is the deformation pattern around the Sicily block which took place from the Messinian to the middle Pliocene, in particular the occurrence of extensional tectonics in the Siracusa escarpment and the Sicily Channel.

In the last extensional phase (upper Pliocene to present), crustal stretching only occurred in the southernmost Tyrrhenian (Marsili basin), with a SEward extensional trend, in connection with downward bending of the Ionian lithosphere. Eastward migration of the rift-thrust belt-foredeep system had instead ceased in the Southern Apennines. It is not clear how the roll-back mechanism can explain the different flexural behaviors of the Adriatic and Ionian forelands. Furthermore, as for the previous phase, it seems hard to reconcile the passive sinking of the Ionian lithosphere with the presence of a SSW-NNE compressional regime in the Calabrian Arc. This zone, in addition, has been affected by a fast uplifting since the last million years. Such type of deformation cannot be easily attributed to the passive sinking of the Ionian margin. Even if one can find a plausible answer to this problem, it remains to be understood why uplifting has also occurred in the Southern Apennines where the roll-back mechanism has ceased since about 2 million years ago.

More in general, the main problem of the passive sinking hypothesis appears to be the difficulty in explaining why this phenomenon has occurred at certain places and certain times. In our opinion, this difficulty is due to the fact that the above hypothesis is a partial concept, which only recognises the second stage of the real complete mechanism. The sinking of continental lithosphere cannot occur if it is not preceded by another tectonic process, i.e. the decoupling between the upper and lower crust, which considerably decreases its buoyancy (see, e.g., Molnar and Gray, 1979). Consequently, the distribution in space and time of lithospheric sinking is mainly controlled by the occurrence of the collisional patterns in the shallow structures which turn a buoyant continental plate in a "subductable" lithosphere. Thus, to understand why and where downward bending of the Adriatic margin occurred, it is first necessary to recognize the mechanism which led the Alpine-Apenninic belt to collide with the adjacent Adriatic foreland. Royden (1993) classifies the Adriatic-Apenninic system as a "retreating plate boundary" and states: "subduction occurs in this type of boundary only when the downgoing plate has a sufficiently high density that gravity can drive the subduction process. An important consequence of this is that subduction at retreating plate boundaries generally ceases shortly after the entry of thick bouyont continental crust into the subduction zone". This view, however, contrasts with the fact that a considerable part of the Adriatic foreland which disappeared beneath Southern Apennines during the Messinian-Lower Pliocene times had most probably a continental-like structural character (Casero et al., 1988; Patacca et al., 1990). The above difficulty could be removed if one modifies the concept described by Royden (1993) as follows:"Subduction at retreating plate boundaries ceases when the collisional pattern in the trench zone, and thus the scratching of the descending lithosphere, is interrupted". However, it appears imprudent to adopt a general interpretative scheme for this kind of phenomena. For instance, the cessation of thrusting in the Apennines around the middle-upper Pliocene might have been also influenced by the previous flexural behavior of the southern Adriatic foreland, as argued earlier in this work.

Several recent papers have pointed out that to fully understand deformations in a continental collision zone, such as the Mediterranean one, the classical plate tectonic approach is not completely adequate and it is necessary to use the concepts of continuum mechanics (see, e.g., Tapponier et al., 1982; England and McKenzie, 1982; Cohen and Morgan, 1987; Cobbold and Davy, 1988; Peltzer and Tapponier, 1988; Boccaletti and Nur, Eds., 1990; Ratschbacher et al., 1991a,b). The arguments reported in this work strongly support this point of view. In addition, it is suggested that the geodynamic evolution of the central Mediterranean region has been considerably conditioned by extrusion tectonics processes.and by the consumption of continental lithosphere, previously separated by its buoyant shallowest part.

We have tried to delineate a detailed evolutionary model, also through speculative considerations, in order to provide all necessary elements to allow numerical and experimental modeling. The proposed tentative reconstruction of deep structural evolution could also allow quantitative checks, through the elaboration of thermal models of the lithosphere-astenosphere system based on tomographic data (De Jonge and Wortel, 1990).

AKNOWLEDGMENTS

We are very grateful to Prof. R.Sartori for reviewing the manuscript. He gave us fruitful suggestions to improve the interpretation of geological evidence, the exposition of tectonic arguments and the english in the text.

We would also like to thank Mrs. F.Falciani, Mr. G.Vannucchi and Mr. R.Galgano (ING) for their collaboration in editing the text and drawing the figures. This work has been financially supported by Italian Research Council (CNR) and the Ministry of University and Scientific Research.

5. References

Albarello, D., Mantovani, E., Babbucci, D. and Tamburelli, C. (1993) 'Africa-Eurasia kinematics in the Mediterranean: an alternative hypothesis', This volume.

Alvarez, W., Cocozza, T. and Wezel, C.F. (1974) 'Fragmentation of the Alpine orogenic belt by microplate dispersal', Nature 248, 309-314.

Amato, A. and Alessandrini, B. (1991) 'Tomografia simica per la regione italiana', Le Scienze 59, 70-72.

Amato, A., Cimini, G.B. and Alessandrini, B. (1993) 'P wave teleseismic tomography: contribution to the delineation of the upper mantle structure in Italy', This volume.

Amodio-Morelli, L., Bonardi, G., Colonna, B., Dietrich, D., Giunta, G., Ippolito, F., Liguori, D., Lorenzoni, S., Paglionico, A., Perrone, V., Piccarreta, G., Russo, M., Scandone, P., Zanettin-Lorenzoni, E. and Zuppetta, A. (1976) 'L'Arco Calabro-Peloritano nell'Orogene Appenninico-Maghrebide', Mem.Soc.Geol.It. 17, 1-60.

Anderson, H. and Jackson, J. (1987) 'The deep seismicity of the Tyrrhenian Sea', Geophys.J.R.astr.Soc. 91, 613-637.

Argnani, A. (1990) 'The Strait of Sicily rift zone: foreland deformation related to the evolution of a back-arc basin', J. Geodyn. 12, 311-332.

Argnani, A.(1993) 'Neogene basins in the Straits of Sicily (Central Mediterranean): tectonic settings and geodynamic implications', This volume.

Argnani, A., Cornini, S., Torelli, L. and Zitellini, N. (1986) 'Neogene-Quaternary foredeep system in the Strait of Sicily', Mem.Soc.Geol.It. 36, 123-130.

Aubouin, J. (1984) 'Mediterraneenne (Aire)', Encyclopedia Universalis, 2nd Ed. 11, pp. 1023-1030.

Auroux, C., Mascle, J., Campredon, R., Mascle, G. and Rossi, S. (1985) 'Cadre géodynamique et évolution récente de la Dorsale Apulienne et de ses bourders', Giorn.Geol. 47, 101-107.

Barbano, M.S., Carrozzo, M.T., Caverni, P., Cosentino, M., Fonte, G., Ghisetti, F., Lanzafame, G., Lombardo, G., Patané, G., Riuscetti, M., Tortorici, L. and Vezzani, L. (1978) 'Elementi per una carta sismotettonica della Sicilia e della Calabria meridionale', Mem.Soc.Geol.It. 19, 681-688.

Barberi, F. and Innocenti, F. (1980) 'Volcanisme Neogene et Quaternaire. Guide a l'excursion 122A', Soc.It. Miner. Petrol., 99-104.

Barone, A., Fabbri, A., Rossi, S. and Sartori, R. (1982) 'Geological structure and Evolution of the Marine Areas Adjacent to the Calabrian Arc' in E. Mantovani and R. Sartori (eds), Structure, evolution and present dynamics of the Calabrian Arc, Earth Evolut.Sci. 3, 207-221.

Barrier, E. (1992) 'Tectonic analysis of a flexed foreland: the Ragusa Platform', Tectonophysics 206, 91-111.

Bartolini, C., Bernini, M., Carloni, G.C., Costantini, A., Federici, P.R., Gasperi, G., Lazzarotto, A., Marchetti, G., Mazzanti, R., Papani, G., Pranzini G., Rau, A., Sandrelli, F., Vercesi, P.L., Castaldini, D. and Francavilla, F. (1983) 'Carta neotettonica dell'Appennino settentrionale. Note illustrative', Boll. Soc.Geol.It. 101, 523-549.

Beccaluva, L., Rossi, P.L., and Serri, G. (1982) 'Neogene to Recent Volcanism of the Southern Tyrrhenian-Sicilian Area: Implications for the Geodynamic Evolution of the Calabrian Arc', in E.Mantovani and R.Sartori (eds), Structure,Evolution and Present Dynamics of the Calabrian Arc, Earth Evolut.Sci. 3, 222-238.

Beccaluva, L., Gabbianelli, G., Lucchini, F., Rossi, P.L. and Savelli, C. (1985a) 'Petrology and K/Ar ages of volcanics dredged from the Aeolian seamounts: implications for geodynamic evolution of the southern Tyrrhenian basin', Earth Planet.Sci.Lett. 74, 187-208.

Beccaluva, L., Di Girolamo, P. and Serri G. (1985b) 'High-K calc-alkaline, shoshonitic and leucitic volcanism of Campania (Roman Province, southern Italy): trace elements constraints on the genesis of an orogenic volcanism in a post-collisional, exstensional setting', IAVCEI Sci.Ass.Potassic Volcanism - Etna Volcano, Giardini Naxos, Italy.

Beccaluva, L., Brotzu, P., Macciotta, G., Morbidelli, R., Serri, G. and Traversa, G. (1989) 'Cainozoic Tectonic Magmatic Evolution and Inferred Mantle Source in the Sardo-Tyrrhenian Area', in A.Boriani, M. Bonafede, G.B.Piccardo, G.B.Vai (eds), The Lithosphere in Italy, Atti Acc. Naz. Lincei, Roma, 80, pp. 229-248.

Beccaluva, L., Di Girolamo, P. and Serri, G. (1991) 'Petrogenesis and tectonic setting of the Roman Volcanic Province, Italy', Lithos 26, 191-221.

Bellon, H. (1981) 'Chronologie radiométrique (K-Ar) des manifestations magmatiques autour de la Méditerranée occidentale entre 33 et 1 MA', in C.F. Wezel (ed), Sedimentary Basins of Mediterranean Margins, Tecnoprint, Bologna, pp.341-360.

Biju-Duval, B., Dercourt, J. and Le Pichon, X. (1977) 'From the Tethys ocean to the Mediterranean sea: a plate tectonic model of the evolution of the western Alpine system', in B. Biju-Duval and L. Montadert (eds), The structural hystory of the Mediterranean basin, Editions Technip, Paris, pp.143-164.

Boccaletti, M., Coli, M., Decandia, F.A., Giannini, E. and Lazzarotto, A. (1980) 'Evoluzione dell'Appennino settentrionale secondo un nuovo modello strutturale', Mem.Soc.Geol.It. 21, 359-373.

Boccaletti, M., Cello, G. and Tortorici, L. (1987) Transtensional tectonics in the Sicily Channel', Journal struct.Geol. 9, 869-876.

Boccaletti, M., Cello, G. and Tortorici, L. (1990) 'Strike-slip deformation as a fundamental process during the Neogene-Quaternary evolution of the Tunisian-Pelagian area', Annales Tectonicae 4, 104-119.

Boccaletti, M. and Nur, A. (1990) Active and Recent strike-slip tectonics, Annales Tectonicae, 4.

Bousquet, J.C. and Philip, H. (1986) ' Neotectonics of the Calabrian Arc and Apennines (Italy): an example of plio-Quaternary evolution from island arcs to collisional stages', in F.C. Wezel (ed), The origin of arcs, Elsevier, Amsterdam, 19, pp.305-326.

Burchfiel, B.C. (1980) 'Eastern European Alpine system and the Carpathian orocline as an example of collision tectonics', Tectonophysics 63, 31-61.

Calamita, F. and Deiana, G. (1988) 'The arcuate shape of the Umbria-Marche-Sabina Apennines (Central Italy)', Tectonophysics 146, 139-147.

Calanchi, N., Colantoni, D., Rossi, P.L., Saitta, M. and Serri, G. (1989) 'The Strait of Sicily continental rift system : physiography and petrochemistry of the submarine volcanic centres', Mar.Geol. 87, 55-83.

Carbone, S., Cosentino, M., Grasso, M., Lentini, F., Lombardo, G. and Patané, G. (1982) 'Elementi per una prima valutazione dei caratteri sismotettonici dell'avampaese ibleo (Sicilia sud-orientale)', Mem.Soc. Geol.It. 24, 507-520.

Casero, P., Roure, F., Endignoux, L., Moretti, I., Muller, C., Sage, L. and Vially, R. (1988) 'Neogene geodynamic evolution of the Southern Apennines', Mem.Soc.Geol.It. 41, 109-120.

Casnedi, R., Crescenti, V. and Tonna, M. (1982) 'Evoluzione dell'avanfossa adriatica meridionale nel Plio-Pleistocene, sulla base di dati del sottosuolo', Mem.Soc.Geol.It. 24, 243-260.

Castellarin, A. (1979) 'Il problema dei raccorciamenti crostali nel Sudalpino', Rend.Soc.Geol.It. 1, 21-23.

Castellarin, A. (1984) 'Schema delle deformazioni tettoniche sudalpine', Boll.Oceanologia Teor.Appl. 2, 105-114.

Castellarin, A., Colacicchi, R., and Praturlon, A. (1978) 'Fasi distensive, trascorrenze e sovrascorrimenti lungo la "Linea Ancona-Anzio", dal Lias medio al Pliocene', Geol. Rom. 17, 161-189.

Castellarin, A., Colacicchi, R., Praturlon, A. and Cantelli, C. (1982) 'The Jurassic-Lower Pliocene history of the Ancona-Anzio line (Central Italy)', Mem.Soc.Geol.It. 24, 325-336.

Castellarin, A. and Sartori, R. (1986) 'Il sistema tettonico delle Giudicarie, della Val Trompia e del sottosuolo dell'alta pianura lombarda', Mem.Soc.Geol.It. 26, 31-37.

Castellarin, A. and Vai, G.B. (1986) 'Southalpine versus Po Plain Apenninic Arcs', in F.C. Wezel (ed), The origin of arcs, Elsevier, Amsterdam, 19, pp.253-280.

Castellarin, A., Cantelli, L., Fesce, A.M., Mercier, J.L., Picotti, V., Pini, G.A., Prosser, G. and Selli, L. (1992) 'Alpine compressional tectonics in the Southern Alps. Relationship with the N-Apennines' Annales Tectonicae, 1, pp 62-94.

Catalano, R. and D'Argenio, B. (1982) 'Schema geologico della Sicilia, Guida alla geologia della Sicilia occidentale' Soc.Geol.It., pp. 9-41.

Catalano, R., D'Argenio, B. and Torelli, L. (1989) 'From Sardinia Channel to Sicily Straits. A Geologic Section Based on Seismic and Field Data', in A. Boriani, M. Bonafede, G.B. Piccardo, G.B. Vai (eds), The Lithosphere in Italy, Atti Acc. Naz. Lincei, Roma, 80, pp.110-128.

Cello, G. (1987) 'Structure and deformation processes in the Strait of Sicily rift zone', Tectonophysics 141, 237-247.

Channell, J.E.T. and Horvath, F. (1976) 'The Africa-Adriatic promontory as a paleogeographical premise for Alpine orogeny and plate movements in the Carpatho-Balkan region', Tectonophysics 35, 71-101.

Ciaranfi, N., Guida, M., Iaccarino, G., Pescatore, T., Pieri, P., Rapisardi, L., Ricchetti, G., Sgrosso, I., Torre, M., Tortorici, L., Turco, E., Scarpa, R., Cuscito, M., Guerra, I., Iannaccone, G., Panza, G.F. and Scandone, P. (1983) 'Elementi sismotettonici dell'Appennino Meridionale', Boll.Soc.Geol.It. 102, 201-222.

Civetta, L., Francalanci, L., Manetti, P. and Peccerillo, A. (1989) 'Petrological and geochemical variations across the Roman Comagmatic Province: inference on magma genesis and crust-mantle evolution', in A. Boriani, M. Bonafede, G.B. Piccardo, G.B.Vai (eds), The Lithosphere in Italy, Atti Acc. Naz. Lincei Roma, 80, pp.249-27.

CNR-PFG: Bigi, G., Castellarin, A., Catalano, R., Coli, M., Cosentino, D., Dal Piaz, G.V., Lentini, F., Parotto, M., Patacca, E., Praturlon, A., Salvini, F., Sartori, R., Scandone, P. and Vai, G.B. (1989) Synthetic structural-kinematic map of Italy - Scale 1:2.000.000, C.N.R., Roma.

Cobbold, P.R. and Davy, P.H. (1988) 'Indentation tectonics in nature and experiment, 2 , Central Asia', Bull.Geol.Inst., Univ. Uppsala, New Ser. 14, 143-162.

Cohen, C.R. (1980) 'Plate tectonic model for the Oligo-Miocene evolution of the Western Mediterranean', Tectonophysics 68, 283-311.

Cohen, S.C. and Morgan, R.C. (1987) 'Intraplate deformation due to continental collision : A numerical study of deformation in a thin viscous sheet', Tectonophysics 132, 247-260.

Conticelli, S., Manetti, P., Peccerillo, A. and Santo, A. (1986) 'Caratteri petrologici delle vulcaniti potassiche italiane: considerazioni genetiche e geodinamiche', Mem.Soc.Geol.It. 35, 775-783.

De Jonge, M.R. and Wortel, M.J.R. (1990) 'The thermal structure of the Mediterranean upper mantle: a forward modelling approach', Terra Nova 2, 609-616.

Dercourt, J., Zonenshain, L.P., Ricou, L.E., Kazmin, V.G., Le Pichon, X., Knipper, A.L., Grandjacquet, C., Sbortshikov, I.M., Geyssant, J., Lepvrier, C., Pechersky, D.H., Boulin, J., Sibuet, J.C., Savostin, L.A., Sorokhtin, O., Westphal, M., Bazchenov, M.L., Lauer, J.P. and Biju-Duval, B. (1986) 'Geological evolution of the Tethys belt from Atlantic to the Pamirs since the Lias', in J. Aubouin, X. Le Pichon and A.S. Monin (eds), Evolution of the Tethys, Tectonophysics 123, 241-315.

Dewey, J.F. and Sengor, A.M.C. (1979) 'Aegean and surrounding regions: complex multiplate and continuum tectonics in a convergent zone', Geol.Soc.Am.Bull. 90, 89-92.

Di Girolamo, P. (1978) 'Geotectonic setting of Miocene-Quaternary volcanism in and around the eastern Tyrrhenian Sea border (Italy) as deduced from major elements geochemistry', Bull.Volcanol. 41, 1-22.

Di Girolamo, P., Morra D., Ortolani, F. and Pagliuca, S. (1988) 'Osservazioni petrologiche e geodinamiche sul magmatismo "orogenico transizionale" della Campania nell'evoluzione della fascia Tirrenica della catena Appenninica', Boll.Soc.Geol.It. 107, 561-578.

Di Nocera, S., Ortolani, F. and Torre, M. (1976) 'La tettonica messiniana nell'evoluzione della catena Appenninica. Meeting on: Il significato geodinamico della crisi di salinità del Miocene Terminale del Mediterraneo, CRN-PFG, Firenze, pp.29-47.

Elter, P., Giglia, G., Tongiorgi, M. and Trevisan, L. (1975) 'Tensional and compressional areas in the recent (Tortonian to Present) evolution of the Northern Apennines', Boll.Geof.Teor. Appl. 65, 3-18.

England, P.C. and Mc.Kenzie, D. (1982) 'A thin viscous sheet model for continental deformation', Geophys.J.R. astr. Soc. 70, 295-321.
Fabbri, A. and Curzi, P. (1979) 'The Messinian of the Tyrrhenian Sea: seismic evidence and dynamic implication', Giorn.Geol. 43, 215-248.
Finetti, I. (1984) 'Geophysical study of the Sicily Channel Rift Zone', Boll.Geof.Teor.Appl. 26, 101-328.
Finetti, I. and Del Ben, A. (1986) 'Geophysical study of the Tyrrhenian opening', Boll.Geof.Teor.App. 110, 75-156.
Ghisetti, F. and Vezzani, L. (1982) 'The Recent Deformation Mechanism of the Calabrian Arc', in E.Mantovani and R.Sartori (eds), Structure, Evolution and present Dynamics of the Calabrian Arc, Earth Evolut.Sci. 3, 197-206.
Ghisetti, F. and Vezzani, L. (1984) 'Thin-skinned deformations of the western Sicily thrust belt and relationships with crustal shortening: mesostructural data on the Mt. Kumeta-Alcantara fault zone and related structures', Boll.Soc.Geol.It. 103, 129-157.
Ghisetti, F. and Vezzani, L. (1986) 'Assetto geometrico ed evoluzione strutturale della catena del Gran Sasso tra Vado di Sella e Vado di Corno', Boll.Soc.Geol.It. 105, 131-171.
Giardini, D. and Velonà, M. (1991) 'The deep seismicity of the Tyrrhenian Sea', Terra Nova 3, 57-64.
Grasso, M. and Lentini, F. (1982) 'Sedimentary and tectonic evolution of the eastern Hyblean Plateau (southeastern Sicily) during Late Cretaceous to Quaternary time', Palaeogeogr. Palaeoclimatol. Palaeoecol. 39, 261-280.
Grasso, M., De Dominicis, A. and Mazzoldi, G. (1990) 'Structure and tectonic setting of the western margin of the Hyblean-Malta shelf, Central Mediterranean', Annales Tectonicae 2, 140-154.
Grundmann, G. and Morteani, G. (1985) 'The young uplift and thermal history of the central Eastern Alps (Austria/Italy), evidence from apatite fission track ages', Jb.Geol.Bundesanst. 128, 2, 197-216.
Horvath, F. (1984) 'Neotectonics of the Pannonian basin and the surrounding mountain belts: Alps, Carpathians and Dinarides', Annales Geophys. 2, 147-154.
Illies, J. (1981) 'Graben formation in the Maltese Islands: a case history', Tectonophysics 73, 151-168.
Jongsma, D., Woodside, J.M., King, G.C.P. and Van Hinte, J.E. (1987) 'The Medina Wrench: a key to the kinematics of the central and eastern Mediterranean over the past 5 My', Earth planet. Sci.Lett. 82, 87-106.
Kastens, K.A., Mascle, J. et al. (1987) 'Proc.Init.Repts. ODP 107', (Pt.A), pp.1013.
Kastens, K.A. et al. (1988) 'ODP Leg 107 in the Tyrrhenian Sea: insights into passive margin and backarc basin evolution', Geol.Soc.Am.Bull. 100, 1140-1156.
Laubscher, H.P. (1983) 'The late Alpine (periAdriatic) intrusions and the Insubric line', Mem.Soc.Geol.It. 26, 21-30.
Laubscher, H.P. (1988) 'Material balance in Alpine orogeny', Geol.Soc.Amer.Bull. 100, 9, 1313-1328.
Lavecchia, G. (1988) 'The Tyrrhenian-Apennines system: structural setting and

seismotectogenesis', Tectonophysics 147, 263-296.

Lavecchia, G., Minelli, G. and Pialli, G. (1988) 'The Umbria-Marche arcuate fold belt', Tectonophysics 146, 125-138.

Lentini, F. (1982) 'The geology of the Mt.Etna basement', Mem.Soc.Geol.It. 23, 7-25.

Malinverno, A. and Ryan, W.B.F. (1986) 'Extension in the Tyrrhenian sea and shortening in the Apennines as result of arc migration driven by sinking of the lithosphere', Tectonics 5, 227-245.

Mantovani, E., Albarello, D., Babbucci, D. and Tamburelli, C. (1992) Recent geodynamic evolution of the Central Mediterranean area (Tortonian to Present), Tipografia Senese, Siena, Italy.

Marinelli, G. (1975) 'Magma evolution in Italy', in C.H. Squyres (ed), Geology of Italy, The Earth Sci.Soc. of Lybian A.R., Tripoli, pp.165-219.

Marton, E. (1987) 'Paleomagnetism and tectonics in the Mediterranean region', J.Geodyn. 7, 33-57.

Marton, E. (1993) 'Paleomagnetism in the Mediterranean from Spain to the Aegean: a review of data relevant to Cenozoic movements', This volume.

Massari, F. (1990) 'The foredeeps of the northern Adriatic margin: evidence of diachroneity in deformation of the Southern Alps', Riv.It.Paleont.Strat. 96, 2/3, 351-380.

Mascle, J. and Rehault, J.P. (1990) 'A revised seismic stratigraphy of the Tyrrhenian Sea: implications for the basin evolution', in K.A. Kastens, J. Mascle et al. (eds), Proc.ODP Sci.Results, 107. College Station, Tx, pp.617-636.

Mercier, J., Carey, E., Philip, H. and Sorel, D. (1976) 'La Néotectonique plio-quaternaire de l'Arc Egéen externe et de la Mer Egée et ses relations avec la seismicité, Bull.Soc.Géol. Fr. 7, 355-372.

Mercier, J., Delibassis, N., Gauthier, A., Jarrige, J., Lemeille, F., Philip, H., Sebrier, M. and Sorel, D. (1979) 'La Néotectonique de l'Arc Egéen', Rev.Géol. Dyn.Géogr.Phys., Spec.Publ. 21, 67-92.

Mercier, J., Sorel, D., Vergely, P. and Simeakis, K. (1989) 'Extensional tectonic regimes in the Aegean basins during the Cenozoic', Basin Research 2, 49-71.

Molnar, P. and Gray, D. (1979) 'Subduction of continental lithosphere: some constraints and concertainties', Geology 7, 58-62.

Montigny, R., Edel, J.B. and Thuizat, R. (1981) 'Oligo-Miocene rotation of Sardinia: K/Ar ages and paleomagnetic data of Tertiary volcanics', Earth Planet.Sci.Lett. 54, 261-271.

Moretti, I. and Royden, L. (1988) 'Deflection, gravity anomalies and tectonics of doubly subducted continental lithosphere: Adriatic and Ionian Seas', Tectonics 7, 875-893.

Mostardini, F. and Merlini, S. (1986) 'Appennino centro-meridionale. Sezioni geologiche e proposta di modello strutturale', Mem.Soc.Geol.It. 35,177-202.

Moussat, E., Rehault, J.P. and Fabbri, A. (1986) 'Rifting et evolution tectono-sedimentaire du Bassin tyrrhenien au cours du Neogene et du Quaternaire', Giorn.Geol. 48, 41-62.

Nur, A., Dvorkin, J., Mavko G. and Ben-Avraham, Z. (1991) 'Deformation of back arc basins', Terra abstracts 3, 43.

Ortolani, F. (1979) 'Alcune considerazioni sulle fasi tettoniche mioceniche e plioceniche dell'Appennino meridionale', Boll.Soc.Geol.It. 97, 609-616.

Ortolani, F. and Aprile, F. (1977) 'Struttura profonda dell'Irpinia centrale (Appennino campano)', Boll.Soc.Geol.It. 95, 903-921.

Parotto, M. and Praturlon, A. (1975) 'Geological summary of central Apennines', in L. Ogniben, M. Parotto and A. Praturlon (eds), Structural model of Italy, Quad.Ric.Sci. 90, 257-311.

Patacca, E. and Scandone, P. (1989) 'Post-Tortonian mountain building in the Apennines. The role of the passive sinking of a relict lithospheric slab', in A. Boriani, M. Bonafede, G.B. Piccardo, G.B. Vai (eds), The Lithosphere in Italy, Atti Acc. Naz. Lincei, Roma, 80, pp.157-176.

Patacca, E., Sartori, R. and Scandone, P. (1990) 'Tyrrhenian basin and Apenninic arcs: kinematic relations since Late Tortonian times', Mem.Soc.Geol.It., in press.

Peccerillo, A. and Manetti, P. (1985) 'The Potassium alkaline volcanism of central-southern Italy: a review of the data relevant to petrogenesis and geodynamic significance', Trans.Geol.Soc.S.Afr. 88, 379-394.

Peltzer, G. and Tapponnier, P. (1988) 'Formation and evolution of strike-slip faults, rifts, and basins during the India. Asia collision: An experimental approach', J.Geophys.Res. 93, 15085-15117.

Philip, H. (1987) 'Plio-Quaternary evolution of the stress field in Mediterranean zones of subduction and collision', Annales Geophys. 5B, 301-320.

Pieri, M. and Groppi, G. (1981) 'Subsurface geological structure of the Po Plain', Pubbl.414, PFG-C.N.R., pp. 23.

Ratschbacher, L., Merle, O., Davy, P. and Cobbold, P. (1991a) 'Lateral extrusion in the Eastern Alps, Part 1: boundary conditions and experiments scaled for gravity', Tectonics 10, 245-256.

Ratschbacher, L., Frisch, W., Linzer, H.G. and Merle, O. (1991b) 'Lateral extrusion in the Eastern Alps, Part 2: structural analysis', Tectonics 10, 257-271.

Rehault, J.P., Moussat, E., Mascle, J. and Sartori, R. (1985) 'Geodynamic evolution of the Tyrrhenian Sea. New data and drilling objectives', 8th Congr.Reg.Comm.Medit. Neogene Strat., Budapest (abstract).

Rehault, J.P., Moussat, E. and Fabbri, A. (1987) 'Structural evolution of the Tyrrhenian back-arc basin', Mar.Geol. 74, 123-150.

Reuther, C.D. (1987) 'Extensional tectonic within central Mediterranean segment of the Afro-European zone of convergence', Mem.Soc.Geol.It. 38, 69-80.

Reuther, C.D. (1990) 'Strike-slip generated rifting and recent tectonic stresses on the African foreland (central Mediterranean region)', Annales Tectonicae 4, 120-130.

Rossi, S. and Sartori, R. (1981) 'A seismic reflection study of the External Calabrian Arc in the Northern Ionian Sea (Eastern Mediterranean)', Mar.Geophys.Res. 4, 403-426.

Rossi, S. and Zarudski, E.F.K. (1978) 'Medina e Cirene: Montagne sottomarine del mar Ionio', Boll.Geof.Teor.Appl. 77, 61-67.

Royden, L. (1985) 'The Vienna basin: a thin-skinned pull-apart basin', in K.T. Biddle and N.C. Blick (Eds), Strike-slip deformation, basin formation and sedimentation, Soc.Econ.Paleont.Miner., Spec.Publ. 37, Tulsa, pp. 227-264.

Royden, L., Horvath, F. and Rumpler, J. (1983).'Evolution of the Pannonian basin system, 1: Tectonics', Tectonics 2, 63-90.

Royden, L., Patacca, E. and Scandone, P. (1987) 'Segmentation and configuration of subducted lithosphere in Italy: An important control on thrust-belt and foredeep-basin evolution', Geology 15, 714-717.

Royden, L. (1993) 'The evolution of retreating subduction boundaries formed during continental collision', Tectonics, in press.

Sartori, R. (1986) 'Notes on the geology of the acoustic basement in the Tyrrhenian Sea', Mem.Soc.Geol.It. 36, 99-108.

Sartori, R. (1989) 'Evoluzione neogenico-recente del bacino tirrenico ed i suoi rapporti con la geologia delle aree circostanti', Giorn.Geol., ser. 3, 51/2, 1-39.

Sartori, R. and ODP Leg 107 Scientific Staff (1989) 'Drillings of ODP Leg 107 in the Tyrrhenian Sea: tentative basin evolution compared to deformation in the surrounding chains', in A.Boriani, M.Bonafede,G.B.Piccardo, G.B.Vai (eds), The lithosphere in Italy, Atti Acc.Naz.Lincei 80, pp.139-156.

Sartori, R., Calalonga, M.L., Gabbianelli, G., Bonazzi, C., Carobene, S., Curzi, P.V., Evangelisti, D., Grasso, M., Lentini, F., Rossi, S. and Selli, L. (1992) 'Note stratigrafiche e tettoniche sul "Rise di Messina" (Ionio nord-occidentale)', Giorn. Geol. 53, (in press).

Savelli, C., Beccaluva, L., Deriu, M., Macciotta, G.B. and Maccioni, L. (1979) 'K/Ar geochronology and evolution of the Tertiary "calk-alcalik" volcanism of Sardinia (Italy)', J.Volc.geotherm Res. 5, 257-269.

Scandone, P. (1979) 'Origin of the Tyrrhenian Sea and Calabrian Arc', Boll.Soc.Geol.It. 98, 27-34.

Scandone, P. (1982) 'Structure and Evolution of the Calabrian Arc', in E. Mantovani and R. Sartori (Eds), Structure, evolution and present dynamics of the Calabrian Arc, Earth Evolut.Sci. 3, 172-179.

Scandone, P., Patacca, E., Radoicic, R., Ryan, W.B.F., Cita, M.B., Rawson, M., Chezar, H., Miller, E., McKenzie, J. and Rossi, S. (1981) 'Mesozoic and Cenozoic Rocks from the Malta Escarpment (Central Mediterranean)', Am.Assoc.Pet.Geol.Bull. 65, 1299-1313.

Schmid, S.M., Aebli, H.R., Heller, F. and Zingg, A. (1989) 'The role of the Periadriatic Line in the tectonic evolution of the Alps', in M.P. Coward, D. Dietrich and R.G. Park (eds), Alpine tectonics, Geol.Soc.Spec.Publ. 45, pp.153-171.

Semenza, E. (1974) 'La fase Giudicariense, nel quadro di una nuova ipotesi sull'Orogenesi Alpina nell'area Italo-Dinarica', Mem.Soc.Geol.It. 13, 187-226.

Serri, G., (1990) 'Neogene-Quaternary magmatism of the Tyrrhenian region: characterization of the magma sources and geodynamic implications', Mem.Soc.Geol.It. 41, 219-242.

Serri, G., Innocenti, F., Manetti, P., Tonarini, S. and Ferrara, G. (1991) 'Il magmatismo neogenico-quaternario dell'area tosco-laziale-umbra: implicazioni sui modelli di evoluzione geodinamica dell'Appennino settentrionale', Studi Geologici Camerti, Spec.Vol., 429-463.

Spakman, W. (1990) 'Tomographic images of the upper mantle below central Europe and the Mediterranean', Terra Nova 2, 512-553

Tapponier, P., Peltzer, G., Le Dain, A.Y., Armijo, R. and Cobbold, P.R. (1982) 'Propagating extrusion tectonics in Asia: New insights from simple experiments with plasticine', Geology 10, 611-616.

Torre, M., Di Nocera, S. and Ortolani, F. (1988) 'Evoluzione Post-Tortoniana dell'Appennino meridionale', Mem.Soc. Geol.It., 41, 47-56.

Vai, G.B. (1987) 'Migrazione complessa del sistema fronte deformativo-avanfossa-Cercine periferico: il caso dell'Appennino settentrionale', Mem.Soc.Geol.It., 38, 95-105.

Van Dijk, J.P. and Okkes, M. (1991) 'Neogene tectonostratigraphy and kinematics of Calabrian basins; implications for the geodynamics of the Central Mediterranean', Tectonophysics 196, 23-60.

White, R. and McKenzie, D. (1989) 'Magmatism at rift zones: the generation of volcanic continental margins and flood basalts', J.Geophys.Res. 94, 7685-7729.

Zitellini, N., Trincardi, F., Marani, M and Fabbri, A. (1986) 'Neogene tectonics of the Northern Tyrrhenian Sea', Giorn.Geol. 48, 25-40.

AFRICA-EURASIA KINEMATICS IN THE MEDITERRANEAN: AN ALTERNATIVE HYPOTHESIS

D. ALBARELLO, E.MANTOVANI, D.BABBUCCI, C.TAMBURELLI
University of Siena
Dept. of Earth Sciences
Via Banchi di Sotto, 55
53100 Siena - Italy

ABSTRACT. The post-Tortonian deformation pattern in the Central Mediterranean can be satisfactorily interpreted in the framework of a SSW-NNE to SW-NE convergence between Africa and Eurasia (Mantovani et al., 1992 and this volume). However, this hypothesis is not in line with the most widely used Africa-Eurasia kinematic models, based on the analysis of North Atlantic kinematic data, which predict a SE-NW to S-N motion between Africa and Eurasia. Here it is argued that these models might be not reliable, since are based on the assumption that Eurasia is a unique coherent block from the Atlantic ridges to the Pacific trenches, which can hardly account for the significant intraplate deformation occurring in Western Europe, and in particular, in the Iberian peninsula and surrounding regions. This evidence would imply that the kinematic data observed along the Mid-Atlantic ridges just North of Azores and those on the Azores-Gibraltar belt cannot be used to constrain the relative motions of Eurasia with respect to North America and Africa respectively. It is shown by quantitative computations that the remaining kinematic data in the North Atlantic can be reconciled, within errors, with a NNE to NEward motion of Africa with respect to Eurasia, in the Central Mediterranean. This kinematic pattern does not involve any significant difficulty in explaining the tectonic pattern in the other sectors of the Mediterranean area and along the boundaries of the Eurasian continent.

1. Introduction

As known, the most straightforward way to determine the relative motion between two adjacent plates is the analysis of kinematic evidence along their common boundary, which, in the case of the Africa-Eurasia plate system, corresponds to the Mediterranean region. However, along this border the African and Eurasian forelands are separated by a large deformed zone which includes microplates, fragmented orogenic belts and thinned zones.

To overcome this problem, the history of the Africa-Eurasia kinematics has been indirectly determined from the relative motion of these two plates with respect to North

America, through the analysis of North Atlantic magnetic lineations (see, e.g., Pitman and Talwani, 1972; Dewey et al., 1973; Le Pichon et al., 1977; Savostin et al., 1986; Dewey et al., 1989). These attempts have suggested that in the Central Mediterranean Africa and Eurasia have been converging along a SE-NW or S-N direction during the last 9-10 My, in line with the results obtained by the reconstructions of present- day plate motions (see, e.g., Minster and Jordan, 1978; Chase, 1978; Argus et al., 1989; DeMets et al., 1990).

However, these kinematic models are not easily reconcilable with the post-Tortonian deformation pattern in the Central Mediterranean, as discussed by Mantovani et al. (1992 and this volume) and also pointed out by Dewey et al. (1989). In this work, we propose a possible solution of the above controversy. We advance the hypothesis that the Africa-Eurasia kinematic models so far proposed are not reliable, since they are based on the assumption that Eurasia behaves as a unique rigid block from the Atlantic ridges to the Pacific trenches. In our opinion, this condition is not fulfilled, because in the Ibero-Maghrebian region there is clear evidence of intraplate deformation (see, e.g., Philip, 1987; Buforn et al., 1988a,b; Cabral, 1989; Nicolas et al., 1990; Rebai et al., 1992; Royden, 1993) which could absorbe part of the spreading activity on the Mid-Atlantic ridges lying just North of Azores (Fig.1). As a consequence, we suggest that the spreading rates observed along the Mid-Atlantic ridges comprised between the Azores and latitude 45°N and the kinematic data observed along Azores-Gibraltar belt must not be used to constrain the Africa-North America and Africa-Eurasia relative motions respectively.

This is a crucial point, since it is possible to demonstrate that the remaining kinematic data in the North Atlantic can be reconciled, within experimental errors, with the trend of the Africa-Eurasia relative motion (SSW-NNE) suggested by the central Mediterranean deformation pattern. This statement is supported by quantitative computations in the next section.

2. Kinematic analysis

Fig.1 shows the distribution of the available kinematic data along the North Atlantic ridges and the Azores-Gibraltar belt which are used in this analysis. The values of spreading rates, transform fault azimuths and earthquake slip vector azimuths are reported in Tab.1. A detailed description of these observations and respective uncertainties is given by Argus et al. (1989) and DeMets et al. (1990).

The main difference between the plate mosaic used in this analysis and the ones adopted in previous kinematic reconstructions is given by the fact that here the Azores-Gibraltar belt is not assumed as a sector of the Africa-Eurasia border, since we suppose that the oceanic zone lying North of the Azores-Gibraltar belt does not move in close connection with Eurasia. We do not known how and where the decoupling between this mobile zone and stable Eurasia may occur. In the kinematic analysis, it has been assumed, for simplicity, a schematic plate boundary corresponding to the dashed line shown in Fig.1, which is very tentatively located along an old Africa-Eurasia border (Klitgord and Schouten, 1986; Whitmarsh et al., 1982; Srivastava et al., 1990) and an alignment of minor shocks (Buforn et al., 1988b).

However, it must be pointed out that the results of the following kinematic analysis are not significantly influenced by the precise position of the Eurasia-Iberia boundary, due to the scarce constraining power of the kinematic data lying on the Mid-Atlantic ridges which are comprised between Iceland and the Iberian block.

The purpose of this analysis is not that of finding the best kinematic solution of the three-plate system Africa-Eurasia-North America, but rather that of verifying the existence of solutions which can account for North Atlantic data and also predict a SSW-NNE motion of Africa with respect to Eurasia in the Central Mediterranean.

The kinematic pattern of the plate system considered is fully described by six parameters, given by the components of the rotation vectors of Africa and Eurasia with respect to North America. The remaining vector, related to the Africa-Eurasia system, is obtained from the composition of the previous ones by imposing the closure condition.

Figure 1 - Location of the data used in the kinematic analysis. Equal area projection (after Argus et al., 1989, modified). The zone comprised between the Azores-Gibraltar active belt and the dashed line (here called Iberian block) is allowed to move independently from Eurasia and Africa. The northern border of this block has been tentatively drawn through an Africa-Eurasia paleoborder (see, e.g., Klitgord and Shouten, 1986) which also corresponds to an alignment of few minor earthquakes (Buforn et al., 1988 b).

Through the Monte Carlo procedure, a number of Eurasia-North America rotation vectors, able to account within experimental errors for kinematic data along the Mid-Atlantic ridges lying North of latitude 45°N, has been found. By the same procedure, we obtained a number of Africa-North America vectors compatible with the kinematic data South of Azores. From the above two sets of Eurasia-North America and Africa-North America vectors, a number of Africa-Eurasia vectors has been obtained by imposing the closure condition. Some of the vectors obtained in such a way are compatible with a SSW-NNE Africa-Eurasia convergence in the Central Mediterranean. This result demonstrates that, in the framework of the plate mosaic here assumed, it is possible to reconcile the kinematic implications of Mediterranean deformation pattern with North

TABLE 1. Kinematic data in the North Atlantic ocean which have been used in the kinematic analysis. Data are grouped according to the plate boundary to which they refer, within the four-plate mosaic shown in Fig.1. The first four columns respectively give the position of each datum, the observed value and its uncertainty (from Argus et al., 1989). Spreading rates are given in mm/year along the direction normal to the ridge. Azimuths of slip vectors (Az.) and transform faults are given in degrees (measured clockwise). The fifth column reports the values computed on the basis of our kinematic model (Tab. 2). The last column gives the difference (D) between the observed and computed values.

Africa - North America

Spreading rates

Lat.	Lon.	Rate	Error	Rate	D
36.8	-33.2	21.5	2.0	21.8	-0.3
36.8	-33.2	20.5	2.0	21.8	-1.3
36.5	-33.7	22.0	3.0	21.4	0.6
36.0	-34.1	20.0	3.0	21.5	-1.5
35.0	-36.5	21.0	4.0	22.7	-1.7
34.3	-37.0	21.0	3.0	22.2	-1.2
31.9	-40.5	23.0	4.0	22.6	0.4
30.9	-41.7	23.0	4.0	22.9	0.1
30.5	-41.9	22.0	3.0	23.4	-1.4
29.6	-43.0	23.0	3.0	23.8	-0.8
27.5	-44.2	24.0	3.0	23.7	0.3
26.9	-44.5	26.0	4.0	24.1	1.9
26.2	-44.8	22.0	3.0	23.0	-1.0
25.7	-45.0	24.0	4.0	23.7	0.3
25.3	-45.4	22.5	2.0	23.8	-1.3
25.1	-45.4	24.5	2.0	23.8	0.7
24.5	-46.1	23.0	4.0	23.9	-0.9
24.2	-46.3	24.5	2.0	24.4	0.1
23.0	-45.0	25.0	4.0	24.6	0.4
22.8	-45.0	25.0	2.0	24.5	0.5

Transform faults

Lat.	Lon.	Az.	Error	Az.	D
35.2	-35.6	105.	2.0	104	1.
33.7	-38.7	105.	2.0	103	2.
30.0	-42.4	102.	3.0	102	0.
23.7	-45.7	98.	2.0	101	-3.

Slip vectors

Lat.	Lon.	Az.	Error	Az.	D
35.4	-36.0	102.	20.	104.	-2.
35.4	-36.0	101.	10.	104.	-3.
35.4	-36.1	100.	10.	104.	-4.
35.1	-35.5	101.	15.	104.	-3.
33.8	-38.6	101.	10.	103.	-2.
33.8	-38.5	102.	15.	103.	-1.
33.7	-38.6	103.	15.	103.	0.
28.7	-43.6	91.	20.	101.	-10.
23.9	-45.6	100.	10.	101.	-1.
23.8	-45.9	100.	10.	101.	-1.
23.8	-45.4	106.	15.	101.	5.
23.7	-45.2	102.	15.	101.	1.

Eurasia - North America

Spreading rates

Lat.	Lon.	Rate	Error	Rate	D
86.5	43.0	12.0	3.0	13.2	-1.2
84.9	7.5	13.0	3.0	14.2	-1.2
84.1	0.0	13.0	2.0	14.5	-1.5
83.4	-4.5	15.0	3.0	14.7	0.3
73.7	8.5	17.0	4.0	15.9	1.1
72.5	3.0	15.0	4.0	15.1	-0.1
71.8	-2.5	14.0	3.0	13.7	0.3
69.6	-16.0	17.0	2.0	17.6	-0.6
69.3	-16.0	17.5	2.0	17.6	-0.1
68.5	-18.0	18.0	2.0	17.8	0.2
67.9	-18.5	18.0	2.0	17.9	0.1
61.6	-27.0	19.0	2.0	17.1	1.9
60.2	-29.1	19.0	2.0	17.0	2.0

Transform faults

Lat.	Lon.	Az.	Error	Az.	D
80.0	1.0	126	5.0	126	0.
78.8	5.0	127	10.0	129	2.
71.3	-9.0	114	3.0	115	1.
52.6	-33.2	96	3.0	96	0.
52.1	-30.9	96	2.0	97	-1.

Slip vectors

Lat.	Lon.	Az.	Error	Az.	D
80.3	-1.9	125.	20.	124	1.
80.2	-0.7	130.	20.	125	5.
79.8	2.9	134.	20.	128	6.
79.8	2.9	139.	20.	128	11.
71.0	-6.9	116.	20.	116	0.
71.2	-8.0	113.	15.	116	-3.
71.2	-8.2	110.	20.	115	-5.
71.5	-10.4	106.	20.	114	-8.
71.6	-11.5	111.	15.	113	-2.
52.8	-34.3	98	10.	95	3.
52.8	-34.2	101	20.	95	6.
52.7	-33.3	100	20.	96	4.
52.7	-32.0	98	10.	97	1.
52.5	-31.9	103	20.	97	6.

Africa - Iberia

Transform faults

Lat.	Lon.	Az.	Error	Az.	D
36.9	-23.5	257.	5.	259.	-2.
37.0	-22.6	265.	3.	263.	2.
37.1	-21.7	265.	3.	266.	-1.
37.1	-20.5	-90.	7.	-90.	0.

Slip vectors

Lat.	Lon.	Az.	Error	Az.	D
37.8	-17.3	-89.	25.	-78.	-11.
37.2	-14.9	-50.	25.	-70.	20.
37.0	-11.8	267.	25.	300.	-33.
36.0	-10.6	-35.	25.	-54.	19.
36.0	-10.3	-60.	25.	-53.	-7.
36.2	-7.6	-35.	25.	-47.	12.

Iberia-North America

Trasform faults

Lat.	Lon.	Az.	Error	Az.	D
44.5	-28.2	25.0	4.0	24.4	0.6
43.8	-28.5	24.0	3.0	24.7	-0.7
43.3	-29.0	23.0	3.0	24.7	-1.7
42.9	-29.3	25.5	2.0	24.7	0.8
42.7	-29.3	23.0	2.0	24.7	-1.7
42.3	-29.3	23.5	2.0	24.3	-0.8
41.7	-29.2	24.5	3.0	24.7	-0.2

Atlantic kinematic data. Tab.2 reports the Africa-North America, Eurasia-North America and Africa-Eurasia poles and related rotation rates which are characterized by the lowest residuals among those explored.

In order to elaborate a more complete kinematic model, able to account for all data available (Tab.1), we also tried to determine a kinematic pattern of the Iberian block (Fig.1) which can account for the data along the adjacent Mid-Atlantic ridges and the Azores-Gibraltar belt and also satisfy the closure conditions of all the three-plate circuits involved: North America-Africa-Iberia, Africa-Eurasia-Iberia and North America-Eurasia-Iberia. By the Monte Carlo procedure, a set of North America-Iberia rotation poles, compatible with spreading rates in the Mid-Atlantic ridges lying between Azores and latitude 45°N, has been identified. The combination of each one of these vectors with the Africa-North America vector reported in Tab.2 has provided a set of Africa-Iberia vectors, from which we selected the one that best fits the kinematic data along the Azores-Gibraltar belt. By imposing the closure condition to the plate circuit Africa-Iberia-Eurasia we also determined an Iberia-Eurasia vector. The North America-Iberia, Africa-Iberia, Eurasia-Iberia vectors obtained through this procedure are reported in Tab.2.

The theoretical predictions of the overall kinematic solution proposed in Tab.2 are compared with observed values in Tab.1.

A statistical check of the reliability of the adopted solution has been carried out through a chi-squared test (Gordon et al., 1987), which indicates that the level of data fitting is reliable at a high confidence level (greater than 99%).

The kinematic solution reported in Tab.2 cannot be considered the best fitting one in a least squares sense. The best-fit solution which was obtained by Argus et al. (1989) for the three-plate mosaic constituted by Africa, North America and Eurasia (this last taken as a unique coherent plate from the Atlantic ridges to the Pacific trenches) is characterized by slightly smaller residuals with respect to those reported in Tab.1. However, this does not necessary mean that the solution of Argus et al. (1989) should be preferred to ours, given that the respective fitting levels are not significantly different, as indicated by an F-test (Gordon et al., 1987).

TABLE 2. Proposed set of rotation vectors for the 4-plate system shown in Fig.1. This kinematic solution predicts a SSW-NNE motion between Africa and Eurasia in the central Mediterranean and can account, within errors, for the spreading rates and the azimuths of transform faults and earthquake slip vectors in the Atlantic ridges and Azores-Gibraltar belt (see Tab.1). The rotation rates refer to the motion of the first plate with respect to the second. See text for other explanations.

PLATES	EULERIAN POLE		RATE
	Lat.	Lon.	(°/My)
Africa - North America	74.0	93.4	0.225
Eurasia - North America	49.1	138.3	0.183
Iberia - North America	49.3	136.7	0.225
Africa - Eurasia	41.6	-11.8	0.117
Africa - Iberia	22.7	-20.6	0.119
Iberia - Eurasia	59.9	129.6	0.040

3. Discussion and Conclusions

The procedure which has been used to determine the current Africa-Eurasia kinematic models (e.g., DeMets et al., 1990) is correct and can satisfactorily account for all the available kinematic data in the North Atlantic. In spite of this, we suspect that the result of this approach, indicating a SE-NW to S-N motion of Africa with respect to Eurasia, in the Central Mediterranean is not reliable. This because we believe that a SW-NE to SSW-NNE Africa - Eurasia motion is the best kinematic framework to explain the post-Tortonian deformation pattern in the Central Mediterranean region (Mantovani et al., 1992 and this volume). This work shows that the hypothesis that Africa moves Northward or NWward with respect to Eurasia is mainly conditioned by the kinematic data observed along the Azores-Gibraltar belt and on the Mid-Atlantic ridges lying just North of Azores (see Tab.1). It is demonstrated that if one does not take into account these data, the remaining observations in the North Atlantic can be reconciled, within errors, with a SW-NE to SSW-NNE motion of Africa relative to Eurasia in the Central Mediterranean.

At this point, the main problem is to understand whether the hypothesis that just North of the Azores-Gibraltar belt there is a zone which does not move in close connection with Eurasia can be reconciled, or at least does not contrast with significant geological and geophysical evidence. The border between the hypothesized mobile zone and Africa can be reasonably assumed in correspondence of the Azores-Gibraltar belt. It seems, however, much more problematic to identify the border between the mobile zone and stable Eurasia. Mantovani et al. (1992) has tentatively assumed that this boundary corresponds to the tectonic belt running through the Rhine graben system and North Sea up to Iceland. This working hypothesis, however, would imply a relative motion of about 4 mm/year along the above discontinuity, which appears slightly greater than the maximum deformation rates estimated on the Rhine graben system. In this work, we have explored another hypothesis. We have assumed that only the region which is identified as Iberian block in Fig.1 moves independently from Eurasia. In this case, the decoupling between the Iberian mobile zone and Eurasia should be taken up by a discontinuity which goes from the Pyrenees towards the Atlantic ridges, such as the one shown in Fig.1. Some moderate seismicity occurs in the sector of the hypothesized discontinuity which lies close to the Iberian peninsula (Buforn et al., 1988b; Udias and Buforn, 1991), but this activity cannot account for the Iberia-Eurasia relative motion of about 4 mm/y predicted by the proposed kinematic solution (see Fig.2 and Tab.3). One might suppose that deformation in this zone is only partially taken up by seismicity. Another possible hypothesis is that the decoupling between the mobile zone and the stable part of Eurasia is taken up by several minor deformations in the zone comprised between the Rhine graben-North Sea-Iceland alignment and the Azores-Gibraltar belt.

As discussed earlier, the Africa-Eurasia motion predicted by the proposed kinematic solution is compatible with the deformation pattern in the Central Mediterranean. It is also opportune to make some comments about the compatibility, or at least the non incompatibility, of the proposed Africa-Eurasia kinematics with the major tectonic evidence in the western and eastern Mediterranean regions.

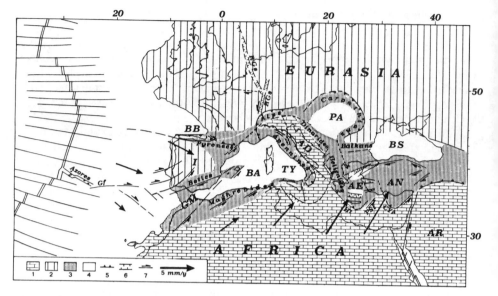

Figure 2 - Relative motions of the African and Iberian blocks relative to Eurasia and major tectonic features in the Mediterranean region. 1) African domain 2) Eurasian domain 3) Orogenic belts 4) Oceanic zones and neogenic basins 5,6,7) Main compressional, tensional and transcurrent features. AD: Adriatic block; AE: Aegea; AN: Anatolia; AR: Arabia; BA: Balearic basin; BB: Bay of Biscay; BS: Black Sea; CYA: Cyprus arc; Gf: Gloria fault; GM: Gibraltar-Morocco block; Ht: Hellenic trench; I: Iberia; PA: Pannonian basin; PSt: Pliny and Strabo trenches; RGs: Rhine Graben system; TY: Tyrrhenian basin. The arrows in Africa and the Iberian block indicate the velocity fields with respect to Eurasia in accord with the rotation poles respectively located at 41.6°N, 11.8°W with a rotation rate of 0.12°/My, and 59.9°N, 129.6°E with a rotation rate of 0.04°/My, (see Tab.2).

Iberia - Africa

The proposed kinematic model (Fig.2) implies that the African and Iberian blocks converge along a roughly SE-NW direction in the western Mediterranean region (see Tab.3). Deformation along this plate boundary involves a wide zone comprising the Iberian peninsula, the Balearic basin, the Maghrebian belt and the Atlantic zone lying offshore Iberia and Morocco (Fig.2). This zone, due to the SE-NW compressional regime between the Iberian block and Africa, has undergone a great fracturation, with the formation of several crustal wedges bounded by transpressional fault systems, as the Gibraltar-Morocco block (see Fig.2). The lateral extrusion of these wedges towards weakly constrained margins might have accommodated the plate convergence along this boundary (see, e.g., Tapponier, 1977; Philip, 1987; Buforn et al., 1988a,b; Rebai et al., 1992; Royden, 1993).

TABLE 3. Relative motion rates (mm/year) and displacement azimuths (degrees clockwise from North) of Africa and Iberian blocks with respect to Eurasia (see Fig.2) and of the Iberian block relative to Africa in a number of points close to the respective plate boundaries.

Africa - Eurasia				Iberia - Eurasia			
Lat.	Lon.	Az.	Rates	Lat.	Lon.	Az.	Rates
35	-20	132	2.1	40	-20	109	4.4
35	-5	53	1.9	43	-10	115	4.4
30	0	54	3.4				
31	5	43	3.8				
35	9	28	3.9				
32	11	34	4.6	Iberia - Africa			
33	13	31	4.8				
30	20	33	6.1	Lat.	Lon.	Az.	Rates
32	22	29	6.2				
29	26	33	7.1	39	-8	127	4.4
31	29	30	7.3	40	-2	138	5.3

A significant part of SE-NW shortening between Africa and Iberia is also absorbed by the reactivation, with thrust mechanisms, of old normal faults in the western Mediterranean basin (see, e.g., Mauffret et al., 1981) and by SE-NW ocean-ocean collisional processes in the area lying off-shore the Southern Portugal (see, e.g., Lynnes and Ruff, 1985; Grimison and Chen, 1988).

At the Gloria fault, the motions of the Iberian block and Africa are almost parallel (see Fig.2), in line with the dominant transcurrent mechanisms observed along this sector of the Azores-Gibraltar belt (Buforn et al., 1988b).

It is interesting to note that in northeastern Algeria, fault mechanisms with a dominant sinistral shear and NNE trending compressional axis have been observed (Bounif et al., 1987). Given that this sector of the belt marks the boundary between Africa and the Corsica-Sardinia massif, which does not show any appreciable movement with respect to Eurasia, the observed sinistral shear in eastern Algeria might indicate a left lateral motion of Africa relative to Eurasia, in accord with the kinematic model shown in Fig.2.

Africa - Eurasia

The interaction zone between Africa and Eurasia involves a wide tectonic belt comprising relatively rigid blocks, such as the Adriatic one, several sectors of the Alpine belt, such as the Alps, the Carpathians, the Balkans, the Dinarides, the Hellenides, the Aegean Arc and the Anatolian belts and internal basins, such as the Tyrrhenian, the Aegean, the Pannonian and the Black Sea. In this structural context, it appears rather hard to find significant evidence about the relative motion between the two converging continents. One can only remark that no clear evidence exists which may contradict a NE to NNE motion of Africa with respect to Eurasia in the Central-Eastern Mediterranean. On the contrary, the dominant trends of compressional fronts (mainly SE-NW at the Hellenic and Cyprus Arcs) and of transcurrent fault systems (mainly SW-NE at the Pliny-Strabo trenches and the minor transversal faults in the Hellenic Arc) observed along the northern African margin (McKenzie, 1978; Le Pichon and Angelier, 1979; Auroux et al., 1984) can easily be reconciled with the Africa-Eurasia kinematic model here proposed (Fig.2).

Arabia - Eurasia

Using the Africa-Eurasia rotation vector reported in Tab. 2 and the Arabia-Africa vector which best fits the kinematic evidence along the Red sea (DeMets et al., 1990), we obtained an Arabia-Eurasia pole located at 28.4°N, 17.0°E and a rotation rate of 0.51 °/My. This vector provides a relative plate motion in Northern Arabia, which is roughly oriented N20°W and has a rate of about 20 mm/y. These predictions are very similar to the ones provided by the kinematic model proposed by DeMets et al. (1990) and are compatible with recent deformation styles and stress field data observed along the Arabia-Eurasia boundary (Jackson, 1992 and this volume).

The kinematic solution proposed in Tab.2 predicts a motion of Eurasia with respect to North America which is significantly different from that obtained by global plate motion reconstructions (see, e.g., DeMets et al., 1990). In the previous sections, we have discussed the major consequences that this new motion of Eurasia implies in the Atlantic and Mediterranean regions. It is not possible to extend this discussion to the other plate boundaries of Eurasia, with the Indian and Pacific plates, since no reliable kinematic data are available in those regions. The only constraints which are imposed to the motion of Eurasia in global plate motion reconstructions are located along the borders of this block with Africa and North America (DeMets et al., 1990).

AKNOWLEDGMENTS

We are very grateful to Prof. J.Jackson for reviewing the manuscript and for the fruitful discussions during the Erice Meeting.
We also thank Mrs. F.Falciani, Mr. G.Vannucchi and Mr. R.Galgano (ING) for their collaboration in editing the text and drawing the figures. This work has been financially supported by Italian Research Council (CNR) and the Ministry of University and Scientific Research.

4. References

Auroux, C., Mascle, J. and Rossi, S. (1984) 'Geologia del margine ionico dalle isole Strofadi a Corfù (estremità settentrionale dell'Arco Ellenico)', Mem.Soc.Geol.It. 27, 267-286.

Argus, D.F., Gordon, R.G., DeMets, C. and Stein, S. (1989) 'Closure of the Africa-North America plate motion circuit and tectonics of the Gloria fault', J.Geophys.Res. 94, 5585-5602.

Bounif, A., Haessler, H. and Meghraoui, M. (1987) 'The Constantine (northeast Algeria) earthquake of October 27, 1985: surface ruptures and aftershock study', Earth Planet.Sci.Lett. 85, 451-460.

Buforn, E., Udias, A. and Mezcua, J. (1988a) 'Seismicity and focal mechanisms in south Spain', Bull.Seism.Soc.Am. 78, 2008-2024.

Buforn, E., Udias, A. and Colombas, M.A. (1988b) 'Seismicity, source mechanism and tectonics of the Azores-Gibraltar plate boundary', Tectonophysics 152, 89-118.

Cabral, S. (1989) 'An example of intraplate neotectonic activity, Vilarica Basin, Northeast Portugal', Tectonics 8, 285-303.

Chase, C.G. (1978) 'Plate kinematics: the Americas, East Africa, and the rest of the world', Earth Planet.Sci.Lett. 37, 355-368.

DeMets, C., Gordon, R.G., Argus, D.F. and Stein, S. (1990) 'Current Plate Motions', Geophys. J. 101, 425-478.

Dewey, J.F., Pitman, W.C., Ryan, W.B.F. and Bonnin, J. (1973) 'Plate tectonics and the evolution of the Alpine system', Geol. Soc. Am. Bull. 84, 3137-3180.

Dewey, J.F., Helman, M.L., Turco, E., Hutton, D.H.W. and Knott, S.D. (1989) 'Kinematics of the western Mediterranean', in M.P. Coward, D. Dietrich and R.G. Park (eds), Alpine Tectonics, Geological Society Special Publ. 45, 265-283.

Gordon, R.G., Stein, S., DeMets, C. and Argus, D.F (1987) 'Statistical tests for closure of plate motion circuit', Geophys.Res.Lett. 14, 587-590.

Grimison, N.L. and Chen, W.P. (1988) 'Source mechanisms of four recent earthquakes along the Azores-Gibraltar plate boundary', Geophys. J. 92, 391-401.

Jackson, J. (1992).'Partitioning of strike-slip and convergent motion between Eurasia and Arabia in Eastern Turkey and the Caucasus', J.Geophys.Res. 97, 12,471-12,479.

Jackson, J. (1993).'Rates of active deformation in the Eastern Mediterranean', this volume.

Klitgord, K.D. and Schouten, H. (1986) 'Plate kinematics of the central Atlantic' in P.R. Vogt and B.E. Tucholke (eds), The Geology of the North America, Vol. M, The Western North Atlantic Region., Geol.Soc.Am. 351-378.

Le Pichon, X., Sibuet, J.C. and Francheteau, J. (1977) 'The fit of the continents around the North Atlantic Ocean', Tectonophysics 38, 169-209.

Le Pichon, X. and Angelier, J. (1979) 'The Hellenic arc and trench system: a key to the neotectonic evolution of the eastern Mediterranean area', Tectonophysics 60, 1-42.

Lynnes, C.S. and Ruff, L.J. (1985) 'Source process and tectonic implications of the great 1975 North Atlantic earthquake', Geophys. J.R. astr. Soc. 82, 497-510.

Mantovani, E., Albarello, D., Babbucci, D. and Tamburelli, C. (1992) Recent geodynamic evolution of the Central Mediterranean Region (Tortonian to Present), Tipografia Senese, Siena.

Mantovani, E., Albarello, D., Babbucci, D. and Tamburelli, C. (1993) 'Post-Tortonian deformation pattern in the Central Mediterranean: a result of extrusion tectonic processes driven by the Africa-Eurasia convergence' , in this volume.

Mauffret, A., Rehault, J.P., Gennesseaux, M., Bellaiche, G., Labarbarie, M. and Lefebvre, D. (1981) 'Western Mediterranean basin evolution: from a distensive to a compressive regime', in F.C. Wezel (ed), Sedimentary Basins of Mediterranean margins. Tecnoprint, Bologna, pp.67-82.

Mc Kenzie, D. (1978) 'Active tectonics of the Alpine-Himalayan belt: the Aegean sea and surrounding regions (Tectonics of Aegean region)', Geophys.J.R.astr.Soc. 55, 217-254.

Minster, B.J. and Jordan, T.H. (1978) 'Present-day plate motions', J. Geophys. Res. 83, 5331-5354.

Nicolas, M., Santoire, J.P. and Delpech, P.Y. (1990) 'Intraplate seismicity: new seismotectonic data in Western Europe', Tectonophysics 179, 27-53.

Philip, H. (1987) 'Plio-Quaternary evolution of the stress field in Mediterranean zones of subduction and collision', Annales Geophys. 5B, 301-320.

Pitman, W.C. and Talwani, M. (1972) 'Seafloor spreading in the North Atlantic', Geol.Soc.Am.Bull. 83, 619-646.

Rebai, S., Philip, H. and Taboada, A. (1992) 'Modern tectonic stress field in the Mediterranean region: evidence for variation in stress directions at different scales', Geophys.J.Int. 110,106-140.

Royden, L.H. (1993) 'The evolution of retreating subduction boundaries formed during continental collision', Tectonics, in press.

Savostin, L.A., Sibuet, J.C., Zonenshain, L.P., Le Pichon, X. and Roulet, M.J. (1986) 'Kinematic evolution of the Tethys belt from the Atlantic ocean to the Pamirs since the Triassic', in J. Aubouin, X., Le Pichon and A.S. Monnin (eds), Evolution of the Tethys, Tectonophysics 123, 1-35.

Srivastava, S.P., Raest, W.R., Kovacs, L.C., Oakey, G., Lévesque, S., Verhoef, J. and Macuab, R. (1990) 'Motion of Iberia since the Late Jurassic: results from detailed aeromagnetic measurements in the Newfoundland Basin', Tectonophysics 184, 229-260.

Tapponier, P. (1977) 'Evolution tectonique du système alpin en Méditerranée: poinçonnement et écrasement rigide-plastique', Bull.Soc.Géol.Fr. 7, 437-460.

Udias, A. and Buforn, E. (1991) 'Regional stresses along the Eurasia-Africa plate boundary derived from focal mechanism of large earthquakes', Pageoph 136,4,433-448.

Whitmarsh, R.B., Ginzburg, A. and Searle, R.C. (1982) 'The structural and origin of the Azores-Biscay Rise, North-east Atlantic Ocean', Geophys.J.R.astr.Soc. 70, 79-107.

TECTONIC EVOLUTION OF THE NORTHERN ARABIAN PLATE IN WESTERN SYRIA

MUAWIA BARAZANGI, DOGAN SEBER, THOMAS CHAIMOV,
JOHN BEST, and ROBERT LITAK
Institute for the Study of the Continents
Snee Hall
Cornell University
Ithaca, New York 14853-1504 , USA

DAMEN AL-SAAD and TARIF SAWAF
Syrian Petroleum Company
Ministry of Petroleum and Mineral Resources
Damascus, Syrian Arab Republic

ABSTRACT. The primary geologic structures of the northern Arabian plate in western Syria include the intracontinental Palmyride mountain belt and the interplate boundary of the Dead Sea transform fault system. The Palmyride belt strikes NE and is sandwiched between two relatively stable crustal blocks of the Arabian platform: the Aleppo plateau in the north and the Rutbah uplift in the south. The Palmyrides were the site of an early Mesozoic aulacogen-type depression that was linked to the Levantine rifted continental margin in the eastern Mediterranean. The location of this postulated aulacogen may be genetically associated with a crustal zone of weakness, possibly a Proterozoic suture and/or shear zone, between the Aleppo and Rutbah crustal blocks. Uplift of the intraplate Palmyride depression initiated in the Late Cretaceous, penecontemporaneous with emplacement of ophiolites along the nearby Arabian plate boundaries in southern Turkey and western Iran. More intense episodes of shortening during the Cenozoic also appear to be temporally related to collision along nearby plate boundaries, implying that stresses have been transmitted hundreds of kilometers across the northern Arabian platform. The style and intensity of the inversion process vary considerably along the strike of the Palmyrides and involves both shortening by folding and reverse faulting as well as translation and rotation along numerous strike-slip faults. Such folds and faults clearly define at least three structurally distinct small crustal blocks within the Palmyrides. Shortening of about 20% in the southwest Palmyrides near Lebanon gradually dies out to the northeast near the intersection of the Palmyrides with the NW-trending Euphrates depression. Depth to metamorphic basement beneath the Palmyra mountain belt increases from 9 km in the northeast to 11 km in the southwest, compared with a basement depth of about 6-8 km in the adjacent Arabian platform, indicating that shortening along the Palmyrides has been insufficient to invert the previously extended basement morphology. Finally, slip measurements along the Dead Sea fault and estimates of crustal shortening in the Palmyride belt indicate that the northern segment of the seismogenic active Dead Sea fault in Lebanon and Syria is considerably younger (Pliocene) than the southern part (Miocene).

1. Introduction

In this paper, we argue that the present tectonic provinces of the northern Arabian platform in Syria are the result of the interaction of Cenozoic Arabian plate boundaries with older intraplate

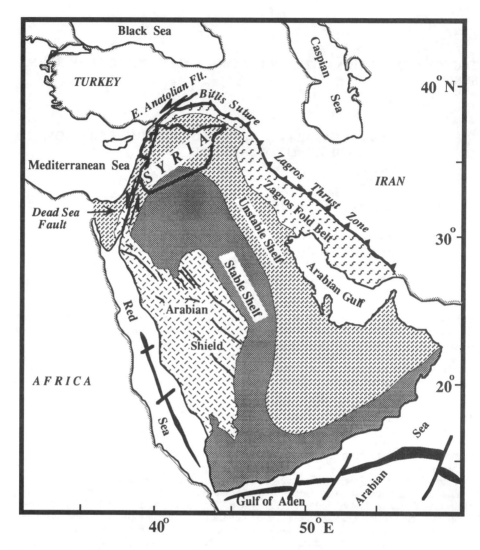

Figure 1. Generalized tectonic map of the Arabian plate, and location of Syria relative to the major plate boundaries.

geologic structures. The Arabian plate was part of the much larger African plate until about mid-Cenozoic time (Figures 1 and 2), when the initiation of sea-floor spreading along the Red Sea and the subsequent development of the Dead Sea transform fault system clearly defined the western boundaries of the Arabian plate (e.g., Dubertret, 1970; Freund et al., 1970; Garfunkel, 1981; Cochran, 1983; Hempton, 1987). The northern boundary is along the Bitlis suture and the East Anatolian fault zone in southern Turkey, which represents the site of convergence and collision of the Arabian and Eurasian plates and the subsequent attempt of the Anatolian

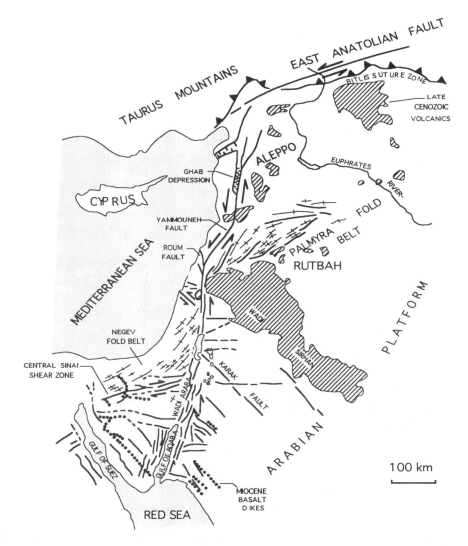

Figure 2. Summary tectonic map of the Dead Sea transform fault system and nearby plate boundaries, including intraplate Palmyride fold- thrust belt within northern Arabian platform.

subplate to escape this collision (e.g., Rigo de Righi and Cortesini, 1964; Sengör et al., 1985; Karig and Kozlu, 1990). Finally, the northeast and east boundary is located along the Zagros collision zone in Iraq and Iran (e.g., Berberian and Berberian, 1981; Barazangi, 1983 and 1989; Ni and Barazangi, 1986). The Arabian plate boundaries are relatively accessible, and abundant geological and geophysical data are available in Syria as a result of extensive hydrocarbon exploration, including seismic reflection and refraction profiles, potential fields, well logs, historical earthquake records, Landsat imagery, digital topography, geologic field studies and maps. These circumstances make it possible to study the structure of these plate boundaries and intraplate geologic features in Syria (especially the Palmyra mountain belt and the Euphrates graben/fault system) as well as to relate the deformation history of the intraplate structures to specific geologic episodes that occurred along the interplate boundaries.

This study is part of an ongoing joint cooperative research project between Cornell researchers and their colleagues from the Syrian Petroleum Company (SPC). The project represents a multidisciplinary program that integrates diverse data sets and sharply focuses on geologic problems rather than on technique. Results presented here are a brief summary of many detailed research papers and/or studies at different stages of development (McBride et al., 1990; Best et al., 1990 and 1993; Chaimov et al., 1990, 1992, and 1993; Al-Saad et al., 1991 and 1992; Seber et al., 1992 and 1993; Barazangi et al., 1992; Sawaf et al., 1993). Hence, the reader should consider this paper mainly as a review of progress to date on the Cornell-SPC Syria Project.

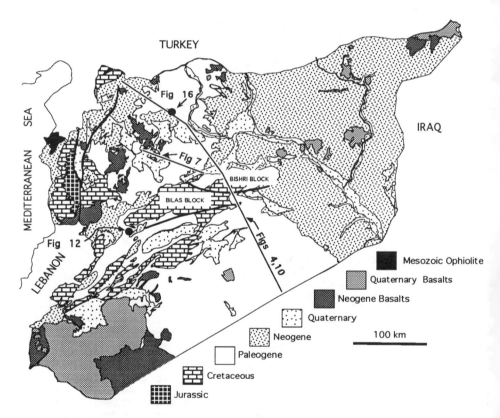

Figure 3. Simplified geological map of Syria. Locations of some of the following figures are also marked on the map.

2. Pre-Cenozoic Evolution of the Northern Arabian Platform

The Arabian shield in western Saudi Arabia and Jordan was assembled during the late Proterozoic (between about 720 Ma and 620 Ma) by accretion of a series of island arcs and continental microplates (e.g., Stoesser and Camp, 1985; Pallister et al., 1987; Johnson et al., 1987). However, the rest of the Arabian plate, i.e., the Arabian platform, is covered by a relatively thick Phanerozoic section (e.g., Beydoun, 1977; Murris, 1980; Husseini, 1989; Ben-Avraham and Ginzburg, 1990). Hence, geophysical observations must be used to infer the possible presence and distribution of other accreted terranes throughout the Arabian platform.

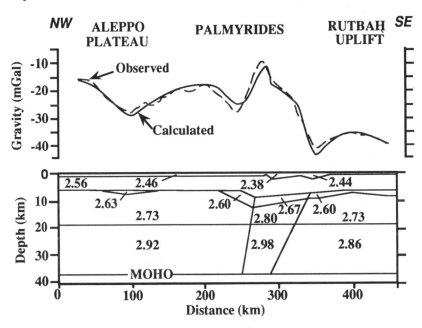

Figure 4. Best fit model to the observed Bouguer gravity data along a north-south transect (see Figure 3 for location). Density values are in g cm^{-3}. Note that the density of the crust beneath the Aleppo plateau is higher than that beneath the Rutbah uplift (assuming constant crustal thickness). Alternatively, the Aleppo may have a thinner crust relative to the Rutbah uplift.

In western Syria, the Aleppo plateau and Rutbah uplift are separated by the Palmyride mountain belt (Figures 2 and 3). Bouguer gravity anomalies in central Syria indicate that there is a fundamental difference in the crusts of the Rutbah and the Aleppo regions: the Aleppo crust is denser and/or thinner than the Rutbah crust (Best et al., 1990; Al-Saad et al., 1991 and 1992). The gravity modeling is well constrained by seismic reflection profiles and drillhole information in the upper part of the crust (less then 8 km); therefore, the observed gravity variations must be the result of deeper crustal anomalies (Figure 4). The above documented difference could be interpreted, though not uniquely, to suggest that the Palmyra mountain belt is the location of a possible Precambrian (Proterozoic?) suture and/or strike-slip fault zone. A prominent deep

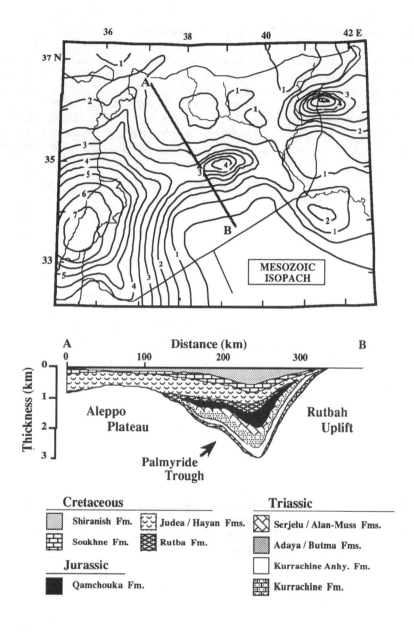

Figure 5. Mesozoic isopach map of Syria (top). Note the narrow trough with thicker section in central Syria (Palmyrides) and the major depocenter on the Levantine margin in the eastern Mediterranean Sea. Contours outside of Syria are extrapolations from Syria data. Contour interval equals 500 meters. (Bottom) a cross section (A-B) through the Mesozoic section showing lithologies and thicknesses.

(about 3 seconds) reflection correlated with the Middle Cambrian Burj limestone is observed in both the Aleppo plateau and in the Rutbah uplift. For this pervasive Burj limestone to have been deposited regionally, the Rutbah and Aleppo crusts must have been joined prior to the Middle Cambrian, and by analogy with the sutures of the Arabian shield to the south a Proterozoic time of accretion is inferred. If this interpretation is valid, then it may explain the subsequent development of the Palmyrides in early Mesozoic time as a reactivation of a zone of crustal weakness along the postulated Proterozoic suture zone.

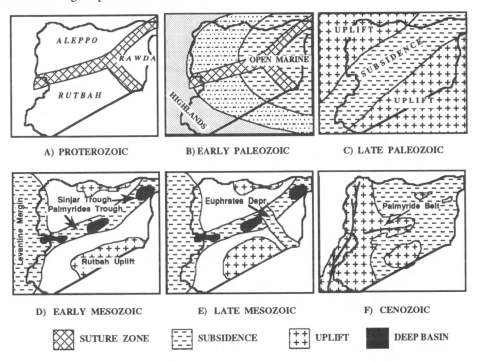

Figure 6. Schematic maps showing tectonic evolution history of the northern Arabian platform in Syria. (A) Proposed suturing of at least three distinct crustal blocks in the Proterozoic, followed by Early Cambrian extension; (B) Early Paleozoic epicontinental region of the northern Gondwana margin; (C) Development of Late Paleozoic depocenter in central Syria, possible precursor to later Mesozoic rifting; (D) Early Mesozoic development of the Levantine margin and the Palmyride rift ; (E) Second phase of extension marked by the development of the Euphrates depression; (F) Cenozoic transpressive uplift of the Palmyride intracontinental mountain belt and subsidence in northeast Syria due to loading and subsequent bending of the Arabian plate along the Zagros orogenic belt.

Based on formation tops from approximately 200 hydrocarbon exploration wells in Syria that are tied to over 100,000 km of seismic reflection profiles, the Paleozoic and Mesozoic depositional settings and history are inferred and mapped (Sawaf et al., 1988; Bebeshev et al.,

1988; Leonov et al., 1989; Best et al., 1993). The Paleozoic depositional setting was an east-facing, clastic dominated continental margin on the northern rim of Gondwanaland (Best et al., 1993). A wide, broad shallow marine basin is identified in western Syria which deepens eastward into marine shales with cherts. Mesozoic rifting dramatically affected the northern Arabian platform and transformed the east-facing Paleozoic margin into a west-facing margin. This transformation is evidenced by the development of the Levantine margin and the associated, onshore continuation of the intracontinental Palmyra rift (i.e., aulacogen) that are characterized by major sediment accumulation throughout the Mesozoic (Figure 5). Since Miocene time, the Levantine margin has been sheared along the Dead Sea fault system (e.g., Quennell, 1958 and 1984; Freund et al., 1970; Darkal et al., 1990; Girdler, 1990). Timing of the margin transformation is inferred to be Triassic in age and is closely associated with the development of the Neo-Tethys (Best et al., 1993). About 9 km of sedimentary section is present along the Levantine margin in and near the eastern Mediterranean Sea. Also, a very thick (over 9 km) sedimentary section trends cratonward approximately perpendicular to the Levantine margin along the Palmyra rift. The Aleppo plateau subsided during the Mesozoic while the Rutbah uplift was a dome-like structure (Figure 6).

In summary, the Precambrian-Paleozoic development of the northern (e.g., Syria) and southern (e.g., Saudi Arabia) regions of the Arabian plate are strikingly similar in many aspects. However, these regions dramatically differ in the Mesozoic time because of interplate rifting and the creation of the eastern Mediterranean basin during the Mesozoic.

3. Sediment Thickness and Basement Structure

An extremely valuable and unique dataset is produced by a detailed seismic refraction survey that was conducted in the northern Arabian platform in Syria in 1972-1973 by Syrian and Soviet scientists for the upper part of the crust (less than about 15 km). Specifically the survey focused on mapping the detailed geometry and velocity structure of the metamorphic basement of Syria. Reanalysis and reinterpretation of the original seismic records of three of the profiles using modern computer software, including 2-D ray tracing technique, provide a detailed image of the basement's morphology and seismic velocity structure beneath the Palmyra mountain belt and the nearby Aleppo and Rutbah regions (Seber et al., 1992 and 1993).

An example of one of the refraction profiles in the Aleppo plateau is presented in this study (Figure 7). The profile is 92 km in length. The distance between any given shot point and the last geophone averages about 40 km and, hence, only about the upper 15 km of the crust is sampled by seismic rays. Geophone spacing is 100 m, and the average charge size varied from 200-1000 kg. Overlapping forward and reverse shot points provide up to 8-fold data coverage, which significantly contributes to obtaining very reliable and unique velocity-depth models. The first P arrivals on the original photographic paper records were digitized and interpreted via ray-trace modeling (Figure 8).

The final velocity model for this profile has a three-layered sequence in the upper 1 km, with velocities increasing from 2.0 to 3.8 km s^{-1}, corresponding to Cenozoic and Mesozoic sedimentary rocks. This sequence was also constrained by the sonic log of the Khanaser-1 (KH-1) well. Even though different geological units are traversed, no lateral velocity variation is observed. The boundary between this three-layered sequence and the underlying layer with a velocity of about 4.8 km s^{-1} corresponds to the Mesozoic-Paleozoic boundary as inferred from the Khanaser-1 well. Beneath this layer lies another sedimentary layer with a velocity of about

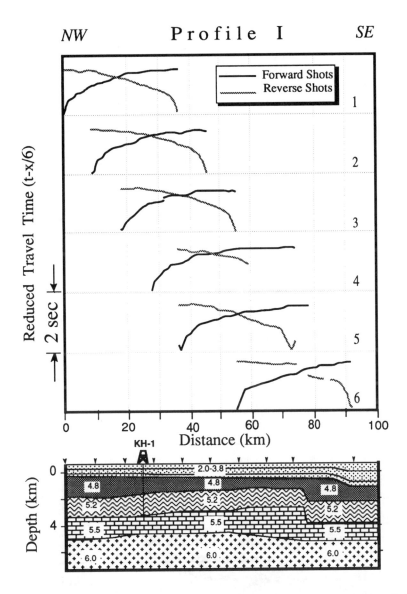

Figure 7. Reduced travel times of first breaks and wide angle reflections (when available) for all shot points on Profile I (see Figure 3 for location). Arrivals from forward and reverse shots are shown as black and gray lines, respectively. Phase correlation was made on the analog records before digitization. In each shot record about 400 picks correspond to arrival times of either first breaks or wide angle reflections. Shot points are numbered 1 through 6. Ray tracing for shot point 1 is shown in Figure 8. Individual shot points are also marked on the ground level of the velocity model. The final velocity- depth model (in km s^{-1}) is shown at the bottom. Zero depth corresponds to sea level.

5.2 km s^{-1}, which corresponds to the lower section of the Paleozoic. The Khanaser-1 well bottoms at about 3800 m in the lower Cambrian. The refraction model gives a velocity boundary at about 4000 m, possibly marking the base of the lower Cambrian in this region. According to the refraction interpretation, approximately 1.5 km lies between this interface and the top of basement, amounting to a total sedimentary thickness of 5.5 km. The lowest sedimentary layer on top of the metamorphic basement has a velocity of 5.5 km s^{-1}, and is interpreted as early Paleozoic and late Precambrian sediments. The deepest boundary on the refraction model is interpreted as the sedimentary-metamorphic basement boundary. The basement velocity is 6.0 km s^{-1}, and the basement depth is about 5.5-6.0 km in this region. An offset in the velocity-depth model beneath the southeastern end of the profile is interpreted as a fault, consistent with seismic reflection and gravity data.

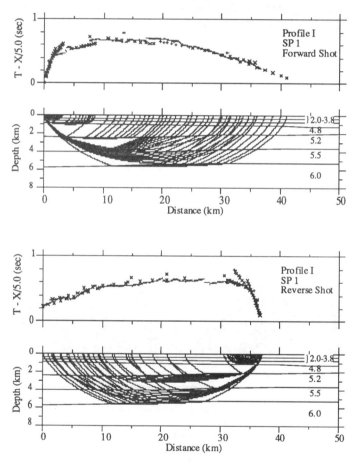

Figure 8. Examples of ray tracings of reversed shot 1 (See Figure 7). Small dots are the actual observations; larger crosses and pluses are calculated arrivals of refraction and wide angle reflections, respectively.

Interpretation of the three refraction profiles shows that the Phanerozoic sedimentary section in the northern Arabian platform varies from 5.5-6.0 km in the Aleppo plateau region to 9.0-11.0 km in the Palmyrides, and to more than about 8.0 km in the Rutbah uplift region (Figure 9). The deepest part of the metamorphic basement (about 11 km) is observed beneath the southwestern segment of the Palmyra mountain belt. This is consistent with geologic and gravity observations that indicate more pronounced extension in the southwestern Palmyrides during Mesozoic time. The mapping of a well-defined rift-related basement trough beneath the Palmyrides further indicates that Cenozoic shortening along the Palmyra belt has been insufficient to invert this trough (Seber et al., 1992 and 1993).

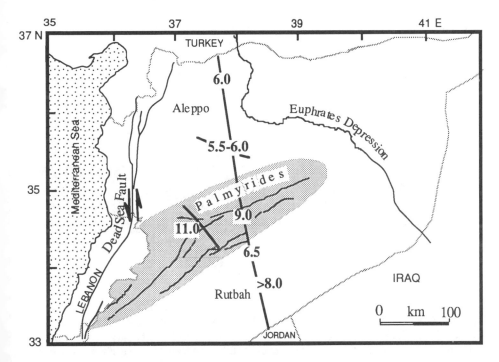

Figure 9. A summary map of refraction results. The three profiles and observed maximum sedimentary rock thicknesses in km beneath the profiles are marked.

4. Palmyra Mountain Belt

The Palmyra fold-thrust belt is a relatively modest intracontinental mountain belt, though it is the most prominent structure in central Syria (e.g., Ponikarov, 1964, 1966, and 1967; Lovelock, 1984; Leonov, 1989). The belt is the northeastern arm of the approximately "S"-shaped Syrian arc that extends from the Negev folds of central Sinai (Moustafa and Khalil, 1989) northward through the Palmyrides and farther northeast until it plunges and vanishes under the Euphrates depression. The Palmyrides, embedded in the northern Arabian platform, are about 400 km in length and about 100 km in width, and strike in a NE-SW direction. The belt approaches an average elevation of about 1.5 km in the southwest, where it merges with the Dead Sea fault system and associated Lebanon and Anti-Lebanon mountain ranges. Toward the northeast the Palmyra belt gradually becomes more subdued with an average elevation that rarely exceeds eight hundred meters, until it intersects with the NW-SE oriented Euphrates graben/fault system (e.g., Sawaf et al., 1993).

As has been discussed in section 2 of this paper, gravity observations suggest the presence of a late Proterozoic zone of weakness, possibly a suture zone, beneath the present-day Palmyrides (Best et al., 1990). This may explain why the location of the Palmyrides preferentially developed as the site of an aulacogen-type depression in early Mesozoic time that was linked to the Levantine rifted continental margin, a part of the Neo-Tethys system. The thick Mesozoic sedimentary section (up to about 5 km), the relative linearity of the depositional axis, and limited volcanism are supporting evidence for the presence of an aulacogen along the Palmyrides during most of the Mesozoic time. Inversion of the Mesozoic depression throughout the Cenozoic, especially in Neogene and Quaternary times, resulted in the present-day Palmyra mountain belt. The inversion process involves both shortening by folding and reverse faulting as well as translation and rotation along numerous strike-slip faults. The Palmyrides can be considered as a type example of an intracontinental transpressive mountain belt (e.g., McBride et al., 1990; Al-Saad et al., 1992; Chaimov et al., 1993).

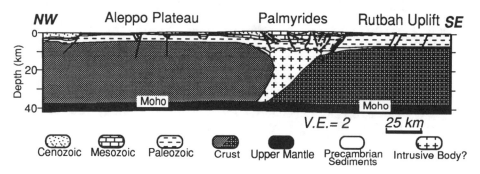

Figure 10. Interpretive cross section to Moho (see Figure 3 for location). The high density intrusive body may be associated with Paleozoic extension. The nature of the difference between The Rutbah and Aleppo crusts is uncertain because of the ambiguity inherent in gravity modeling. The crust of the Aleppo plateau is either thinner or denser than Rutbah crust.

An important conclusion of our ongoing study of the Palmyra mountain belt is that different regions of the belt exhibit considerable variations in the nature of deformation (e.g., Barazangi et al., 1992). To a first approximation the Palmyrides can be divided into northeast and southwest segments that are separated by the E-W oriented Jhar fault, an apparent strike-slip fault zone. The northeast Palmyrides (i.e., the Bishri and Bilas blocks) consist of broad, relatively symmetric anticlines, with reverse faults along the southern and northern flanks of the belt that dip towards the interior of the belt (Figures 10 and 11). The seismic reflection data clearly show that this deformation is thick-skinned and affects the whole Phanerozoic section; i.e., no local or regional detachment is apparent in the Mesozoic section (Al-Saad et al., 1992). A system of broad folds and somewhat ill-defined fault zones (mostly strike-slip faults) that bound the folded regions characterize the northeastern sector of the Palmyrides. The focal

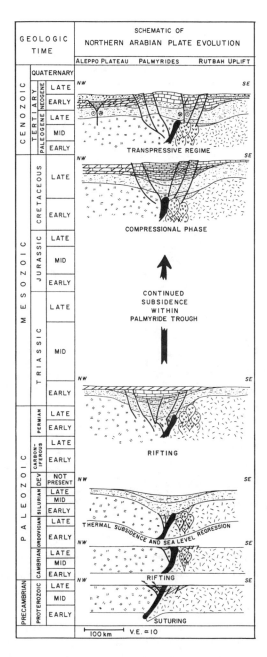

Figure 11. Schematic tectonic evolution of the Palmyrides and adjacent crustal blocks from Precambrian to Ceonozoic.

mechanisms of two moderate-size earthquakes that occurred in 1970 and 1987 in the northeast Palmyrides show right-lateral oblique reverse motion along the faults that separate the Bishri block from the Bilas block (Chaimov et al., 1990). Clearly, strike-slip faulting significantly contributes to the present structure and the geological evolution of the Palmyra mountain belt.

The southwestern Palmyrides are characterized by a system of short-wavelength (about 5-10 km) en echelon, northeast-trending folds that are associated with surface-mapped and/or blind, south-vergent frontal thrust faults with small intermontane basins (e.g., Chaimov et al., 1992). The South Palmyra fault zone approximately defines the southern limits of the southwestern Palmyrides (Figure 3). Seismic reflection profiles clearly image local, sub-horizontal decollement surfaces within the relatively thick Triassic evaporites (Figure 12). However, a regional, pervasive detachment is not observed beneath the southwestern Palmyrides.

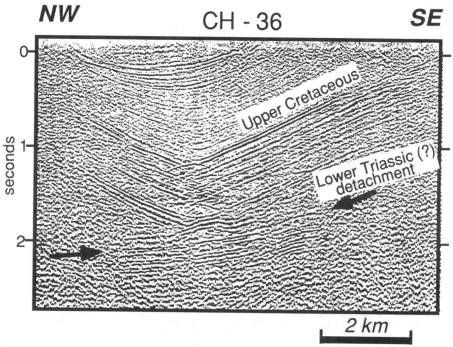

Figure 12. An example of a possible Lower Triassic detachment in the southwestern Palmyrides decoupling relatively undeformed Paleozoic rocks from overlying Mesozoic and Cenozoic strata (see Figure 3 for location).

Seismic stratigraphic evidence indicates that minor Late Cretaceous uplift marked the first inversion phase of the Mesozoic Palmyride trough (Chaimov et al., 1992). This phase had a direct temporal relationship to the emplacement of ophiolites along the northern and eastern Arabian plate margins. Thus, the initiation of inversion in the Palmyrides apparently predates development of the Red Sea/Dead Sea plate boundary and is approximately coeval with the closing of the Neo-Tethys. However, the development of the present-day Palmyrides, mostly in

Neogene and Quaternary times, is synchronous with intense movements documented along nearby Arabian plate boundaries, including the development of the Dead Sea transform fault system, the Bitlis suture and the East Anatolian fault, and the Zagros continental collision zone (Chaimov et al., 1992). The northward movement and anticlockwise rotation of the Arabian plate since Miocene time must have played a critical role in the intense inversion of the Palmyrides depression. The above discussion suggests that plate stresses have been transmitted for hundreds of kilometers from the surrounding Arabian plate boundaries through the stable northern Arabian platform to the Palmyrides in the interior of the plate. It is tempting to speculate based on limited seismic evidence about the presence of a regional master detachment beneath the Aleppo plateau, either along the basement surface or within the basement, that accommodates some of the convergence along the Bitlis suture in southern Turkey and transmits it farther southwards to the Palmyrides in central Syria. Evidence for this scenario is sparse at the present time.

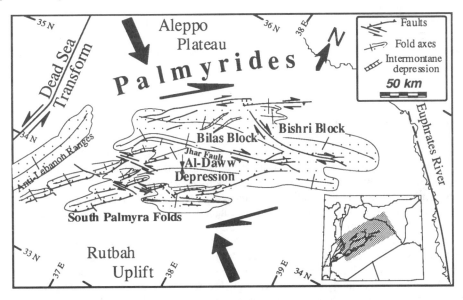

Figure 13. Simplified kinematics of the Palmyrides, with emphasis on known senses of strike-slip component across major fault zones. Proportion of strike-slip vs. compression is not known; however, compression most likely dominates.

Based on the available seismic data it is clear that some of the Mesozoic rift-bounding normal faults have been reactivated into reverse faults in response to the transpressive forces acting on the different blocks of the Palmyrides during Cenozoic time. An interesting observation is that no foreland basin exists along the margins of the intracontinental Palmyride belt. The belt is strictly sandwiched between the two relatively rigid subprovinces of the northern Arabian platform: the Rutbah uplift in the south and the Aleppo plateau in the north. This is in marked contrast to observed foreland basins associated with interplate and collisional mountain belts. Finally, Neogene and Quaternary basaltic volcanism is widespread in the Rutbah and Aleppo regions, but no surface volcanism exists within the Palmyrides. No simple explanation is yet available for the cause of this volcanism.

A summary map (Figure 13) of the most reliably identified and prominent faults along the Palmyra mountain belt, mapped at the surface and/or inferred from seismic reflection profiles, indicates that the whole Palmyride belt is affected by an overall right-lateral sense of shear (Chaimov et al., 1993). Though the main cause of the Palmyrides has been compression, strike-slip faulting significantly contributes to the fabric and tectonic evolution of this mountain belt. However, the amount of strike-slip motion along the different faults is not well-documented, and considerably more future field work combined with satellite imagery and other geophysical observations are required to resolve this problem. It is interesting to note that the above suggested right-lateral shear in the Palmyrides requires that the Aleppo plateau is escaping northeastward from the overall northward motion of the Arabian plate. Modest seismicity, coupled with geologic evidence based on morphotectonic observations, indicate that many of the Palmyride faults are still active (Trifonov et al., 1983; Leonov, 1989).

In conclusion, the evolution of the Palmyrides, an active right-lateral transpressive mountain belt, probably can be traced back as far as the Proterozoic. The present structure of the belt, however, is the result of successive tectonic phases that were dramatically influenced by variations in the sedimentary section and the presence of crustal zones of weakness, and by interactions with nearby plate boundaries.

5. Dead Sea Transform Fault System

The discussion in this section is not intended to present an overall review of the structure of the Dead Sea transform fault system. Rather, we address the issue of the reported paradox concerning the apparent discrepancy in the amount of left-lateral offset between the northern (in Lebanon and Syria) and the southern segments of the Dead Sea fault and its relationship to Cenozoic shortening in the nearby Palmyra mountain belt. But first it should be emphasized that the Palmyra belt is not the result of the restraining bend in the Dead Sea fault system. A comparison between the San Andreas transform fault system in southern California and the Dead Sea system (Figure 14) clearly suggests that the Transverse Ranges along the San Andreas fault, which are a result of a restraining bend, are equivalent and similar in many respects to the Anti-Lebanon Ranges along the Dead Sea fault (Chaimov et al., 1992). The Palmyrides extend more than 300 km farther to the northeast of the restraining bend in the Dead Sea fault system. Moreover, the Palmyra belt considerably predates the development of the Dead Sea fault by many tens of millions of years.

A well-documented 105 km offset along the southern segment (south of Lebanon) of the Dead Sea fault since Miocene time (e.g., Quennell, 1958 and 1984; Freund et al., 1970; Garfunkel, 1981) is contrasted with only about 25 km offset along the northern segment in Lebanon and Syria (see Chaimov et al., 1990). One reported possible explanation for the 80 km difference in the observed offsets is that the Palmyrides absorb this documented difference through shortening. Estimates of crustal shortening along strike of the Palmyrides belt based on balanced cross sections constrained by available seismic reflection profiles, drill holes, and detailed geologic maps give a minimum estimate of about 20 km in the intensely deformed, southwestern segment of the belt (Chaimov et al., 1990). Shortening estimates are much less to the northeast. Clearly this is insufficient to explain the above discussed paradox. Two geologically reasonable models are (1) considerable strike-slip movements exist along the

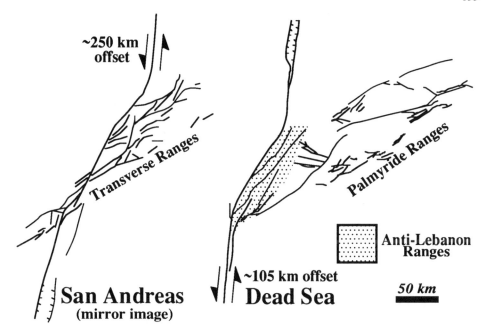

Figure 14. A scale comparison of the Transverse Ranges and central San Andreas fault system of California with the Palmyride ranges and the central Dead Sea fault system of the Middle East. A mirror image of the San Andreas system is shown in order to make the comparison more direct. Only major faults are shown. The Anti-Lebanon range in the Middle East is analogous to the Transverse Ranges of California, both being the result of compression in the region of the restraining bend of a major strike-slip fault. That is, the Palmyrides are not equivalent to the Californian Transverse Ranges, but are a result of the inversion of a preexisting basin.

Palmyrides and possibly within the Aleppo region (Figures 15 and 16), though displacements can not yet be estimated, or (2) the southern Dead Sea fault segment formerly continued in a NW direction towards the Mediterranean (the Roum fault) and only began to propagate northward in Lebanon and Syria (the Yammouneh and Ghab faults) since about the beginning of the Pliocene time (Figure 17). This second postulated model would mean that the 80 km of apparent missing displacement is illusive, and that there really has been a total of only about 45 km of N-S shortening in Syria (Chaimov et al., 1990). The "truth" likely lies in some combination of these two proposed models.

Finally, it is important to report that the segments of the Dead Sea fault in Lebanon (the Yammouneh fault) (e.g., Ron, 1987; Walley, 1988; Khair et al., 1993) and in Syria (the Ghab fault) are historically very active seismogenic faults. Many large earthquakes ($M \approx 7$) occurred along these segments (e.g., Jackson and McKenzie, 1988; Ambraseys and Barazangi, 1989) as well as possibly along nearby subsidiary faults (e.g., Walley, 1988; Girdler, 1990; Khair et al., 1993).

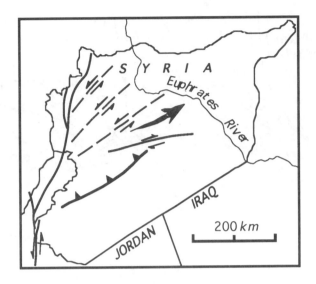

Figure 15. A possible model that invokes distributed strike-slip to accommodate excess shortening. This model requires a total of about 80 km of NE-SW strike-slip motion.

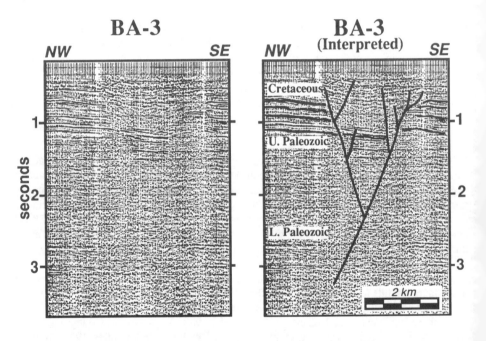

Figure 16. An example from the Aleppo plateau showing interpretation of a typical flower structure such as commonly results from strike-slip motion. (see Figure 3 for location).

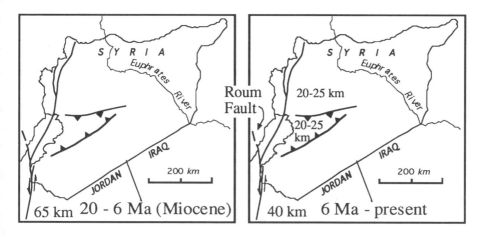

Figure 17. Alternative kinematic model to explain the apparent discrepancy between left-lateral offsets observed on the segments of the northern versus the southern Dead Sea fault system. The Roum fault may have been the main northward continuation of the Dead Sea system during Miocene.

6. Concluding Remarks

The intracontinental Palmyride fold-thrust belt in central Syria was formed along a Mesozoic rift (aulacogen) that was linked to the Levantine rifted continental margin in the eastern Mediterranean. The location of this rift may be associated with a postulated Proterozoic suture or shear zone within the northern Arabian platform. The different documented episodes of uplift of the Palmyrides since Late Cretaceous time can be related to specific deformation episodes along nearby Arabian plate boundaries, implying that stresses have been transmitted hundreds of kilometers across the northern Arabian platform.

The general increase of shortening along the Palmyra mountain belt from the northeast towards the southwest suggests a scissored-type convergence in the Palmyrides with the pivot point located near the intersection of the Palmyrides and the Euphrates depression (Figure 18). In addition, the inferred right-lateral shear along the Palmyrides belt suggests a northwest-oriented maximum horizontal compressive stress across the belt. The above two observations together imply a clockwise rotation of the Rutbah block around the postulated pivot point (Chaimov et al., 1993). Such a regional kinematics may explain the bend in the Dead Sea fault in Lebanon as well as the development of extension along the southern Euphrates system.

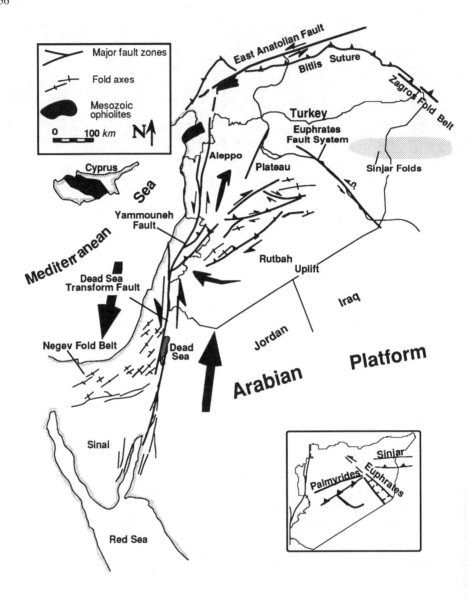

Figure 18. Map showing the regional kinematics of the northern Arabian platform. Right-lateral sense of shear in the Palmyrides requires the extrusion of the Aleppo plateau towards the northeast. Fault arrow sizes are approximately proportional to the amount of slip along fault zones. A local pivot point between the Aleppo plateau and Rutbah uplift is inferred to lie near the Palmyride-Euphrates intersection (inset). The Euphrates faults system may decouple deformation west of the Euphrates (Palmyrides) from that to the east (Sinjar).

7. Acknowledgements

We thank the Syrian Petroleum Company for providing the data base for this joint cooperative research project. The project was sponsored by Amoco, Arco, British Gas, Exxon, Marathon, Mobil, Occidental, and Unocal. Institute for the Study of the Continents contribution No. 188.

8. References

Al-Saad, D., Sawaf, T., Gebran, A., Barazangi, M., Best, J. and Chaimov, T. (1991) 'Northern Arabian platform transect across the Palmyride mountain belt, Syrian Arab Republic', *Global Geoscience Transect* **1**, Copublished by the Inter-Union Commission on the Lithosphere and American Geophysical Union, Washington, D.C.

Al-Saad, D., Sawaf, T., Gebran, A., Barazangi, M., Best, J. and Chaimov, T. (1992) 'Crustal structure of central Syria: The intracontinental Palmyride mountain belt', *Tectonophysics* **207**, 345-358.

Ambraseys, N.N. and Barazangi, M. (1989) 'The 1759 earthquake in the Bekaa valley: Implications for earthquake hazard assessment in the eastern Mediterranean region', *Journal of Geophysical Research* **94**, 4007-4013.

Barazangi, M. (1983) 'A summary of the seismotectonics of the Arab region', in K. Cidlinsky and B. Rouhban (eds.), *Assessment and mitigation of earthquake risk in the Arab region*, UNESCO, Paris, France, pp. 43-58.

Barazangi, M. (1989) 'Continental collision zones: Seismotectonics and crustal structure', in D.E. James (ed.), *The Encyclopedia of Solid Earth Geophysics*, Van Nostrand Reinhold Company, New York, pp. 58-75.

Barazangi, M., Seber, D., Al-Saad, D. and Sawaf, T. (1992) 'Structure of the intracontinental Palmyride mountain belt in Syria and its relationship to nearby Arabian plate boundaries', *Proceedings of 1st International Symposium on Eastern Mediterranean Geology, Bulletin of Earth Sciences* **20**, Cukurova University, Adana, Turkey, 111-118.

Bebeshev, I.I., Dzhalilov, Y.M., Portnyagina, L.A., Yudin, G.T., Mualla, A., Zaza, T. and Jusef, A. (1988) 'Triassic stratigraphy of Syria', *International Geology Review*, 1292-1301.

Ben-Avraham, Z. and Ginzburg, A. (1990) 'Displaced terranes and crustal evolution of the Levant and the Eastern Mediterranean', *Tectonics* **9**, 613-622.

Berberian, F. and Berberian, M. (1981) 'Tectono-plutonic episodes in Iran', in H. Gupta and F. Delany (eds.), *Zagros, Hindu Kush, Himalaya, geodynamic evolution: Geodynamics Series* **3**, American Geophysical Union, Washington, D.C., pp. 5-32.

Best, J.A., Barazangi, M., Al-Saad, D., Sawaf, T. and Gebran, A. (1990) 'Bouguer gravity trends and crustal structure of the Palmyride Mountain belt and surrounding northern Arabian platform in Syria', *Geology* **18**, 1235-1239.

Best, J.A., Barazangi, M., Al-Saad, D., Sawaf, T. and Gebran, A. (1993) 'Continental margin evolution of the northern Arabian platform in Syria', *American Association of Petroleum Geologists Bulletin* **77**, 173-193.

Beydoun, Z.R. (1977) 'The Levantine countries: the geology of Syria and Lebanon (maritime regions)', in A.E.M. Nairn, W.H. Kanes, and F.G. Stehli (eds.), *The Ocean Basins and Margins: Volume 4A: the Eastern Mediterranean*, Plenum Press, New York, pp. 319-353.

Chaimov, T., Barazangi, M., Al-Saad, D., Sawaf, T. and Gebran, A. (1990) 'Crustal shortening in the Palmyride fold belt, Syria, and implications for movement along the Dead Sea fault system', *Tectonics* **9**, 1369-1386.

Chaimov, T., Barazangi, M., Al-Saad, D., Sawaf, T. and Gebran, A. (1992) 'Mesozoic and Cenozoic deformation inferred from seismic stratigraphy in the southwestern intracontinental Palmyride fold-thrust belt, Syria', *Geological Society of America Bulletin* **104**, 704-715.

Chaimov, T., Barazangi, M., Al-Saad, D. and Sawaf, T. (1993) 'Seismic fabric and 3-D upper crustal structure of the southwestern intracontinental Palmyride fold belt, Syria', *American Association of Petroleum Geologists Bulletin*, in press.

Cochran, J.R. (1983) 'Model for the development of the Red Sea', *American Association of Petroleum Geologists Bulletin* **67**, 41-69.

Darkal, A.N., Krauss, M. and Ruske, R. (1990) 'The Levant fault zone: An outline of its structure, evolution and regional relationship', *Zeitschrift Geologische Wissenschaft* **18**, 549-562.

Dubertret, L. (1970) 'Review of the structural geology of the Red Sea and surrounding areas', *Royal Society of London Philosophical Transactions, Series A* **267**, 9-20.

Freund, R., Garfunkel, Z., Zak, I., Goldberg, M., Weissbrod, T. and Derin, B. (1970) 'The shear along the Dead Sea rift', *Philosophical Transactions of the Royal Society of London, Series A* **267**, 107-130.

Garfunkel, Z. (1981) 'Internal structure of the Dead Sea leaky transform (rift) in relation to plate Kinematics', *Tectonophysics* **80**, 81-108.

Girdler, R.W. (1990) 'The Dead Sea transform fault system', *Tectonophysics* **180**, 1-13.

Hempton, M.R. (1987) 'Constraints on Arabian plate motion and extensional history of the Red Sea', *Tectonics* **6**, 687-705.

Husseini, M.I. (1989) 'Tectonic and deposition model of late Precambrian-Cambrian Arabian and adjoining plates', *American Association of Petroleum Geologists Bulletin* **73**, 1117-1131.

Jackson, J. and McKenzie, D. (1988) 'The relationship between plate motions and seismic moment tensors, and the rates of active deformation in the Mediterranean and Middle East', *Geophysical Journal* **93**, 45-73.

Johnson, P.R., Scheibner, E. and Smith E.A. (1987) 'Basement fragments, accreted tectonostratigraphic terranes and overlap sequences: Elements in the tectonic evolution of the Arabian shield' in A. Kroner (ed.), *Proterozoic Lithospheric Evolution: American Geophyscial Union Geodynamics Series* **17**, 323-343.

Karig, D.E. and Kozlu, H. (1990) 'Late Paleogene-Neogene evolution of the triple junction region near Maras, south-central Turkey', *Geological Society of London Journal* **147**, 1023-1034.

Khair, K., Khawlie, M., Haddad, F., Barazangi, M., Seber, D. and Chaimov, T. (1993) 'Bouguer gravity and crustal structure of the Dead Sea transform fault and adjacent mountain belts in Lebanon', submitted to *Geology*.

Leonov, Y.G. (ed.) (1989) *The Tectonic Map of Syria, scale 1:500,000*, Ministry of Petroleum and Mineral Resources of the Syrian Arab Republic and Academy of Sciences of the USSR.

Leonov, Y.G., Sigachev, S.P., Otri, M., Yusef, A., Zaza, T. and Sawaf, T. (1989) 'New data on the Paleozoic complex of the platform cover of Syria', *Geotectonics* (English Edition) **23**, 538-542.

Lovelock, P. (1984) 'A review of the tectonics of the northern Middle East region', *Geological Magazine* **121**, 577-587.

McBride, J.H., Barazangi, M., Best, J., Al-Saad, D., Sawaf, T., Al-Otri, M. and Gebran, A. (1990) 'Seismic reflection structure of intracratonic Palmyride fold-thrust belt and surrounding Arabian platform, Syria', *American Association of Petroleum Geologists Bulletin* **74**, 238-259.

Moustafa, A.R. and Khalil, M.H. (1989) 'North Sinai structures and tectonic evolution', *Middle East Research Center, Ain Shams University, Earth Science Series* **3**, 215-231.

Murris, R.J. (1980) 'Middle East: stratigraphic evolution and oil habitat', *American Association of Petroleum Geologists Bulletin* **64**, 597-618.

Ni, J. and Barazangi, M. (1986) 'Seismotectonics of the Zagros continental collision zone and a comparison with the Himalayas', *Journal of Geophysical Research* **91**, 8205-8218.

Pallister, J.S., Stacey, J.S., Fischer, L.B. and Premo, W.R. (1987) 'Arabian shield ophiolites and late Proterozoic microplate accretion', *Geology* **15**, 320-323.

Ponikarov, V.P. (ed) (1964) 'Tectonic map of Syria: scale 1:1,000,000', *Ministry of Industry*, Damascus, Syrian Arab Republic.

Ponikarov, V.P. (ed) (1966) 'The Geological map of Syria: scale 1:1,000,000', *Ministry of Industry*, Damascus, Syrian Arab Republic.

Ponikarov, V.P. (ed) (1967) 'The geology of Syria: explanatory notes on the geological map of Syria, scale 1:500,000 part I: stratigraphy, igneous rocks and tectonics', *Ministry of Industry*, Damascus, Syrian Arab Republic, 229 pp.

Quennell, A.M. (1958) 'The structural and geomorphic evidence of the Dead Sea Rift', *Quarterly Journal of the Geological Society of London* **114**, 1-24.

Quennell, A.M. (1984) 'The Western Arabian rift system' in J.E. Dixon and A.H.F. Robertson (eds.), *The Geological Evolution of the Eastern Mediterranean, Geological Society of London Special Publication* **17**, Blackwell Scientific Publications, Oxford, 775-778.

Rigo de Righi, M. and Cortesini, A. (1964) 'Gravity tectonics in foothills structure belt of southeast Turkey', *American Association of Petroleum Geologists Bulletin* **48**, 1911-1937.

Ron, H. (1987) 'Deformation along the Yammuneh, the restraining bend of the Dead Sea transform: Paleomagnetic data and kinematic implications', *Tectonics* **6**, 653-666.

Sawaf, T., Zaza, T. and Sarriyah, O. (1988) 'The distribution and litho-stratigraphic base for the sedimentary formations in the Syrian Arab Republic', *Syrian Petroleum Company Unpublished Report*, Damascus, Syria, 89 pp.

Sawaf, T., Al-Saad, D., Gebran, A., Barazangi, M., Best, J.A. and Chaimov, T. (1993) 'Structure and stratigraphy of eastern Syria across the Euphrates depression', *Tectonophysics*, in press.

Seber, D., Barazangi, M., Chaimov, T., Al-Saad, D., Sawaf, T. and Khaddour, M. (1992) 'Geometry and velocity structure of the Palmyride fold-thrust belt and surrounding Arabian platform in Syria', *Proceedings of 1st International Symposium on Eastern Mediterranean Geology, Bulletin of Earth Sciences* **20**, Cukurova University, Adana, Turkey, 103-110.

Seber, D., Barazangi, M., Chaimov, T.A., Al-Saad, D., Sawaf, T. and Khaddour, M. (1993) 'Upper crustal velocity structure and basement morphology beneath the intracontinental Palmyride fold-thrust belt and north Arabian platform in Syria, *Geophysical Journal International*, in press.

Sengör, A.M.C., Gorur, N. and Saroglu, F. (1985) 'Strike-slip faulting and related basin formation in zones of tectonic escape: Turkey as a case study', in K.T. Biddle and N. Christie-Blick (eds.), *Strike-Slip Deformation, Basin Formation, and Sedimentation, Society of Economic Paleontologists and Mineralogists Special Publication* **37**, 227-265.

Stoesser, D.B. and Camp, V.E. (1985) 'Pan-African microplate accretion of the Arabian shield', *Geological Society of America Bulletin* **96**, 817-826.

Trifonov, V., Youssef, A., Al-Khair, Y. and Zaza, T. (1983) 'Using satellite imagery to infer the tectonics and the petroleum geology of Syria' , *Syrian Petroleum Company Unpublished Reports*, Damascus, Syria, 18 pp.

Walley, C.D. (1988) 'A braided strike-slip model for the northern continuation of the Dead Sea fault and its implications for Levantine tectonics', *Tectonophysics* **145**, 63-72.

THE TECTONIC REGIMES ALONG THE CONVERGENT BORDER OF THE AEGEAN ARC FROM THE LATE MIOCENE TO THE PRESENT; SOUTHERN PELOPONNESUS AS AN EXAMPLE

MERCIER J.L., SOREL D. , LALECHOS S. and KERAUDREN B.*
URA (CNRS 1369), Laboratoire de Géophysique et Géodynamique Interne
Paris-Sud University, Bât. 509, 91405 Orsay Cedex
France; *URA (CNRS 184).

ABSTRACT. Field analysis of fault kinematics demonstrates changes in the tectonic regime on the Aegean Arc border. During the Pliocene-Early Pleistocene extensional tectonics were active; extension trended radial to the arc border which suffered subsidence. It is suggested that this resulted from a seaward retreat of a high angle-dipping slab which decreased the magnitude of the push on the Aegean lithosphere. During the Mid-Late Pleistocene compression has been active in the lowlands and extension in the highlands of the arc border, the extensional directions being orthogonal to the compressional directions and parallel to the arc border. It is suggested that this has resulted from a push of a low angle-dipping slab on the Aegean lithosphere. These stress patterns are interpreted at a large wavelength as a result of a balance between the boundary forces acting along the arc border and the body forces due to the thickening of the crust. Rotations of the paleostress directions due to rigid rotations of the material and perturbations of the stress trajectories are also considered in analysing the paleostress maps of the Aegean.

1. Introduction

As a whole, the Aegean Arc exhibits geological and geophysical characteristics comparable to those of the island arcs [see detailed ref. in Mercier (1979) and Mercier and Scholtz (1984)]. As demonstrated by seismicity, the subduction zone dips from the Hellenic trench toward the northeast as deep as 180 km and a calcalkaline volcanic arc outlines its curvature (Fig. 1). The southern part of the Aegean sea is a back-arc basin, north of the volcanic arc, and a fore-arc basin (the southern Aegean trough), south of it. It is a thinned continental crust that underlies these regions, having a mean thickness of 25-30 km and a minimum thickness of 20 km beneath the southern Aegean trough. The continental crust reaches a maximum thickness of 35-46 km beneath the southern Aegean Arc and Peloponnesus [Makris (1975)]. The convergent border of the Aegean Arc displays two different geodynamic situations. In the western and southern parts of the Arc, a subduction zone is active since 12-13 Ma [Le Pichon and Angelier (1979)] and probably since 16-17Ma (Mid Miocene) subsequent to the Langhian compressional event [Sorel (1989)]. To the NW, the transition of this Aegean subduction zone to the Adriatic collision takes place in the Ionian Islands [Sorel (1976); Lyberis and Lallemant (1985)]. There during the Late Miocene the convergent boundary of the arc (the "Ionian Arc") was located in continental Greece, east of these islands. Then, during the Lower Pliocene ($\approx$ 5Ma) it jumped west of the Ionian islands to form the Present day boundary of the Aegean Arc in this northwestern region.

This paper concerns the tectonic regimes along the convergent border of the Aegean Arc during the Late Miocene-Present day period. Many data concerning these tectonic regimes have been already published in two papers [Mercier et al (1987), (1989)]. Field data which concern the tectonic regimes acting in the Aegean from the Late Miocene to the Present may be also found in several synthetic papers [Angelier (1979); Le Pichon and Angelier (1979); Lyberis (1984); Pavlides (1985); Sorel (1989); Caputo (1990)]. Those concerning the seismic activity may be found in Mc Kenzie (1972), (1978), Drakopoulos and Delibassis (1982), Papazachos and al. (1984), Anderson and Jackson (1987), Taymaz et al. (1991). Papers concerning the microseismic activity are also available [Hatzfeld et al. (1990)]. The tectonic regimes have been

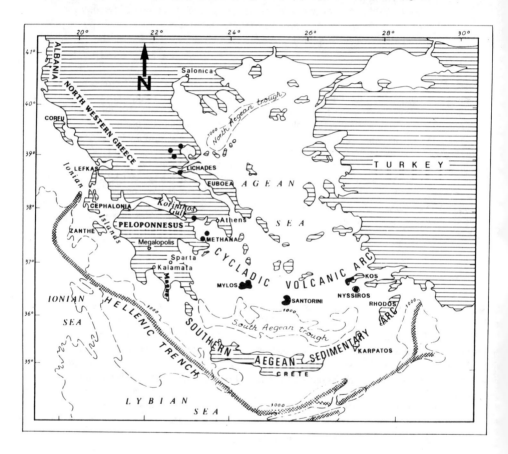

Figure 1. Map showing essential geographic and bathymetric features of the Aegean, Ionian and Lybian seas. Black patches represent Upper Pliocene and Quaternary volcanoes constituting the Cycladic calcalkaline volcanic arc. The arc edge (external Aegean domain) crops out on the Ionian islands of Corfu, Lefkas, Cephalonia, Zanthe and in NW Peloponnesus; towards the southeast it is submerged. The southern Aegean sedimentary arc crops out on the southern Peloponnesus, Crete, Karpathos, Rhodos; it belongs to the internal Aegean domain [after J. Mercier et al. (1987)].

defined from the kinematics of fault populations as shown by striations measured on fault planes in the field or by focal mechanisms. Supposing that sliding occurs in the direction of the shear stress resolved on the fault plane, a mean deviatoric stress tensor is computed within a factor k from a set of striated faults or of focal mechanisms [see Carey (1976), (1979); Carey-Gailhardis and Mercier (1989) and references therein]. In the present paper some examples are choosen in southern Peloponnesus to illustrate the methodology used for separating and analysing the striations sets in the field.

2. Fault Kinematics Analysis : Neotectonic Fauting In Southern Peloponnesus As An Example

Neotectonic faulting has been studied at different scales at numerous sites in Southern Peloponnesus [Lalechos (1992)]. Here we present a brief summary of the results.

2.1. FAULTS AFFECTING DATED SEDIMENTARY DEPOSITS OF LATE PLIOCENE-PLEISTOCENE AGE

Few striated faults are known which affect formations of Mid-Late Pleistocene age. In the vicinity of the major fault of Sparte and in the Magne peninsula some faults affect reddish continental fans and screes attributed to the Mid Pleistocene [Dufaure (1975)]. The pitches of the measured striations (stereonet A3, Fig. 2) show a dextral strike-slip component which clearly demontrates that the direction of extension cannot trend E-W as it has been suggested by Armijo et al. (1992). However, the azimuthal distribution of these faults does not permit us to precisely compute a direction of extension. Therefore we have analysed striated faults affecting formations of Pliocene - Early Pleistocene age.

About 30 km southeast of Kalamata (Fig. 1) near Kardamili, a marine formation of Late Pliocene - Early Pleistocene age lies unconformably on the bedrock. This formation is faulted and some faults show two families of striations. The first family results from a tensional principal stress direction trending NE-SW (stereonet B2, Fig. 2). The second family results from a tensional principal stress direction trending N125° (stereonet B3, Fig. 2) which is in agreement with the kinematics of the faults we have observed affecting Mid Pleistocene fans and screes (stereonet A3, Fig. 2).

2.2. MINOR FAULTS AFFECTING THE BED-ROCK NEAR KARDAMILI

At the same site the bedrock is composed of limestones of the Ionian zone (Mid Trias to Mid Eocene) and of the Tripolis zone (Late Jurassic - Cretaceous). The bedrock is intensively faulted and numerous faults exhibit two, rarely three, families of striations. The third and last family of striations T3 (stereonet C3, Fig. 2) measured at 15 different sites results from extension in the N145 ± 20° direction and are in agreement with the second family of striations (stereonet B3, Fig. 2) affecting the marine formation of Late Pliocene - Early Pleistocene age. They also agree with the kinematics of the faults we have observed affecting continental fans and screes attributed to the Mid-Late Pleistocene (stereonet A3, Fig. 2). Therefore we consider that they are representative of the Mid-Late Pleistocene tectonic regime. It is noteworthy that the computed R ratio [$R = (\sigma_2 - \sigma_1)/(\sigma_3 - \sigma_1)$] of the stress tensors T3 generally has a high value ($R \approx 0.8$). This implies that in southern Peloponnesus the deviatoric value of the intermediate principal stress σ_2 trending SW-NE is also extensional and that $\sigma_2 \approx \sigma_3/2$. The family of striations T2 results from

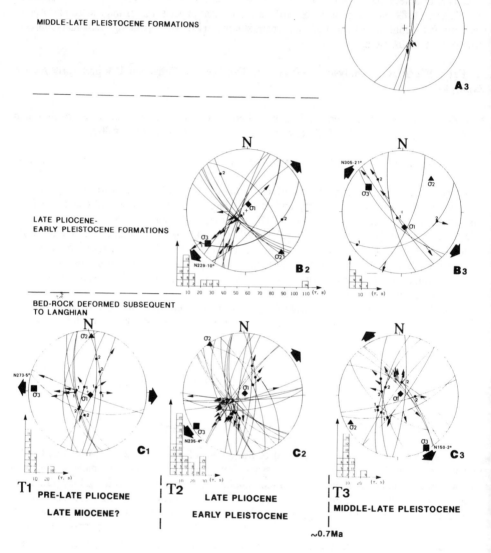

Figure 2

an extension in the N45 ± 20° direction. It is of Late Pliocene -Early Pleistocene age because it affects dated marine deposits of Early Pleistocene age [nannozone NN 19, site M02 in Lallemant (1984)] and predates the Mid Pleistocene. It is not known if this kinematics is older than the Late Pliocene. Finally the first family of striations T1 results from an extension in the N85° ± 20° direction; it predates the Late Pliocene and is probably contemporaneous with the Late Miocene extension demonstrated in Crete by Angelier (1979) and Delrieu (1990).

2.3. DIRECTIONS OF EXTENSION DEFINED BY THE KINEMATICS OF THE MAJOR FAULTS

The Kinematics of major faults, some km to tens of km long, have been measured in the field. These major faults are located on the border of the Megalopolis basin, in the region of Kalamata, along the fault of Sparte and in the Magne Peninsula (Fig. 1). The deviator T1 (Fig. 3) is computed from the first family of striations measured on the bedrock. It demonstrates an extensional tectonic regime with a direction of extension trending N92°. The deviator T2 (Fig. 3) is computed from faults affecting Pliocene deposits and from the family of striations measured on the bedrock subsequent to the first family T1 defined above. It demonstrates an extensional tectonic regime with a direction of extension trending N53°. The deviator T3 (Fig. 3) is computed from the faults affecting Mid -Late Pleistocene deposits or exhibiting striations that postdate to those of the above families. It demonstrates an extensional tectonic regime with a

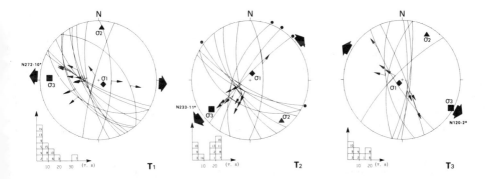

Figure 3. Major normal fault data used to compute the mean (regional) state of stress in southern Peloponnesus. Symbols as in Figure 2.

Figure 2. Minor fault data from southern Peloponnesus. Arrows attached to the fault traces correspond to the measured slip-vectors (Wulff stereonet, lower hemisphere). Histograms show the deviations between the measured and the predicted slip vectors on each fault plane. Divergent black arrows give the azimuths of the minimum σ3 computed principal stress directions. Three families (1, 2, 3) of striations have been separated from faults affecting, (A) Mid-Late Pleistocene formations, (B) Late Pliocene-Early Pleistocene formations and (C) the bedrock. Numbers attached to the filled circles on the fault plane indicate the relative chronology of superimposed striations when observed.

direction of extension trending N120° [more easterly than that (N145 ± 20°) computed from minor faults].

In conclusion, since probably the Late Miocene period, the tectonic regime in southern Peloponnesus has been extensional. The direction of extension trended roughly NW-SE (Fig. 4A) and NE-SW (Fig. 4C) during the Mid -Late Pleistocene and the Late Pliocene - Early Pleistocene respectively. This agrees with the general stress pattern of the border of the Aegean Arc we examine in the following section. The stress pattern of Late Miocene (?) - Lower Pliocene age will be discussed subsequently (section 3.3.2).

3. The Tectonic Regimes Of The Aegean Arc Border During The Late Miocene - Present Day Period

3.1. THE TECTONIC REGIMES OF THE AEGEAN ARC BORDER FROM THE PLIOCENE TO THE PRESENT

Fig. 4A, B and C summarize the main characteristics of the different stress patterns during this period [Mercier et al. (1987), (1989)].

3.1.1. Stress pattern during the Mid Pleistocene - Present day period (Fig. 4A). Along the convergent boundary, focal mechanisms and fault analysis indicate a compressional tectonic regime in Albania and in the Ionian islands; the mean direction of compression trends roughly ENE-WSW to NE-SW. To the southeast the compressional zone extends on the northwestern part of Peloponnesus as shown by some outcrop data [Sorel (1989)] and by microseismic activity [Hatzfeld et al. (1990)]. Further to the south, the compressional zone extends offshore; the direction of compression trends roughly N30° as shown by reverse faulting earthquake focal mechanisms [Mc Kenzie (1972)]. On the internal border of the arc (Northwestern Greece, Peloponnesus, Crete, Karpathos, Rhodes), focal mechanisms and fault analysis show extensional deformation. The mean direction of extension trends roughly N-S in continental Greece and northern Peloponnesus, NW-SE in southern Peloponnesus, ≈N120° in Crete, Karpathos and Rhodos. Thus, during the Mid Pleistocene - Present day period the direction of extension is roughly parallel to the arc curvature and orthogonal to the direction of compression acting along the edge of the arc.

3.1.2. The stress pattern during the Early Pleistocene (proparte) (Fig. 4B). Along the convergent boundary, deformation was strongly compressional in Albania and in the Ionian islands (Zanthe, Cefalonia, Corfu). There, folds, thrusts and reverse faults, several kilometers long, formed along the active Apulian margin. The direction of compression is clearly defined : it trends ENE-WSW [Mercier et al. (1987), Sorel (1989)]. The compressional event is dated as subsequent to the Calabrian and prior to the Mid Pleistocene [between 1 and 0.7 Ma, Sorel et al. (1992)].

In the internal domain, folds and reverse faults of low magnitude suggest a weak compressional tectonic regime. On an E-W section joining the Ionian Islands to Euboea, two directions of compression are observed; they are roughly parallel and orthogonal to the compressional direction acting along the convergent boundary. Few new data have been added to those presented in Mercier et al's (1987) paper. In southern Peloponnesus compressional deformations of Early Pleistocene age have not been evidenced. In eastern Albania, the tectonic regime was extensional with a direction of extension ($\sigma 3$) trending roughly N-S. However the intermediate deviatoric stress axis trending E-W is compressional; this tectonic regime is

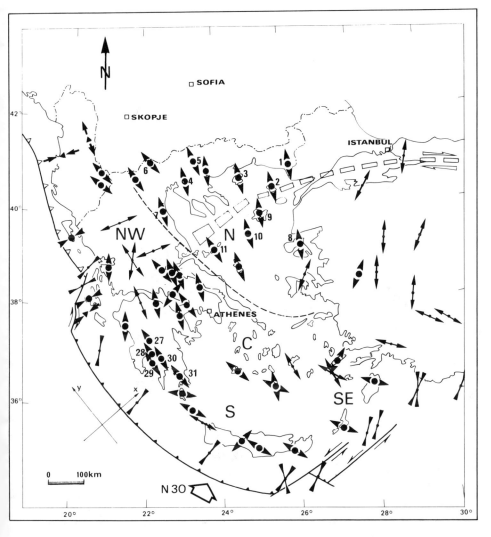

Figure 4A

Figure 4. Stress patterns in the Aegean domain deduced from earthquake focal mechanisms and structural analysis of faults of Pliocene to Quaternary age, for three periods: A - Middle Pleistocene to Present day; B - Early Pleistocene (proparte); C - Early Pliocene (pp) to Early Pleistocene (pp) [modified from Mercier et al. (1987), see detailed references therein].

Double arrows give the directions of the slip vectors on the nodal planes of focal mechanisms of earthquakes (divergent: direction of lengthening, convergent: direction of shortening). Double arrows heads attached to filled circles give the principal stress directions computed from striations measured on fault planes (divergent: σ_3 axis, convergent: σ_1 axis). Small boxes represent the North Anatolian fault system and its continuation into the North Aegean trough.

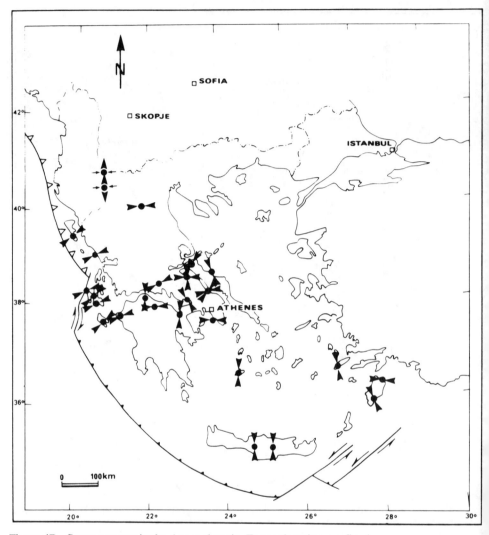

Figure 4B. Stress patterns in the Agean domain. For explanation see fig. 4.

different from the subsequent one of Mid-Recent Pleistocene age [Tagari et al. (1992)].

The regional nature of this compressional regime in the internal domain has been discussed [Jackson et al. (1982); Mercier (1983)]. Although it has been demonstrated at many sites, the directions of compression are not well defined because compressional structures are often subtle. Indeed, such a short-lived compressional event occuring during a long period of extensional tectonic regime in a back-arc domain is not particular to the Aegean. For instance, this has been

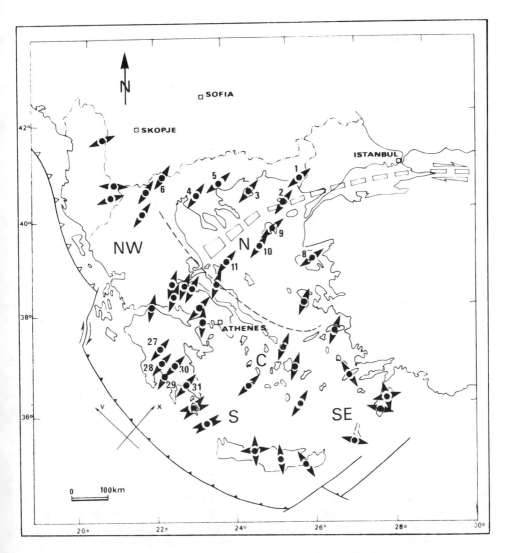

Figure 4C. Stress patterns in the Agean domain. For explanation see fig. 4.

also demonstrated in the High Cordillera above the Andean subduction zone [Sebrier et al. (1985), Mercier et al. (1992)].

3.1.3. *The stress pattern during the Pliocene - Early Pleistocene period (Fig. 4C).* In the Ionian Islands and in the Adriatic depression in front of Albania, the marine sedimentation continued from the Early Pliocene until the Early Pleistocene (Calabrian); subsidence probably reflected a tectonic activity during this period. But no tectonic structures have permitted us to decide whether the tectonic regime was compressional or extensional at that time. No data are available for the region located more to the south as the convergent margin is located offshore. On the internal border of the arc, deformation was clearly extensional. The directions of extension trend roughly E-W in Albania, NE-SW in Peloponnesus i.e. roughly radial to the arc curvature of the western part of the arc. In Eastern Aegean (E. Crete, Karpathos, Rhodos), the tensional directions are more variable but seem to trend toward the deep bathymetric depression of the Lybian sea.

Rotations deduced from paleomagnetic studies are less than 20° in the Ionian branch of the arc since the Late Pliocene ($\approx$ 3.5 Ma), 0° in Crete since the Tortonian ($\approx$ 11 - 6.5 Ma) and 0° in Rhodos Is. since the Early Pliocene [Kissel et al. (1985)]. Thus these small rotations allow us to discuss the Pliocene - Early Pleistocene stress directions in their Present day orientation to a first approximation. The effects of these rotations are discussed in section 3.3.

3.2 REGIONAL CHANGES IN THE STATE OF STRESS ON THE AEGEAN ARC BORDER FROM THE PLIOCENE TO THE PRESENT

The Agean stress pattern changed from the Pliocene - Early Pleistocene (Fig. 4C) to the Mid Pleistocene - Present day period (Fig. 4A). To interpret these changes in the state of stress at a large wavelength (> 100km) we have used a model of compensated topography [see Fleitout and Froidevaux (1982)]. In the island arcs, the topography, the crustal thickness and the deep thermal structure strongly vary laterally and so does the vertical stress value. For different isostatic models, Froidevaux et al. (1987) have computed the average lithospheric vertical stress σ_{zz} in excess to a reference lithostatic stress which has been chosen as that of an ad hoc fore-arc structure shown on Fig. 5. Absolute values of σ_{zz} seen on this figure are not significant but their lateral variations may help to qualitatively explain the spacial changes in the state of stress in the Aegean. For a more detailed analysis the reader may report to Mercier et al's [(1987), (1989)] papers.

Figure 5. Qualitative physical interpretation of the variations of the state of stress in the overriding plate of theAegean subduction zone. Figure at the top, drawn from Froidevaux et al. (1987) shows different isostatic models used to compute average lithospheric vertical stresses σ_{zz} in excess to a reference lithostatic stress chosen as that of the forearc (water depth = 4 km, crustal thickness = 15 km, lithospheric thickness = 80 km). It shows the variation of σ_{zz} with respect to crust thicknesses and deep thermal structures.
A - Middle Pleistocene to Present day period. Mean crustal thicknesses under the forearc, southern Peloponnesus, the southern Aegean trough and the backarc basin are from Makris (1975, 1978).Values of σ_{zz} are those computed by Froidevaux et al. (1987) for models whose crustal thicknesses and deep thermal structures resemble those of these regions; of course, absolute values are not significant. B - Early Pleistocene (p.p.) period. The push σ_{xx} of the slab increases, compression affects the internal domain. C - Early Pliocene (p.p.) - Early Pleistocene (p.p.) period. The push of the slab being weak, or tractional forces acting on the Aegean lithosphere, tension occurs radial to the arc boundary.The sketch shows that large stretching amounts and subsidence have affected the south Aegean trough [from Mercier et al. (1987)].

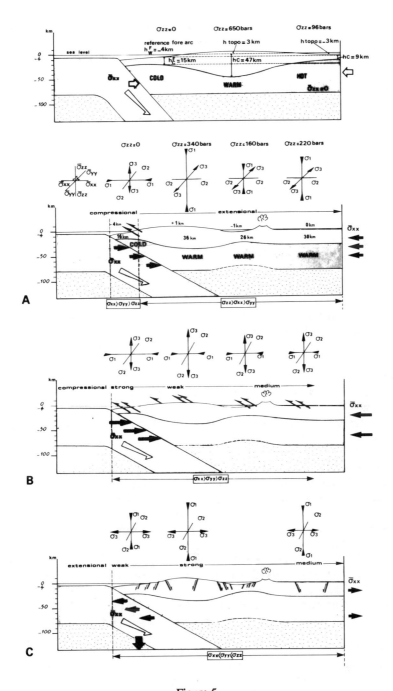

Figure 5

3.2.1. *The state of stress on the Aegean Arc border during the Mid Pleistocene to Present day period.* Fig. 5A is drawn so that the section contains the σ_{xx} (i.e. σ_{Hmax}), and the σ_{zz} axes. Probably due to the curvature of the arc and to the internal extensional deformation of the Aegean, the σ_{xx} azimuth varies along the arc boundary (Fig. 4A). In Albania and in the Ionian Islands, tectonics are compressional: σ_{xx} is σ_1 and σ_{yy} is σ_2; σ_{zz} chosen arbitrarily equal to zero is σ_3 (Fig. 5A). The same situation occurs along the Hellenic trench where the σ_{xx} direction, as shown by the seismic slip vectors on reverse faults [Mc Kenzie (1972), (1978)], trends NE-SW to NNE-SSW roughly parallel to the Aegean/Africa convergence. On the internal border of the arc, the vertical stress σ_{zz} value increases essentially due to the increasing thickness of the continental crust. The vertical stress value σ_{zz} becomes σ_1, then σ_{xx} being σH_{max} is σ_2 and σ_{yy} being σ_{Hmin} becomes σ_3, allowing extension to occur in a roughly N-S to NW-SE - trending direction, orthogonal to the σ_1 direction acting along the convergent boundary (Fig. 4A). In our previous papers [Mercier et al. (1987), (1989)], we had emphasized that the state of stress in Peloponnesus seemed to disagree with the regional state of stress predicted by the model, the already published directions of extension in this region trending ENE-WSW to EW [Sébrier (1977); Angelier (1979); Lallemant (1984)]. As shown in section 2 this E-W extension deduced from striations measured on the bedrock of the major faults (stereonet 1, Fig. 3) predates the Late Pliocene. Indeed the Mid Pleistocene to Present day direction of extension does trend NW-SE (Fig. 4A), in agreement with the model. Thus, at a several hundred kilometers wavelength, the Mid Pleistocene - Present day extensional tectonic regime may be explained by a slab push σ_{xx} acting on the continental Aegean lithosphere plus variations of the vertical stress σ_{zz} value depending on inherited mass heterogeneities.

3.2.2. *The compressional state of stress during the Early Pleistocene period.* During the Early Pleistocene, tectonics were strongly compressional in the Ionian Islands and along the sea-shore of Albania (Fig. 4B): σ_{xx} was σ_1, σ_{yy} was σ_2 and σ_{zz} was σ_3. The compressional direction, in its present-day orientation, is similar to that of the Mid Pleistocene - Present day period but compressional deformations were of much higher magnitude. Assuming that the rheology of the material has not radically changed during this short compressional event, these deformations have to result from an increase of the σ_{xx} value (Fig. 5B). In the internal domain a correlation seems to exist between the magnitude of the compressional deformations and the elevations above sea level i.e. with the corresponding vertical stress value σ_{zz}. During the Early Pleistocene the elevations were lower than today (see section 3), but since this period the crustal thickness in these different regions, has not significantly changed. Indeed deformations are extensional in the high mountains of eastern Albania, with an E-W trending intermediate axis σ_2 having a compressional deviatoric stress value and no compressional deformations have been observed in Peloponnesus; compressional deformations are subtle in Crete [Angelier (1979)] and medium in the lowlands of the Aegean sea.

According to the above model, the increase of the σ_{xx} value implies that the compressional direction in the internal domain is parallel to that acting along the arc boundary (Fig. 5B). This is observed in a section from the Ionian islands to Euboea (Fig. 4B). However other compressional directions are observed, the most frequent trends N-S roughly orthogonal to the previous one. This recalls the two orthogonal directions of compression we have observed in Peru above the subduction zone of the Andes and we have suggested that this might result from an heterogeneous style of deformation [see Mercier et al. (1992)].

3.2.3. *The extensional state of stress during the Pliocene - Early Pleistocene.* A comparison of Fig. 4A and Fig. 4C clearly shows that during the Pliocene - Early Pleistocene, the directions of extension in the internal domain strongly differed from those of the Mid-Late Pleistocene period; they trended roughly NE-SW (Fig. 4C) instead of N-S to NW-SE (Fig. 4A). Using the same notation σ_{xx} for the roughly NE-SW trending stress trajectory as in Fig. 4A, in the internal domain σ_{xx} was σ_3, σ_{yy} was σ_2 and σ_{zz} was σ_1 (Fig. 5C). Thus σ_{xx} was not the maximum but the minimum horizontal stress value. This extensional tectonic regime is not in agreement with a NE-SW to ENE-WSW trending compression in the overriding plate due to a push of the slab. Therefore we have suggested that the slab migrated seaward with respect to the Aegean [Makris (1978), Le Pichon and Angelier (1979)] during the Pliocene so that during this period the slab push was weak, or possibly a traction was applied to the Aegean boundary.

In summary, at a several hundred kilometers wavelength, the spacial and also the temporal changes in the tectonic regime may be satisfactorily explained by a balance between the boundary forces due to the Aegean/African convergence and the body forces due to the inherited mass heterogeneities. Indeed such models do not explain the details of the Aegean stress pattern nor the geometry and the distribution of the Aegean deformations. Local perturbations of the stress directions may occur and possibly some stress directions "frozen" in the rocks have rotated during time as a result of rotational deformations.

3.3 LOCAL PERTURBATIONS AND ROTATIONS OF THE STRESS DIRECTIONS

As an example we again examine the directions of extension obtained in Southern Peloponnesus (Fig. 2 and 3). Each family T1, T2 and T3 of directions of extension obtained from the analysis of faults at numerous sites shows a large scattering around a mean value (see section 2.2). This may be due to local perturbations of the stress trajectories or to rigid rotations of different magnitudes of a regional stress direction "frozen" in the rocks in the form of brittle deformations. Paleomagnetic studies [Kissel et al. (1985)] have demonstrated that the western branch of the Aegean Arc in the Ionian islands and in Western Greece has rotated in the order of 25° since the Early Pliocene ($\approx$ 5Ma) and in the order of 50° since the Mid Miocene (16Ma). Rotations of similar magnitude have probably also affected the Peloponnesus regions as it has been shown that a clockwise rotation in the order of 15° has affected the Pliocene deposits of southern Peloponnesus [Laj et al. (1982)].

3.3.1. *Local perturbations of the stress directions of Mid-Late Pleistocene age.* The directions of extension T3 (Fig. 2) defined from the analysis of minor faults range between N125 and N170°. As the rigid rotation of the western branch of the Aegean Arc has not exceeded 5° during the last 1My [Kissel et al. (1985)], this range of directions must result from local pertubations of the stress trajectories whose mean direction trends $\approx$ N140°. On the other hand the difference between this mean direction of extension T3 (N140°) and the mean direction of extension T2 (N45°) of Pliocene - Early Pleistocene age cannot be explained by the clockwise rotation of the former related to the clockwise rotation of the western branch of the Aegean Arc. The difference of 85° between these two mean directions by far exceeds the maximum 25° clockwise rotation of the western branch of the arc that could have taken place during the last 5My. The two mean directions of extension T2 and T3 are really indicative of two different tectonic regimes.

3.3.2. *Rotations of the tensional stress directions of Late Pliocene - Early Pleistocene age (T2) and of previous Late Pliocene age (T1).* The directions of extension T2 may have been submitted to a maximum 25° clockwise rotation. If the extension T1 started during the Late Miocene, these directions might have been submitted to a maximum 50° clockwise rotation. It is observed that the directions of extension T1 and T2 together range continuously from N25° to N100°. This suggests the possibility that an initially NNE-SSW regional direction of extension "frozen" in the rocks has rotated since the Late Miocene as a consequence of the clockwise rotation of the western branch of the Aegean Arc. At each site where the two T1 and T2 directions of extension have been separated, the deviations between the T1 and T2 directions range between 15° and 60°. Yet, these values may result from both rotations of the material and local perturbations of the stress trajectories. It is not really possible to separate the two effects at each site because it is not known when the striations measured at a given site formed and thus to know how much they may have rotated. For a given event, for example T2, these striations may have formed together during a short event at any time between 3.5Ma and 1Ma or they may have formed successively during all this period. Nevertheless considering the deviation between the mean T1 (N85°) and the mean T2 (N45°) directions, the obtained 40° value is compatible with a clockwise rotation of the western branch of the Aegean Arc during the Late Miocene - Early Pliocene as shown by paleomagnetic data. Thus, the scattering of the directions of extension from N25° to N100° may result from both spacial perturbations of the stress trajectories and subsequent rigid rotations of the striations on the fault planes.

If the E-W direction of extension T1 results from the rotation of an extensional direction trending NNE-SSW to NE-SW, this direction may be compared with the NNE-SSW direction of extension of Late Miocene (Serravalian - Tortonian) age evidenced in Crete [Angelier (1979); Delrieu (1990)] where no rotations have occurred since the Late Miocene. Another E-W to NW-SE direction of extension has been evidenced in Crete [Angelier (1979)]; it has been dated of Uppermost Tortonian to Messinian age (7-5Ma) by Delrieu (1990). This is probably equivalent to the N110-120° extension evidenced in the Northern Aegean [Mercier et al. (1989)] where it affects Late Miocene - ? Early Pliocene deposits. This Messinian extension has not been evidenced in southern Peloponnesus due to the lack of Late Miocene deposits in this region. If this extension is present in Peloponnesus, it affects the bedrock and if it has undergone a 20-30° clockwise rotation, then it is almost impossible to separate the corresponding striations from those of Mid-Upper Pleistocene age.

4. Tectonic Regimes And Vertical Motions On The Aegean Arc Border During The Pliocene - Pleistocene

Supposing that the rheology of the Aegean lithosphere has not significantly changed during the last 5 My, the drastic change in the tectonic regime in the Aegean Arc from the Pliocene - Early Pleistocene to the Mid-Late Pleistocene implies an increase of the boundary forces at the convergent limit of the Aegean Arc during the Early Pleistocene ($\approx$ 1 - 0,7 Ma). It may be suggested that the Aegean Arc has been in a geodynamic situation of collision during the Quaternary [Le Pichon and Angelier (1979)] and that this has to increase the compressional forces applied to the Aegean lithosphere. Yet, similar changes in the tectonic regimes are also observed above the subduction zone of the Central Andes [Sebrier et al. (1985); Mercier et al. (1992)] where no collision has occurred. This suggests that these changes in the tectonic regimes result from the subduction process itself.

On the other hand, these changes in the tectonic regime appear to be roughly contemporaneous with changes in the vertical motions of the Aegean Arc border [Sorel et al. (1988)]. Along this border (Rhodos, Karpathos, Crete, southern Peloponnesus, Ionian Islands, southern Albania) mesozoic limestone massifs exhibit a particular staircase morphology over a few hundred meters high. The steps of this morphology are littoral abrasion platforms. The marine transgressive Late Pliocene deposits which lie on these platforms and synsedimentary faults present evidence that these platforms result from subsidence [Keraudren (1970-72)] which ended during the Early Pleistocene. During this period of subsidence the extensional directions trended radial to the arc (Fig. 4C). This stress pattern involves compressional boundary forces of low magnitude, possibly extensional forces, acting along the arc border (Fig. 5C). Then a strong uplift episode brought these marine submerged platforms back above sea level at elevations which may attain 500 m on Rhodos Is. and along the southern Peloponnesus coast. This uplift occurred during (?) and subsequent to the Early Pleistocene compressional event. During this period tectonics were mostly compressional (Fig. 4B), then compressional along the arc border and extensional in the Highlands (Fig. 4A). The extensional directions were roughly parallel to the arc and orthogonal to the compressional directions in the adjacent lowlands of the arc border. These tectonic regimes involve compressional forces acting along the arc border (Fig. 5B and A).

Sorel et al. (1988) have suggested a common origin for these changes in the vertical motions of the arc border and in the tectonic regimes. The following scenario has been proposed which assumes that the slab-pull force governs the tectonic regimes in subduction zones [see Jarrard (1986)]. A strong pull resulting from a long high angle slab sinking in the asthenosphere decreases the mechanical coupling between the two plates. This decreases the radial stress in the arc (Fig. 5C) and allows extension to be radial (Fig. 4C); the Aegean extension has been already explained in this way [Makris (1978); Le Pichon (1979)]. The length of the slab increasing, as well as the downward pull-force, this causes the subsidence of the arc border during the Pliocene-Early Pleistocene. The self-amplifying mechanism may lead to the rupture of the slab under its own weight [Forsyth and Uyeda (1975)]. This occurred at the end of the Early Pleistocene; the resulting drop in the pull force may explain the uplift of the arc and the strong increase of the radial stress (Fig. 4B and 5B). Later on, subduction slowly restored the slab and this compression weakened (Fig. 5A) so that compression is active in the lowlands of the arc border and extension became longitudinal to the arc border in the highlands (Fig. 4A). Seismic tomography has shown a rupture of the slab [Spakman et al. (1988)] which may have caused the Early Pleistocene compressional event and the uplift of the arc border. If this is correct, the slab rupture would have to be effective all along the Aegean Arc. Older compressional events affected the Aegean arc [see Mercier et al. (1987)], particularly during the Early Pliocene. They are of different origin because they were caused by the jump of the frontal boundary of the arc west of the Ionian Islands whereas no similar jump is known during the Early Pleistocene period.

5. Concluding remarks

5.1. THE AEGEAN STRESS PATTERNS AT A LARGE WAVELENGTH (>100 KM)

A detailed field analysis conducted in the Aegean domain during the last decades and focal mechanisms of earthquakes suggest that two different types of extensional tectonic regimes have been active in the Aegean domain since the Pliocene:

(1) one has the tensional directions σ_3 trending roughly radial to the arc (Figure 4C). This R. type recalls the Marianna type [Uyeda and Kanamori (1979)] or class 1 [Jarrard (1986)] of the

subduction zones. A strong subsidence of the Aegean arc border is contemporaneous with this tectonic regime.

(2) The other has the tensional directions σ_3 trending roughly parallel to the arc and orthogonal to the compressional direction σ_1 acting along the arc boundary (Figure 4A). This O. type recalls the Andean type [Uyeda and Kanamori (1979)] or class 7 [Jarrard (1986)] of the subduction zones. A strong uplift of the Aegean arc border is contemporaneous with this tectonic regime.

(3) A period of compressional tectonics during which compression may invade the internal domain submitted to extension (Figure 4B) separates this two types of extensional tectonic regimes.

At a large wavelength, the spatial changes (Figure 5A) and the temporal changes (Figures 5A, B and C) in the tectonic regimes may be explained satisfactorily by a balance between boundary forces due to the push of the slab on the Aegean lithosphere and body forces due to mass heterogeneities in the Aegean lithosphere. It is suggested that the change in the tectonic regime during the Early Pleistocene may be due to a rupture of a high-angle slab whose upper part subsequently flattened.

Such changes in the tectonic regime possibly occurred during the Late Miocene-Early Pliocene. The previous Late Pliocene extension in Peloponnesus probably trended initially NNE-SSW (section 3.3.2) as the direction of extension of Serravalian-Tortonian age evidenced inCrete which has not rotated since the Tortonian. On the other hand, the direction of extension trended WNW-ESE to NW-SE in Crete during the Uppermost Tortonian-Messinian. The N100-120° trending extension evidenced in northern Agean in regions which have not rotated since this period is probably of the same age. It is recalled that during the Late Miocene the geometry of the arc boundary was more rectilinear than nowadays, trending roughly WNW-ESE. The Serravalian-Tortonian extension was possibly a radial extension. The Uppermost Tortonian-Messinian extension was possibly an extension parallel to the arc border; the change in the tectonic regime is possibly related to the compressional event of Early Pliocene- (?)Messinian age.

This Late Miocene-Present day tectonic history is related to both the right lateral strike-slip motion on the North Anatolian fault [Sengor (1979)] and to the subduction process of the Aegean Arc. Yet, the extensional tectonic regime started earlier in the Aegean. Subsident basins in the Northern Aegean have formed since the end of the Lutetian ($\approx$ 45Ma), subsequent to the main folding and thrusting of Middle Eocene age [Mercier et al., (1989); Papanikolaou and Dermitzakis (1982)]. These basins are related to an older arc, the 'Pelagonian-Pindic Arc' [Mercier et al. (1989)].

5.2. Perturbations of the Stress Trajectories and Rotations of the Stress Directions 'Frozen' in the Material

The analysis of the state of stress at a large wavelength does not explain neither the details of the stress patterns nor the distribution of brittle deformation in the Aegean. This latter also depends on the inherited planes of weakness due to a long tectonic history of Mesozoic and Paleogene age.

Local perturbations of the stress trajectories may be evidenced in southern Peloponnesus (section 3.3.1) and in northern Aegean [Mercier et al. (1989)]. Thus, a statistical analysis of the stress directions is needed when studying the kinematics of minor faults to obtain a satisfaying picture of the regional stress direction.

Rotations of the stress directions 'frozen' in the material in the form of brittle deformations may also occur. Rigid rotations of the western branch of the Aegean Arc probably explain the Late Miocene(?) E-W trending direction of extension in southern Peloponnesus (section 3.3.2). Clockwise rotations related to the dextral strike-slip motion of the North Aegean trough fault zone [Mc Kenzie and Jackson (1983)] probably explain the 25° clockwise rotation of the Late Miocene directions of extension south of the fault zone with respect to those north of the fault zone [Simeakis et al. (1989)]. Thus, paleomagnetic data are necessary when studying fault kinematics in terms of stress.

5.3. DUCTILE AND BRITTLE DEFORMATIONS OF LATE OLIGOCENE-PRESENT DAY AGE IN THE AEGEAN

Extensional directions of Oligocene to Late Miocene age have been deduced from analyses of ductile deformations which affect metamorphic and plutonic bodies in the islands of the Aegean Sea [Gauthier et al. (1990), Faure and Bonneau (1988)]. These extensional directions trend roughly N-S in regions which probably have not been submitted to rotations of large magnitudes. These directions give a good picture of the finite deformation resulting from a roughly N-S lengthening. We suggest that this N-S lengthening essentially formed during the periods of radial extension. Such an analysis is indicative of the Aegean behavior in terms of deformation,yet it can hardly demonstrate subtle changes in the tectonic regime. On the other hand, the analysis of brittle deformations in terms of stress allows us to separate changes in the tectonic regimes which reflect changes in geodynamic conditions i.e. in the balance between the boundary (tectonic) forces and the body forces.

6. References

Anderson, H. and Jackson, J.A. (1987) 'Active tectonics of the Adriatics Region', Geophys. J. R. Astr. Soc. 91, 937-983.

Angelier, J. (1979) 'Néotectonique de l'Arc Egéen', Thèse d'Etat, Soc. Géol. du Nord n°3, Lille.

Armijo, R., Lyon-Caen, H. and Papanastassiou, D. (1992) 'East-West extension and Holocene normal-fault scarps in the Hellenic arc', Geology 20, 491-494.

Caputo, R. (1990) 'Geological and structural study of the recent and active brittle deformation of the Neogene-Quaternary basins of Thessaly (Central Greece)', Sci. Annals of the Geological Department n°12, Aristotle Univ., Thessaloniki.

Carey, E. (1976) 'Analyse numérique d'un modèle mécanique élémentaire appliqué à l'étude d'une population de failles : calcul d'un tenseur moyen des contraintes à partir des stries de glissement', Thèse 3ème Cycle, Univ. Paris-Sud-Orsay, 138 pp.

Carey, E. (1979) 'Recherche des directions principales de contraintes associées au jeu d'une population de failles', Rev. Géol. Dyn. Géogr. Phys. 21(1), 57-65.

Carey-Gailhardis, E. and Mercier, J.L. (1987) 'A numerical method for determining the state of stress using focal mechanisms of earthquake populations: application to Tibetan teleseisms and microseismicity of southern Peru', Earth Planet. Sci. Lett. 82, 165-179.

Delrieu, B. (1990) 'Evolution tectonosédimentaire du Malévisi et du secteur d'Ano Moulia au Miocène supérieur (bassin d'Héraklion, Crète centrale, Grèce)', Mémoire Géologue I.G.A.L. n° 42, 387 pp., Paris.

Dufaure, J.J. (1975) 'Le relief du Péloponnèse', Thèse d'Etat, Univ. de Paris-Sorbonne, 1422 pp.

Drakopoulos, J. and Delibassis, N. (1982) 'The focal mechanisms of earthquakes in the major areas of Greece for the period 1947-1981', Seismological Laboratory n° 2, Univ. of Athens, 72 pp.

Faure, M. and Bonneau, M. (1988) 'Données nouvelles sur l'extension néogène de l'Egée: la déformation ductile du granite miocène de Mykonos (Cyclades, Grèce)', C.R. Acad. Sci. Paris 307, 1553-1559.

Fleitout, L. and Froidevaux, C. (1982) 'Tectonics and topography for a lithosphere containing density heterogeneities', Tectonics 1(1), 21-56.

Forsyth, D. and Uyeda, S. (1975) 'On the relative importance of the driving forces on plate motion', Geophys. J. R. Astr. Soc. 43, 163-200.

Froidevaux, C., Uyeda, S. and Yueshima, M. (1987) 'Island arc tectonics', Tectonophysics 148, 1-9.

Gauthier, P. Ballèvre, M. Brun, J.P. and Jolivet, L. (1990) 'Extension ductile et bassins sédimentaires Mio-Pliocène dans les Cyclades (îles de Naxos et de Paros)', C.R. Acad. Sci. Paris 310, 147-153.

Hatzfeld, D., Pedotti, G., Hatzidimitriou, P. and Makropoulos, K. (1990) 'The strain pattern in the western Hellenic arc deduced from a microearthquake survey', Geophy. J. Int. 101, 181-202.

Jackson, J.A., King, G. and Vita-Finzi, C. (1982) 'The neotectonics of the Aegean, an alternative view', Earth Planet. Sci. Lett. 61, 303-318.

Keraudren, B. (1970, 71, 72) 'Les formations quaternaires marines de la Grèce', Thèse d'Etat, Bull. Mus. Anthrop. Préhist. Monaco 16, 5-153; 17, 87-169; 18, 223-270.

Kissel, C., Laj, C. and Müller, C. (1985) 'Tertiary geodynamical evolution of Northwestern Greece: paleomagnetic results', Earth Planet. Sci. Lett. 72, 190-204.

Laj, C., Jamet, M., Sorel, D. and Valente, J.P. (1982) 'First paleomagnetic results from Mio-Pliocene series of the Hellenic sedimentary Arc', in : X. Le Pichon, S.S. Augustidhis and J. Mascle (eds), Geodynamics of the Hellenic Arc and Trench, Tectonophysics 86, 45-67.

Lalechos, S. (1992) 'Etudes néotectoniques et sismotectoniques dans le Péloponnèse méridonal (Grèce). Relations avec la géodynamique plio-quaternaire de l'Arc Egéen', Thèse de l'Univ. Paris Sud-Orsay, 230 pp.

Lallemant, S. (1984) 'La transversale Nord-Maniote ; étude géologique et aéromagnétique d'une structure transversale à l'arc Egéen externe', Thèse 3ème Cycle, Univ. Paris 6, 175 pp.

Le Pichon, X. (1979) 'Bassins marginaux et collision continentale : exemple de la zone égéenne', C.R. Acad. Sci. Paris 288, 1083-1086.

Le Pichon, X. and Angelier, J. (1979) 'The Hellenic arc and trench system: a key to the neotectonic evolution of the Eastern Mediterranean area, Tectonophysics 60, 1-42.

Lyberis, N. (1984) 'Géodynamique du domaine égéen depuis le Miocène supérieur', Thèse d'Etat, Univ. Paris 6, 367 pp.

Lyberis, N. et Lallemant, S. (1985) 'La transition subduction-collision le long de l'arc égéen externe', C.R. Acad. Sci. Paris 300, sér. 2, 885-890.

Makris, J. (1975) 'Crustal structure of the Aegean sea and the Hellenids obtained from geophysical surveys', J. Geophys. 41, 441-443.

Makris, J. (1978) 'The crust and upper mantle of the Aegean Region from deep seismic soundings', Tectonophysics 46, 269-284.

Mc Kenzie, D.P. (1972) 'Active tectonics of the Mediterranean region', Geophys. J. R. Astr. Soc. 30, 109-185.

Mc Kenzie, D.P. (1978) 'Active tectonics of the Alpine-Himalayan belt: the Aegean sea and surrounding regions', Geophys. J.R. Astr. Soc. 55, 217-254.

Mc Kenzie, D.P. and Jackson, J. (1983) 'The relation between strain rates, crustal thickening, paleomagnetism, finite strain and fault movements within a deforming zone' Earth Planet. Sci. Lett. 65, 182-202.

Mercier, J.L. (1979) 'Signification néotectonique de l'Arc Egéen. Une revue des idées' Rev. Géol. Dyn. Géogr. Phys. 21(1), 5-15.

Mercier, J.L. (1983) 'Some remarks concerning the paper 'The neotectonics of the Aegean: an alternative view' by J.A. Jackson, G. King and C. Vita-Finzi', Earth Planet. Sci. Lett. 66, 321-325.

Mercier, J.L. and Scholtz, C. (1984) 'Fensability of an international experimental site for earthquake prediction in Greece' Report UNESCO/IASPEI work, on a code of practice for earthquake prediction, édit. de l'UNESCO, Paris.

Mercier, J.L., Sébrier, M., Lavenu, A., Cabrera, J., Bellier, O., Dumont, J.F. and Machare, J. (1992) 'Changes in the tectonic regime above a subduction zone of Andean type: the Andes of Peru and Bolivia during the Pliocene-Pleistocene', J. Geophys. Res. 97, 11, 945 - 11, 982.

Mercier, J.L., Simeakis, K., Sorel, D. and Vergely, P. (1989) 'Extensional tectonic regimes in the Aegean basins during the Cenozoic', Basin Research 2, 49-77.

Mercier, J.L., Sorel, D. and Simeakis, K. (1987) 'Changes in the state of stress in the overriding plate of a subduction zone: the Aegean arc from the Pliocene to the present', Annales Tectonicae 1(1), 20-39.

Papanikolaou, D.J. and Dermitzakis, M.D. (1982) 'Major changes from the last stages of the Hellenids to the actual Hellenic arc and trench system', Internat. Symp. on the Hellenic arc (HEAT), Athens 8-10/4/1981, proceedings vol. II, 57-73.

Papazachos, B.C., Kiratzi, A.A., Hatzidimitriou, P.M. and Rocca, A.C. (1984) 'Seismic faults in the Aegean area', Tectonophysics 106, 71-85.

Pavlides, S.B. (1985) 'Neotectonic evolution of the Florina-Vegoritis-Ptolemais basin (W. Macedonia, Greece)', Thesis, Univ. of Thessaloniki, Greece, 265 pp.

Sébrier, M. (1977) 'Tectonique récente d'une transversale à l'Arc Egéen: le golfe de Corinthe et ses régions périphériques', Thèse de 3ème Cycle, Univ. Paris Sud-Orsay, 176 pp.

Sébrier, M., Mercier, J.L., Mégard, F., Laubascher, G. and Carey-Gailhardis, E. (1985) 'Quaternary normal and reverse faulting and the state of stress in Central Andes of South Peru', Tectonics 4(7), 739-780.

Sengor, A.M.C. (1979) 'The North Anatolian transform fault: its age, offset and tectonic significance', J. Geol. Soc. Lond. 136, 269-282.

Simeakis, C., Mercier, J.L., Vergely, P. and Kissel, C. (1989) 'Late Cenozoic rotations along the North Aegean trough: structural constraints', in: C. Kissel and C. Laj (eds), Paleomagnetic Rotations and Continental Deformations, Kluwer Academic Publishers, Dortrecht, NATO ASI Ser., C 254, 131-143.

Sorel, D. (1976) 'Etude néotectonique de l'Arc Egéen externe occidental : les îles ioniennes de Kephallinia et Zakinthos et l'Elide occidentale', Thèse 3ème Cycle, Univ. Paris Sud-Orsay, 196 pp.

Sorel, D. (1989) 'L'évolution structurale de la Grèce nord-occidentale depuis le Miocène dans le cadre géodynamique de l'Arc Egéen', Thèse d'Etat, Univ. Paris Sud-Orsay, 305 pp.

Sorel, D., Bizon, G., Alliaj, S. and Hassani, L. (1992) 'Calage stratigraphique de l'âge et de la durée des phases compressives des Hellénides externes (Grèce nord-occidentale et Albanie) du Miocène à l'actuel', Bull. Soc. Géol. France 163(4), 447-454.

Sorel, D., Mercier, J.L., Keraudren, B. and Cushing, M. (1988) 'Le rôle de la traction de la lithosphère subductée dans l'évolution géodynamique plio-pleistocene de l'Arc égéen : mouvements verticaux alternés et variations du régime tectonique', C.R. Ac. Sci. Paris 307, sér. 2, 1981-1986.

Spakman, W., Wortel, M.J. and Vlaar, N.J. (1988) 'The Hellenic subduction zone: a tomographic image and its geodynamic implications', Geophys. Res. Lett. 15, 60-63.

Tagari, D., Vergely, P. and Aliaj, S. (1993) 'Tectonique polyphasée plio-quaternaire en Albanie orientale (région de Korça - Progradeci - Peshkopia)', Bull. Soc. Géol. France, in press.

Taymaz, T., Jackson, J. and Mc Kenzie, D.P. (1991) 'Active tectonics of the north and central Aegean sea', Geophys. J. Int. 106, 433-490.

Uyeda, S. and Kanamori, H. (1979) 'Back-arc opening and the mode of subduction', J. Geophys. Res. 84, 1049-1061.

TYRRHENIAN BASIN AND APENNINES. KINEMATIC EVOLUTION AND RELATED DYNAMIC CONSTRAINTS

E. PATACCA[1], R. SARTORI [2], P. SCANDONE [1]

[1] Dip. Scienze della Terra Università, v. S. Maria 53, Pisa. - Italy
[2] Dip. Scienze Geologiche Università. v. Zamboni 67, Bologna. - Italy

ABSTRACT. The post-Oligocene kinematic evolution of the central Mediterranean region is synthetically described. The proposed reconstruction is based on a comparative study of tectonically- controlled sedimentary sequences and of coeval tectonic features in areas experiencing stretching (Algero-Provençal Basin, Tyrrhenian Basin) and shortening (Apennines). The derived step by step palinspastic reconstruction at Langhian times states a number of constraints for the geodynamic processes that affected the area during Neogene and Quaternary times. Both continental (often thinned) and oceanic lithosphere were involved in subduction processes. The former prevailed in the northern sector, while both oceanic and continental lithosphere were present in the southern area. The long duration of the subduction processes in the southern sectors may be related to the existence of large portions of sinking oceanic lithosphere which dragged down intervening portions of continental lithosphere.

1. Introduction

Four major geodynamic elements occur in the central Mediterranean region. The Algero-Provençal Basin is a back-arc area that developed between the upper Oligocene and the lower-middle Miocene. The Corsica-Sardinia "Block", is a fragment of normal continental crust which acted as a frontal volcanic arc, translated and rotated, during the opening of the Algero-Provençal Basin. The Tyrrhenian Basin is another back-arc area that started stretching since upper Miocene times. The Apennines are an orogenic system that records shortening and deformations coeval to the development of the previous domains.

Following the "roll-back" model proposed by Malinverno and Ryan (1986), the opening of the extensional basins may be interpreted as a consequence of a fast flexure-retreat of the subducting foreland lithosphere, which exceeded the convergence rate between Africa and Europe. The evolution of the central Mediterranean region followed two main steps.
1. From about 30 to 13 Ma ago the opening of the Algero-Provençal Basin and the drifting/rotation of the Corsica-Sardinia "Block" took place. This interval (during which Sardinia acted as a volcanic arc), is longer than that one proposed in the past (see Montigny et al. 1981); nevertheless, new paleomagnetic data (Vigliotti and Langenheim, 1992) suggest that rotation of Sardinia was quite complex and at least in some areas it was not completed in Langhian times. The Algero-Provençal basin developed inside an area previously consisting of normal continental lithosphere and representing the European foreland of the Cretaceous-Paleogene Alpine chain. The newly individuated Corsica-Sardinia "Block" acted, during its Neogene evolution up to late Tortonian times, as the hinterland of an eastward-migrating arc-

trench system. The evolution of this arc-trench system is recorded in the Apennine units piled up during the drifting/rotation of the "Block". Some of these units were also affected, during shortening, by greenschist metamorphism.
2. From about 8 Ma ago to present, the Tyrrhenian area underwent back-arc-type extension. Large portions of the previous mountain chain were involved in the extensional processes; new compressional deformations (post-Tortonian Apennines) were in turn produced, following the further outward migration of the arc-trench system.

The Apenninic chain as a whole is a complex pile of thrust and nappe units, transported towards the Padan-Adriatic-Ionian- Hyblean foreland starting from Late Oligocene times. As a result of the post-Tortonian evolution, the Apenninic chain (including Calabria and Sicily) is presently split into two major arcuate features: the northern and the southern Apenninic arcs (Fig. 1). They merge in Central Italy along a marked feature, called Ortona-Roccamonfina Line, which acted as dextral strike-slip transfer during Quaternary times (Patacca et al. 1991). In parallel, also the Tyrrhenian Basin can be subdivided into two sectors, respectively N and S of a prominent lineament running W-E from northern Sardinia to Campania (41° N Line in Fig. 1). North and south of this late Neogene transfer, extension rates were markedly different. Only to the south, extension was so high to generate oceanic lithosphere in two small districts.

Patacca et al. (1990) proposed a palinspastic restoration of the Apenninic paleogeographic domains in late Tortonian times and reconstructed a possible kinematic evolution of the Tyrrhenian-Apennine couple. The major tectonic events were defined by the recognition of the regional structural features (listric faults in the extensional areas, thrust fronts and lateral ramps in the compressional ones) and by a careful stratigraphic analysis of the sedimentary sequences deposited under tectonic control. The investigation was focused on three different structural settings:
- extensional areas, where syn-rift clastic wedges are widespread in half-graben basins produced by listric faults;
- mountain belt, where piggy-back basins have developed on top of the advancing thrust sheets;
- flexure zones, where siliciclastic flysch deposits were filling active foredeep basins produced by the progressive flexure retreat of the subducting foreland lithosphere.

We refer to Patacca et al. (1990) for most of the references which are not reported here for the sake of readability. In this paper, we tentatively extend the palinspastic restoration of the central Mediterranean domains to Langhian times, in order to compare the early-middle Miocene tectonic evolution of the area (when the Algero-Provençal Basin, Corsica and Sardinia were still active elements) with the late Miocene-Quaternary one. We shall also discuss some major implications concerning the subduction processes and the nature of the consumed foreland lithosphere.

Figure 1. Structural sketch of the Apennines and Tyrrhenian Basin. The base-of-Pliocene/Quaternary isobaths (1, in kilometres) show the deformation pattern of the foreland areas. In the northern Apenninic arc, the foreland appears still sinking beneath the thrust belt. The Ortona-Roccamonfina Line (OR Line) acts as lateral ramp of the northern arc. The front of the thrust belt (2), major post-Tortonian thrusts (3), normal faults (4), strike-slip faults (5), antiforms (6) and synforms (7) show the regional structural trends. Areas with widespread magmatic activity (8 volcanites, 9 intrusive bodies) are also indicated.

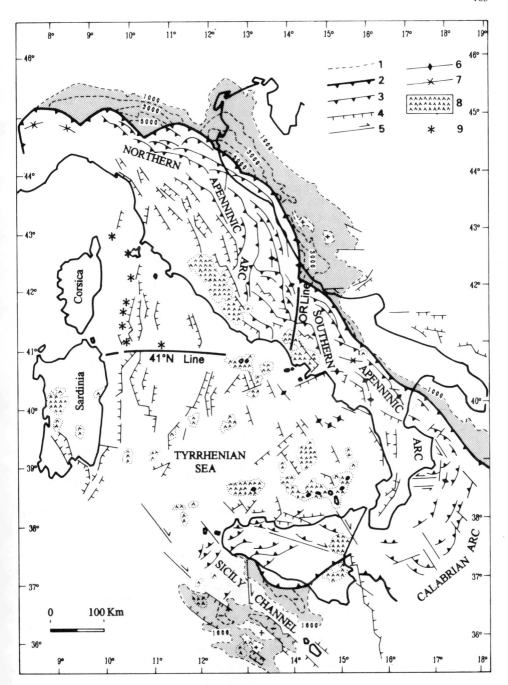

Figure 1

2. Palinspastic restoration and kinematic evolution.

Fig. 2 is a palinspastic sketch of the study region reconstructed at Langhian times (about 16 Ma), when several south-Apenninic domains where supplied by Africa-deriving Numidian sands (see Patacca et al. 1992). At that time, the mountain chain included the Ligurian and Calabrian nappes carrying a series of piggy-back basins (e.g. Stilo-Capo d'Orlando, Albidona, Ranzano- Bismantova sedimentary sequences, Bonardi et al. 1980, 1985; Ricci Lucchi 1986). In the south, the compression front had not reached yet the Sicilide domain, where in fact Langhian Numidian sandstones were conformably deposited over Aquitanian-Burdigalian sediments rich in volcaniclastic material. The Numidian input also reached the San Donato domain, when the latter began to be incorporated in the foredeep basin (Patacca et al. 1992). Greenschist metamorphism affecting both the Verbicaro and the San Donato units points (Amodio Morelli et al. 1976) to a collision between the Sardinia "Block" and the Apulian continental margin post-dating the Langhian (age of the Numidian sandstones). In the Northern Apennines, on the contrary, the Langhian mountain chain already included the equivalent of the Sicilide units (Canetolo unit) and the equivalent of the Verbicaro and San Donato units (Tuscan nappe and Tuscan metamorphic units). Greenschist metamorphism in the Apuane-Montagnola Senese units (Carmignani et al. 1978; Kligfield et al. 1986) and crustal doubling in Tuscany (Roeder 1990) suggest a pre-Langhian collision between the Corsica "Block" and the Padan-Adriatic continental margin, with development of a deep-seated shear zone within the Padan-Adriatic crust. These facts strongly support the idea that Sardinia had not yet completed counterclockwise rotation during the Langhian (Vigliotti and Langenheim, 1992), whilst Corsica had already reached, in that time, the present- day position.

In Fig. 2, the Verbicaro and the San Donato domains have been linked to the western Apenninic platform, the Verbicaro domain representing the original inner slope of the shallow platform. The connection between the western Apenninic platform and the Panormide domain are still questionable and the more internal position of the Panormide carbonate domain suggested in Fig. 2 is merely hypothetical.

Another open problem is represented by the original position of the Lagonegro basin, which is usually considered the southern continuation of the Molise basin, located between the western platform and the Inner Apulia one (e.g. Mostardini and Merlini, 1986; Casero et al. 1988). Nevertheless, some tectonic features, together with stratigraphic considerations, suggest a more internal relocation of the Lagonegro basin. An important tectonic feature is the occurrence of Ligurian and Sicilide slices tectonically sandwiched between the upper and the lower Lagonegro units, which prove the out-of-sequence nature of the tectonic doubling.

Figure 2. Palinspastic sketch of the central Mediterranean region in Langhian times. According to the picture, Sardinia did not complete rotation and a calc-alkaline volcanic arc was still active in the western part of the island. Within the Apenninic domains, basinal areas possibly floored by oceanic crust (Sicilide realm, South Molise-Ionian basin) and areas occupied by persistent shallow platforms (Panormide realm, western platform, Inner Apulia platform and Apulia platform s.str.) are roughly delimited. The front of the thrust belt in the Alps and Apennines, as well as the Insubric Line are merely geographic references. The hatched area between the Po-Adriatic foreland and the Southern Alps-Dinarides represents the amount of shortening related to the Europa-Africa convergence from Langhian times. This shortening is quite small, compared to the Apennine shortening related to the roll-back of the Padan-Adriatic-Ionian lithosphere.

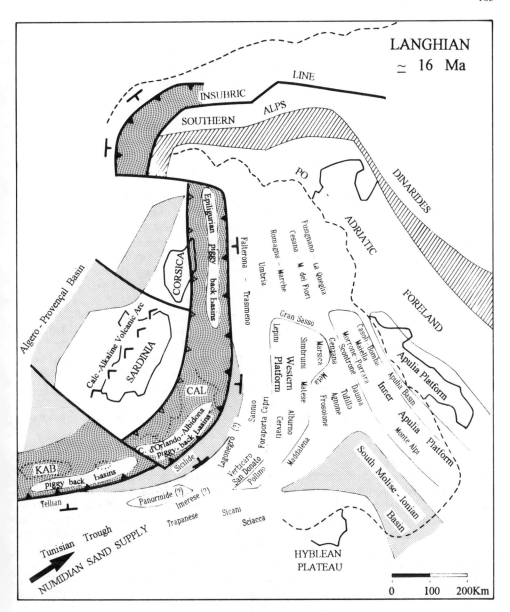

Figure 2

In addition, field evidences clearly show that the tectonic superposition of the western-platform carbonates (together with the Sannio and the Sicilide nappes) over the Lagonegro units is systematically an out-of-sequence thrust contact, not older than Messinian, with the footwall units previously doubled. Two stratigraphic considerations further stress the proposed paleogeographic attribution. The most important one is that no post-Langhian deposits occur in the Lagonegro sequences. In the Molise units, on the contrary, post-Langhian sequences are preserved everywhere, with their upper part represented by thick flysch deposits of Messinian age. The second indication is that no rocks older than middle-upper Liassic are known in the Molise units, whilst the Lagonegro basin became a deep-water furrow already in Middle Triassic times. This matches the observation that no extensional tectonics older than middle-upper Liassic has been found so far either along the eastern margin of the western platform or along the western margin of the Inner Apulia platform.

The last questionable relocation concerns the Sannio domain. The Sannio unit has been often confused, following Ortolani et al. (1975), with the Lagonegro units in spite of their different geometric position (the Sannio unit tectonically overlies whilst the Lagonegro units tectonically underlie the carbonate thrust sheets) and in spite of some different characteristics of the respective sedimentary sequences (see Patacca et al., 1992). The tectonic position of the Sannio unit and the occurrence of Serravallian siliciclastic flysch deposits in the sedimentary sequence univocally fix its original position west of the western Apenninic platform, since the latter was incorporated in the foredeep basin during upper Tortonian times.

A relocation of the San Donato and Verbicaro domains along the inner margin of the western Apenninic platform brings as a consequence that the collision between the Sardinia "Block" and the Apulia continental margin took place after the deposition of the youngest deposits of the Sannio unit, that is not before the early Serravallian. This age is not in conflict with the recent paleomagnetic results on Sardinia of Vigliotti and Langenheim (1992) and concides with the end of the calc-alkaline volcanic activity in that island at about 13 Ma (Savelli et al. 1979).

Referring to Fig. 2, we are going now to describe the main steps of the regional kinematic evolution.

Upper Oligocene - Langhian
Starting from kinematic conditions of a neutral arc, where convergence rate and rate of flexure retreat of the subducting lithosphere were approximately equal, an increase in the rate of foreland flexure-retreat caused back-arc opening in the Algero-Provençal Basin, drifting of Corsica-Sardinia, and forward migration of the thrust belt-foredeep system. Due to the spatial distribution of the paleogeographic domains, Corsica collided with the Tuscan realms and the foredeep shifted east of the shear zone, reaching the Falterona-Trasimeno domain. From this times onwards, no oceanic crust was available in the northern part of the study region.

Wide portions of oceanic or strongly-thinned continental crust (Sicilide, Sannio and possibly Lagonegro domains) were still available in the south, so that spreading in the Algero-Provençal Basin, drifting of Sardinia and progressive outward migration of the thrust belt-foredeep system continued, the accretionary wedge being constituted of rather thin units derived from the sedimentary cover detached from the subducting lithosphere.

Serravallian - Lower Tortonian
Around 13 Ma Sardinia collided with the Verbicaro and San Donato domains, causing deep-seated shear zones and greenschist metamorphism. No evidences of lower Tortonian foredeep basins have

been found up to now in the Southern Apennines.

In the northern sectors, convergence rate and rate of flexure retreat had to be roughly equal, as it is suggested by the progressive eastward migration of the thrust belt-foredeep system (from the Falterona-Trasimeno to the Umbria-western Romagna domains) without clear evidences of back-arc extension between Corsica and the Apennines. Piggy-back basins developed (see, e.g. Ponsano and Bismantova sedimentary sequences, Mazzei et al. 1980; Ricci Lucchi 1986) on top of the advancing thrust sheets.

Upper Tortonian - Lower Messinian

At the present state of the art, we are not able to evaluate how long time elapsed between the incorporation into the foredeep of the platform domains not yet reached by the Serravallian compression and the enucleation of the earliest extensional features in the Southern Tyrrhenian area. In any case, during late Tortonian times the rate of flexure retreat of the foreland lithosphere had to exceed the convergence rate both in the Northern an Southern Apennines, since back-arc extension began at that time in the whole Tyrrhenian region. Here, the 41° N lineament acted as an important transfer fault which allowed for larger amounts of back-arc extension, flexure retreat, and forward migration of the compression fronts in the southern areas. In the Southern Tyrrhenian basin, limited portions of oceanic lithosphere may have been emplaced near the end of this interval.

Upper Messinian - Lower Pliocene

Severe rifting occurred in the southern Tyrrhenian area across the Central Fault, followed by oceanic crust emplacement in the central bathyal plain. North of the still active N 41° N lineament, moderate extension occurred in the northern Tyrrhenian Sea and Southern Tuscany accompanied, in the mountain chain, by eastward migration of the compressional fronts. In the central Apennines, arcuate features, corresponding to out-of-sequence thrusts, widely developed from the Sibillini mountains to the Gran Sasso-Genzana area. These features, which sometimes form high angles with the strike of the previous compressional structures, produced block rotations in the concave portions of the arcs (Mattei et al., 1991). In the Southern Apennines and Calabria, data on the buried foredeep deposits are very poor, due to the geometry of the thrust belt (duplex system) and to the insufficient subsurface information. Some indications regarding the compression front migration are supplied by the piggy-back- basin deposits unconformably overlying the roof units of the duplex.

Upper Pliocene

The upper Pliocene kinematic evolution of the study region does not markedly differ from the lower Pliocene one, with the main variant represented by a southward migration of the arcuate features in the thrust belt (see Patacca et al. 1991). Near the end of the Pliocene, the Apennine compression front reached the present-day position in the Ortona-Lucera segment, suggesting the end of the Adriatic-Apulia flexure retreat in the area.

Quaternary

During the Pleistocene, the differentiation of the two Apenninic arcs became more and more pronounced. In the northern Apenninic arc, extension and compression still followed SW-NE vectors, with dextral transpression along the Ortona-Roccamonfina Line (eastern wing of the arc). In the southern arc, on the contrary, flexure retreat and thrust belt-foredeep migration progressively ceased from north to south (near the Pliocene-Quaternary boundary in the Molise-Daunia segment; about 1 Ma ago in the Campania- Lucania segment, see Cinque et al. 1992).

Flexure retreat of the Ionian lithosphere and migration of the Calabrian Arc were still active when the previous south-Apenninic segments were already undergoing "isostatic" rebound. We are not able, however, to establish whether roll-back processes are still working in the Calabrian Arc or they have ceased in late Pleistocene-Holocene times. It is interesting to point out that the rhombic Marsili basin, floored by young oceanic crust, opened just back of the Calabrian Arc. This belt was still moving toward the Ionian Sea when the Campania-Lucania Apenninic front was already near to the present-day position.

3. DISCUSSION AND CONCLUSIONS

As already described, the post-Tortonian evolution of the area had different kinematics in the northern and southern sectors. The northern Tyrrhenian Basin experienced modest rifting and crustal thinning, whilst the southern basin was severely stretched, with emplacement of oceanic crust. In parallel, the Apenninic chain evolved into two arcs, the northern arc having been affected by an amount of transport quite smaller than the southern one.

The proposed palinspastic restoration at Langhian times (Fig. 2) shows that a N versus S differentiation had also to exist during the opening of the Algero-Provençal Basin and the drifting/rotation of Corsica-Sardinia. For instance, collisional greenschist metamorphism affected the north-Apenninic units (Tuscan units) in Lower Miocene times, whilst the south-Apenninic units (San Donato and Verbicaro p.p.) were metamorphosed in post-Langhian times, probably during the early Serravallian. In addition, arc-type volcanism was absent in Corsica, while it was widespread in Sardinia.

As regards the lithosphere subducting under the northern Apennines, it had to be of thinned-continental type and rather homogeneous laterally. This lithosphere originally floored the Umbria-Marche basinal domains which were parts of the Jurassic passive margin of Tethys.

In the southern sector, markedly-different crustal and lithospheric elements had to be involved in the roll-back processes. In Langhian times, we can recognize from W to E (Fig. 2):
- domains with oceanic or extremely-thinned continental crust (Sicilide, Sannio and possibly Lagonegro domains);
- domains with thinned to normal continental crust (carbonate platforms and their margins);
- domains with thinned continental crust (north-Molise) and domains with possible oceanic crust (south-Molise - Ionian Basin).

These lithosphere inhomogeneities favoured segmentation and differential sinking accomodated by lithospheric tear faults acting as free boudaries (see Royden et al. 1987; Mantovani et al. 1992). These lateral variations across the southern sector may also account for the puzzling time/space distribution of the arc-type volcanism. Calc-alkaline volcanism was active in Sardinia from about 29 to 13 Ma (Savelli et al. 1979) and had petrochemical characters indicating NW-dipping subduction of normal oceanic lithosphere (Coulon, 1977). After a long interval of quiescence, it resumed about 2 Ma ago in Campania (Beccaluva et al. 1984), becoming widespread and important starting with 1.3 Ma in peninsular Italy and in the Aeolian Arc. The Sardinia volcanism may have reflected the subduction of the ocean-type lithosphere of the Liguride and Sicilide domains. The stop in arc-type volcanism corresponds temporally to the Sardinia collision against the western platform domains. No calc-alkaline volcanism has been found coeval with the subsequent subduction of the continental lithosphere underlying the western Apenninic platform. Arc-type magmatism resumed only when the thinned Molise crust, and the oceanic Ionian crust were involved in the subduction. These domains were located in the southern part of the study area (see Fig. 2). According to Serri (1990) and Serri et al. (1991), the Pleistocene volcanism of Italy has typical arc-type characters related to oceanic subduction only in the Aeolian Islands and Campania, whilst it becomes more and more reminiscent of continental influences moving northwards, in Latium and Tuscany.

The southern and northern domains also show marked differences in earthquake patterns. In the Northern Apennines only shallow and possibly intermediate hypocenters (≤ 90 km) occur (Selvaggi and Amato 1992). The deeper hypocenters seem to depict a faint seismogenic zone deepening from the Adriatic foreland beneath the chain, in the same sense as the deflection of the continental lithosphere should have occurred. A true, though complex, Wadati-Benioff zone occurs in the Southern Tyrrhenian Sea back of Calabria, with hypocenters exceeding 450 km of depth (Anderson and Jackson, 1987). This well-delineated and narrow seismogenic slab is about 700 kilometres in lenght. According to the deformation rates calculated by Patacca et al. (1990) and taking into account Fig. 2, the slab should result from subduction processes which started before the early opening of the Tyrrhenian basin. Such a long duration may be related to the presence of wide sectors of oceanic lithosphere which dragged down intervening sectors of continental lithosphere.

Acknowledgments
We are grateful to Patrizia Pantani and Simonetta Ruberti for data base organization and for drawings.

REFERENCES

Amodio-Morelli L., Bonardi G., Colonna V., Dietrich D., Giunta G., Ippolito F., Liguori V., Lorenzoni S., Paglionico A., Perrone V., Piccarreta G., Russo M., Scandone P., Zanettin-Lorenzoni E., Zuppetta A. (1976) 'L'Arco calabro-peloritano nell'orogene appenninico-maghrebide', Mem.Soc.geol.ital., 17, 1- 60.

Anderson H., Jackson J. (1987) 'The deep seismicity of the Tyrrhenian Sea' Geophys.J.r.astron.Soc., 91, 613-637.

Beccaluva L., Di Girolamo P., Serri G. (1984) 'High-K calcalkalic shoshonitic and leucitic volcanism of Campania (roman province, southern Italy): trace elements constraints on the genesis of an orogenic volcanism in a post-collisional extensional setting' Ed. Tipografia C. Cursi - Pisa, 47 pp.

Bonardi G., Ciampo G., Perrone V. (1985) 'La formazione di Albidona nell'Appennino calabro-lucano: ulteriori dati stratigrafici e relazioni con le unità esterne appenniniche' Boll.Soc.geol.ital., 104, 539-549.

Bonardi G., Giunta G., Perrone V., Russo M., Zuppetta A., Ciampo G. (1980) 'Osservazioni sull'evoluzione dell'arco calabro- peloritano nel Miocene inferiore: la formazione di Stilo-Capo d'Orlando' Boll.Soc.geol.ital., 99, 365-393.

Carmignani L., Giglia G., Kligfield R. (1978) 'Structural evolution of the Apuane Alps: an example of continental margin deformation in the Northern Apennines, Italy' J.Geol., 86, 487-504.

Casero P., Roure F., Moretti I., Müller C., Sage L., Vially R. (1988) 'Evoluzione geodinamica neogenica dell'Appennino Meridionale' Mem.Soc.geol.ital., 41, 109-120.

Cinque A., Patacca E., Scandone P., Tozzi M. (1991) 'Quaternary kinematic evolution of the Southern Apennines. Possible relationships between surface geological features and deep lithospheric structures' Intern.School of Solid Earth Geophysics. 7th Course "Modes of crustal deformation: from the brittle upper crust through detachments to the ductile lower crust", (Erice, 18-24 november 1991). In press.

Coulon C. (1977) 'Le volcanisme calco-alcalin cénozoique de Sardaigne (Italie). Pétrographie, géochimie et genèse des laves andésitiques et des ignimbrites - signification géodynamique' Thése Doct. 3e cycle Univ. Aix-Marseille III, 288 pp.

Kligfield R., Hunziker J., Dallmeyer R.D., Schamel S. (1986) 'Dating of deformation phases using K-Ar and $^{40}Ar/^{39}Ar$ techniques: results from the Northern Apennines' J.Struct.Geol., 8, 781-798.

Malinverno A., Ryan W.B.F. (1986) 'Extension in the Tyrrhenian Sea and shortening in the Apennines as a result of arc migration driven by sinking of the lithosphere' Tectonics, 5, 227-245.

Mantovani E., Albarello D., Babbucci D., Tamburelli C. (1992) 'Recent geodynamic evolution of the central Mediterranean region - Tortonian to Present' Tipografia Senese - Siena, 88 pp.

Mattei M., Funiciello R., Kissel C., Laj C. (1991) 'Rotazioni di blocchi crostali neogenici nell'Appennino centrale: analisi paleomagnetiche e di anisotropia della suscettività magnetica (AMS)' In: M. Tozzi, G.P. Cavinato and M. Parotto (eds.), Studi preliminari all'acquisizione dati del profilo CROP 11 Civitavecchia-Vasto, AGIP-CNR-ENEL, Stud.geol.Camerti, vol.spec. 1991-2, 221-229.

Mazzei R., Pasini M., Salvatorini G., Sandrelli F. (1980) 'L'età dell' "arenaria di Ponsano" della zona di Castellina Scalo (Siena)' Mem.Soc.geol.ital., 21, 63-72.

Montigny R., Edel J.B., Thuizat R. (1981) 'Oligo-Miocene rotation of Sardinia: K-Ar ages and paleomagnetic data of Tertiary volcanics' Earth and planet.Sci.Lett., 54, 261-271.

Mostardini F., Merlini S. (1986) 'Appennino centro-meridionale. Sezioni geologiche e proposta di modello strutturale' Mem.Soc.geol.ital., 35, 177-202.

Ortolani F., Narciso G., Sanzò A. (1975) 'Prime considerazioni sulla presenza del flysch numidico nell'Appennino sannita' Boll.Soc.natur. Napoli, 84, 31-44.

Patacca E., Sartori R., Scandone P. (1990) 'Tyrrhenian basin and Apenninic arcs: kinematic relation since Late Tortonian times' Mem.Soc.geol.ital., 45, In corso di stampa.

Patacca E., Scandone P., Bellatalla M., Perilli N., Santini U. (1991) 'La zona di giunzione tra l'arco appenninico settentrionale e l'arco appenninico meridionale nell'Abruzzo e nel Molise' In: M. Tozzi, G.P. Cavinato and M. Parotto (eds.), Studi preliminari all'acquisizione dati del profilo CROP 11 Civitavecchia- Vasto, AGIP-CNR-ENEL, Stud.geol.Camerti, vol.spec. 1991-2, 417- 441.

Patacca E., Scandone P., Bellatalla M., Perilli N., Santini U. (1992) 'The Numidian-sand event in the Southern Apennines' Mem.Sci.geol. già Mem.Ist.Geol.Mineral.Univ. Padova, allegato al 43, 297-337.

Ricci Lucchi F. (1986) 'The foreland basin system of the Northern Apennines and related clastic wedges: a preliminary outline' G.Geol., s. 3, 48, 165-185.

Roeder D. (1990) 'Crustal structure and kinematics of Ligurian and west-Alpine regions' In: R Freeman, P. Giese and St. Müller (eds.), The European Geotraverse. Integrative Studies, Eur Sci.Found., 311-326.

Royden L., Patacca E., Scandone P. (1987) 'Segmentation and configuration of subducted lithosphere in Italy: An important control on thrust-belt and foredeep-basin evolution' Geology 15, 714-717.

Savelli C., Beccaluva L., Deriu M., Macciotta G., Maccioni L. (1979) 'K/Ar geochronology and evolution of the Tertiary "calc-alkalic" volcanism of Sardinia (Italy)' J.Volcanol. Geotherm.Res., 5, 257-269.
Selvaggi G., Amato A. (1992) 'Subcrustal earthquakes in the northern Apennines (Italy): evidence for a still active subduction?' Geophys.Res.Lett., 19, 2127-2130.
Serri G. (1990) 'Neogene-Quaternary magmatism of the Tyrrhenian region - characterization of the magma sources and geodynamic implications' Mem.Soc.geol.ital., 41, 219-242.
Serri G., Innocenti F., Manetti P., Tonarini S., Ferrara G. (1991) 'Il magmatismo neogenico-quaternario dell'area tosco- laziale-umbra: implicazioni sui modelli di evoluzione geodinamica dell'Appennino settentrionale' In: G. Pialli, M. Barchi, M. Menichetti (eds.), Studi preliminari sull'acquisizione dati del profilo CROP 03 Punta Ala-Gabicce, Stud. Geol.Camerti, Vol.Spec. 1991/1, 429-463.
Vigliotti L., Langenheim V.E. (1992) 'When did Sardinia stop rotating? New paleomagnetic results' EOS Trans.AGU, 73, Fall Meeting Suppl., 147, (Abstracts).

NEOGENE BASINS IN THE STRAIT OF SICILY (CENTRAL MEDITERRANEAN): TECTONIC SETTINGS AND GEODYNAMIC IMPLICATIONS

A. ARGNANI
Istituto per la Geologia Marina - CNR
Via Zamboni 65,
40127 Bologna,
Italy

ABSTRACT. Two groups of sedimentary basins have been recognised in the upper Miocene-Quaternary of the Strait of Sicily: basins in one group have a Tortonian sedimentary fill, while in the others the filling is Plio-Quaternary. Tortonian basins are the Adventure foredeep and some narrow grabens, trending NE-SW, present in the foreland. Afterwards, the foredeep was incorporated into the fold-and-thrust belt and the grabens were inverted. Plio-Quaternary basins are the Gela foredeep and the troughs of the Strait of Sicily. Data concerning both the Tortonian and the Plio-Quaternary indicate that extension in the grabens and shortening in the fold-and-thrust belt occurred along the same orientation and that this orientation rotated in time from NW-SE in the Tortonian to about N-S in the Plio-Quaternary. These facts can be accounted for by processes like subducted slab roll-back or lithospheric mantle delamination, occurring at the scale of the Central Mediterranean, which may have originated the Tyrrhenian basin.

Introduction

The Strait of Sicily is a shallow sea area (Fig. 1) located between Sicily and Tunisia., Water depth is greater than 1000 m only in three NW-SE-elongated troughs. Geological and geophysical evidence indicates that it belongs to the African plate as part of the foreland of the Maghrebian fold-and-thrust belt that runs along north Africa and Sicily (Barberi et al., 1974; Catalano and D'Argenio, 1982; Argnani, 1990). The front of this fold-and-thrust belt can be followed across the northern part of the Strait of Sicily (Fig. 2) where it is linked to two foredeep basins (Fig. 4) of different age, one being Tortonian and the other being Plio-Quaternary (Argnani et al., 1987).

The troughs of the Strait of Sicily are major topographic features that have been known since a long time (Finetti and Morelli, 1972; Zarudzki, 1972), although their tectonic interpretation is still non-univocal. Some Authors (Colantoni, 1975; Winnock, 1981; Finetti, 1984; Argnani, 1990) interpret these troughs as rift basins while Others (Jongsma et al., 1985; Boccaletti et al., 1987) favour a strike slip setting.

As a matter of fact, in the Strait of Sicily there are several sedimentary basins of different tectonic origin that formed from Tortonian to Present. During the same time span most of the Central Mediterranean area was shaped to its present form.

The aims of this paper are to unravel the tectonic history of the Strait of Sicily and to relate such a history to the geodynamic evolution of the Central Mediterranean.

Geologic Setting

The crustal structure of the Strait of Sicily has been investigated by refraction studies (Colombi et al., 1973; Boccaletti et al., 1984) that show a thinning of the crust from 30-35 km onshore Sicily, typical of continental crust, to about 20 km below the trough system. A rather high heat flow, of about 110-130 mWm^{-2}, has been observed in the areas of thin crust (Della Vedova and Pellis, 1985).

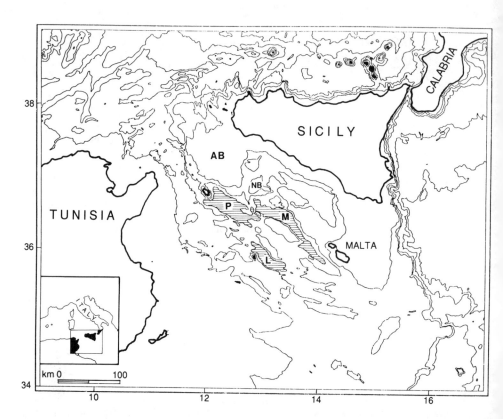

Fugure 1. Bathymetric map of the Strait of Sicily showing location of the main troughs (P: Pantelleria; M: Malta; L: Linosa). The Adventure Bank (AB) and Nameless Bank (NB) are also indicated. Simplified after the International Bathymetric Chart of the Mediterranean (I.O.C., 1987)

Exploration wells in the shelf offshore Sicily allow outlining the stratigraphic evolution of the area (Jongsma et al., 1985; Argnani et al., 1986 ; Antonelli et al., 1988). They display a passive margin succession typical of the Tethyan domain with platform carbonates of Triassic age followed by deeper water sediments that span the rest of the Mesozoic. An exception to this trend is observed in the Maltese region where shallow water conditions persisted until the present. Deposition of deep water carbonates continued until the Oligocene when siliciclastic sediments made their appearence. These clastics reflect a shallowing upward trend continuing until the Messinian and mark the onset of the contraction that affected this continental margin during the convergence between Africa and Europe. The effects of such a contraction become evident in the north-western part of the Strait of Sicily where a small foredeep basin of Tortonian age is present (Argnani et al., 1987). After the Messinian salinity crisis, that exposed most of the Strait of Sicily, fairly deep water sediments were deposited in the early Pliocene. The main depocentres of Plio-Quaternary sediments occur in the troughs and in the Gela foredeep basin.

Volcanic activity is widespread in the Strait of Sicily and in the islands of Pantelleria and Linosa (Fig.2). On these islands the oldest volcanic rocks are Quaternary (Civetta et al., 1984), but volcanites 10 Ma old have been dredged near Nameless Bank (Beccaluva et al., 1981). Affinity of volcanic rocks is alkaline in the majority of cases. In the island of Pantelleria, where the crust is thinner, more transitional products are present. The distribution of volcanic centres shown in Fig. 2 is based on magnetic anomalies and on characteristic facies on seismic profiles. For some of them the volcanic nature has been confirmed by direct sampling (Calanchi et al., 1989). Volcanic centres tend to align along a roughly N-S-trending belt that is also reflected in the bathymetry (Fig. 1).

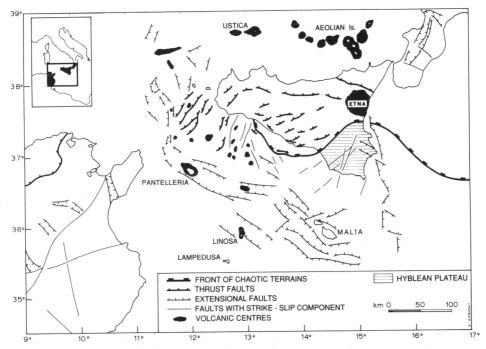

Figure 2. Structural sketch of the Strait of Sicily. The main tectonic elements here represented are Tortonian to Present.

The major source of information on the shallow crustal features and on the sediment depositional patterns in the Strait of Sicily is provided by seismic reflection profiles. The grid over the study area (Fig. 3) is mainly composed of multichannel seismic reflection profiles belonging to the Italian Commercial Zone "G" and "C" plus single channel 30-kJ sparker profiles covering the trough system.

Previous works based on seismic profiles (Colantoni, 1975; Finetti, 1984; Argnani et al., 1987; Antonelli et al., 1988; Argnani, 1990) have shown that main structural elements in the Strait of Sicily are: i) the trough system in the centre of the channel; ii) the foredeep system in the northern part of the channel; and iii) the N-S-trending belt that interferes both with the foredeep and the trough systems. These elements will be briefly described in the following and their possible geodynamic implications will be discussed.

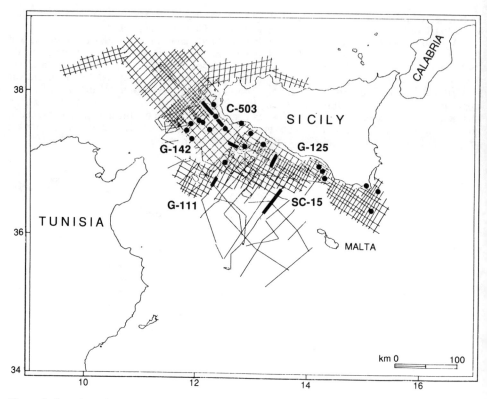

Figure 3. Location of commercial wells and seismic profiles over the study area. Bold lines indicate profiles shown in the paper.

Foredeep System

Along the front of the Maghrebian fold-and-thrust belt, that runs just south of Sicily, two distinct foredeep basins are present, the Adventure and Gela foredeeps (Fig. 4).

The NE-SW-trending Adventure foredeep is filled with a sequence of fine grained clastics of Tortonian-Messinian age (Terravecchia Fm.) up to 3000 m thick that overlies a carbonate substrate. This foredeep originated in the early Tortonian, in response to the SE-ward emplacement of thrust sheets. Subsequent shortening is documented by the folded sedimentary fill and by the general uplift of the basin (Fig. 5). The onlap of Plio-Quaternary sediments marks the end of deformation.

The Gela foredeep is much younger than the Adventure foredeep as its sedimentary fill is composed of Plio-Quaternary fine grained clastics (Fig. 6). Part of this basin is concealed beneath a large feature known as the Gela Nappe, whose emplacement was completed by the early Pleistocene (Argnani, 1987). A comparable age of emplacement of the Gela Nappe has been reported from a study carried out in SE Sicily (Butler et al., 1992). The arcuate front of the Gela Nappe (Fig. 2) suggests a roughly N-S contraction.

The different age of the two foredeeps and the different shortening orientations that can be deduced from the structural data, indicate that the stress field rotated through time from

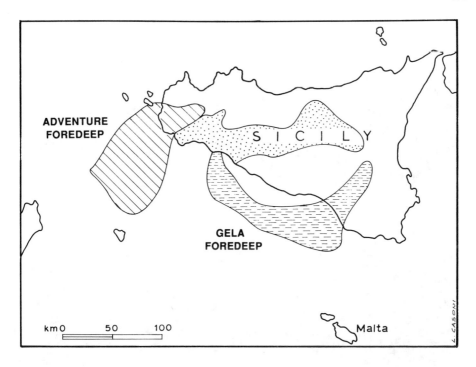

Figure 4. Location of foredeep sediments not or little deformed. Filling of the Adventure foredeep is mainly Tortonian while sediments in the Gela foredeep are Plio-Quaternary. The stippled area indicates the depocentre of Messinian evaporites onshore Sicily.

NW-SE in early Tortonian time to N-S in the early Pleistocene. This is in agreement with the results of a paleomagnetic study carried out in the thrust sheets of the Maghrebian belt outcropping in western Sicily (Oldow et al., 1990). Inferred shortening direction in this belt rotated clockwise from NW-SE in late Miocene to N-S in the Pleistocene.

Trough System

Three major troughs comprise this system, namely Pantelleria, Linosa and Malta (Fig. 1). The Pantelleria trough is separated from the Malta and Linosa troughs by a roughly N-S-trending belt (Fig. 2). As mentioned in the Introduction, two quite different hypotheses, strike-slip vs dip-slip, have been put forward to explain the origin of these troughs. The latter one is here preferred; however, a more thorough description of the reasoning in favour of this hypothesis can be found elsewhere (Argnani, 1990).

Two profiles are used to illustrate some features of the troughs that can help understand their structural evolution. The first one is located on the northern shoulder of the Pantelleria trough (Fig. 7), while the other crosses the Malta trough (Fig. 8).

A set of faults bounding rotated blocks (Fig. 7) affect the M-reflector, that represents the base of the Plio-Quaternary, and create wedge-like depocentres filled with Pliocene sediments. Subsequent sediments are not affected by faulting and the reflections of the most recent unit deposited within the basin onlap the basin margin.

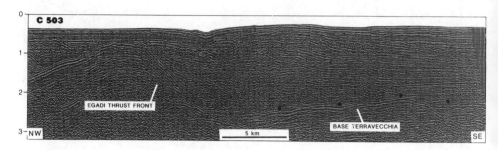

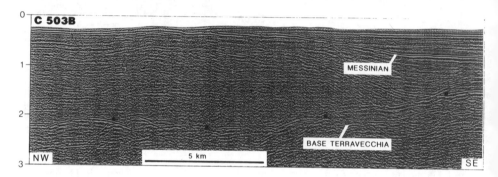

Figure 5. Profile C-503 crossing the Adventure foredeep and showing the Tortonian sedimentary fill (Terravecchia Fm.) slightly deformed and uplifted. Location in Figure 3.

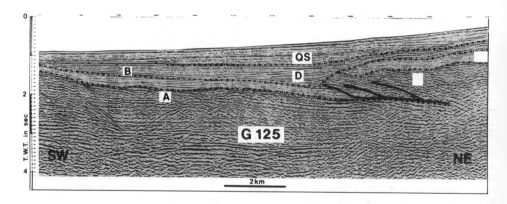

Figure 6. Profile G-125 across the Gela basin. "A" marks the base of the Plio-Quaternary and "B" the interpreted base of the Pleistocene. Thin-skinned thrusting can be observed at the Gela nappe front (deformed sediments "M" and sedimentary cover "C") with a downlapping unit ("D") that marks the end of thrusting. "QS" is a late Quaternary submarine slide. For location see Figure 3.

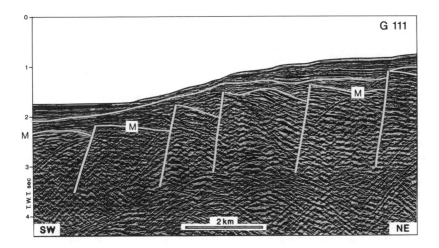

Figure 7. Profile G-111 located on the northern shoulder of the Pantelleria trough. Planar faults affect the reflector "M" that marks the base of Plio-Quaternary. Reflections within the basin onlap the trough margin. Location in Figure 3.

The profile crossing the Malta trough (Fig. 8) shows an upper unit almost undeformed and with reflections onlapping the basin margins. The underlying reflections, in contrast, display wedging and drag folding suggesting syn-depositional deformation.

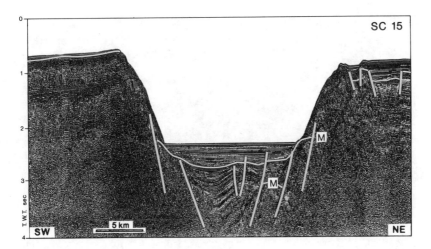

Figure 8. Profile SC-15 that crosses the Malta trough. The upper unit is not affected by faulting and onlaps the basin margins. The lower part of the sedimentary fill shows convergence of reflections and drag folds. "M" marks the base of Plio-Quaternary. Location in Figure 3.

A slight asymmetry of the basin fill indicates that the master fault is on the southern border where there is a large shoulder uplift. The Linosa trough gives a very similar picture (Colantoni, 1975; Argnani, 1990). On plan view, the Malta trough has an arcuate shape (Fig. 1) which is indicative of a NE-directed extension.

The troughs appear to have had a similar history of filling with a syn-rift unit followed by a post-rift unit. Dating these two basinal units is not easy due to the impossibility to correlate reflections from the adjacent shelf areas where wells are available. From the northern shoulder of Pantelleria (Fig. 7) a Pliocene (possibly early Pliocene) age for the extensional faulting, and therefore for the syn-rift unit, can be inferred. Given that the post-rift unit is up to 400 m thick, and assuming normal sedimentation rates (< 100 m/Ma), it seem likely that this unit spans most of the Quaternary. It seems therefore, that the trough system developed mainly during Pliocene and that it was hardly affected by any subsequent tectonics.

As to the origin of the troughs, the coverage of the seismic grid is not ideal to clearly discriminate between a strike-slip and a dip-slip mechanism. However, the consistency of the extensional structures at the basin margins, the lack of characteristic strike-slip features, and the shape in plane view of the troughs suggest that strike-slip motion along the boundary faults, if present, is not particularly important.

North-South Belt

This belt separates the Adventure foredeep from the Gela foredeep and the Pantelleria trough the Malta and Linosa, troughs. On the bathymetric map (Fig. 1), this belt appears as an alignment of small depressions and shallow banks. The Linosa volcanic island of Linosa and other volcanic centres are also located along this belt (Fig. 2).

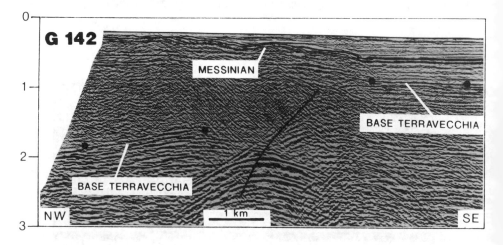

Figure 9. Profile G-142 crossing a the Tortonian grabens. Sedimentary fill (Terravecchia Fm.) has been subsequently uplifted. See Figure 3 for location.

Seismic profiles show the presence of small NE-SW trending grabens filled with Tortonian-Messinian sediments. These grabens have been later tectonically inverted (Fig. 9) and their uplifted depocentres created some of the shallow sea areas observed along the N-S belt. Onlap of Plio-Quaternary sediments (Fig. 9) marks the end of the inversion episode, but elsewhere the deformation lasted troughout the Pliocene. The mechanics of this tectonic inversion, however, is not completely understood.

Discussion

Two groups of sedimentary basins have been recognised on the grounds of their age. A group has a Tortonian-Messinian sedimentary fill, while the other presents a Plio-Quaternary filling.

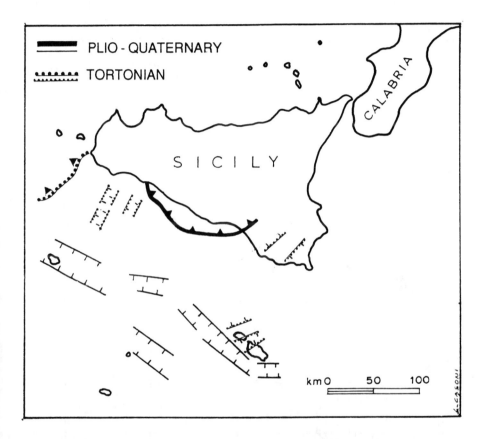

Figure 10. Simplified sketch of the main Tortonian and Plio-Quaternary structural elements. Note the similar orientations of thrust fronts and extensional faults for each time intervals.

Basins of Tortonian-Messinian age are the Adventure foredeep and the grabens located along the N-S belt. Both the foredeep and the grabens have the same NE-SW trend. Furthermore, evidence of faults with a similar trend and active during the Tortonian are widespread in Malta and in the Hyblean plateau (Illies, 1981; Carbone et al., 1982; Reuther and Eisbacher, 1985; Grasso et al., 1990). Therefore, during Tortonian, contraction at the thrust front and extension in the foreland occurred along the same orientation, i.e. NW-SE (Fig. 10).

The Gela foredeep and the troughs of the Strait of Sicily belong to the second group of basins. Here again, contraction in the thrust-and-fold belt and extension in the foreland appear to be roughly along the same orientation (Fig. 10).

This peculiar relationship between contraction and extension can be useful to understand the geodynamic setting on a broader scale. The major event that shaped the Central Mediterranean, i.e. the opening of the Tyrrhenian basin, initiated in the Tortonian (Kastens et al., 1988) and occurred simultaneously with shortening in the surrounding thrust-and-fold belts (Patacca et al., 1990). Models aiming at explaining the opening of the Tyrrhenian basin have also to take into account this shortening in the adjacent mountain ranges. Two models, out of the several so far proposed (Hsu, 1977; Scandone, 1979; Horvath and Berckhemer, 1981; Wezel, 1982; Moussat, 1983), are particularly successful at explaining this geologic evidence. According to the first model, the roll-back of a subducted oceanic, or thinned continental, lithosphere can stretch the overlying lithosphere up to the point of break up (Dewey, 1980; Kincaid and Olson, 1987). This sinking of the lithosphere has been suggested to be typical of the Tyrrhenian subduction (Ritsema, 1979) and can account for the contemporaneous extension and shortening observed (Malinverno and Ryan, 1986). The second model assumes that the opening of the Tyrrhenian basin occurred in the wake of a continental collision zone, as indicated by pieces of geologic evidence (Sartori et al., 1987).

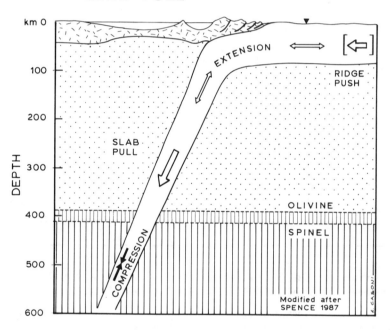

Figure 11. Diagram illustrating the stresses within a subducted lithosphere subjected to slab pull. The Strait of Sicily should be located approximately in the position indicated by the black triangle.

In this setting the lithospheric root can detach because of its gravitational instability (Houseman et al., 1981) and trigger the delamination of the lithospheric mantle. This process of lithospheric delamination appear to be applicable to the southern Tyrrhenian basin and surrounding mountain belts (Channell, 1986). The Strait of Sicily is located in the foreland of the thrust-and-fold belts propagating away from the Tyrrhenian basin. Both of the two models previously described can, to some extent, account for the presence of extensional stresses within the foreland. The slab-pull force acting within a sinking slab (Fig. 11) generates a tensional stress that can propagate updip where it is counteracted by the ridge-push force (Spence, 1987). If the ridge-push force is significantly reduced, extension can propagate within the foreland area. In a setting of locked continental collision such as that of the Central Mediterranean (Mascle et al., 1988), where the northward motion of Africa is mainly absorbed in the Alpine collision, the Strait of Sicily represents an area where the tensional stress due to slab pull can become significant. As to the delamination of lithospheric mantle, numerical modelling (Channell and Mareschal, 1988) has shown that this process can propagate laterally in one direction if the crustal and lithospheric thickening are offset. In this case, the asymmetric mantle flow produced generates a region of compressional stress flanked by two regions of tensional stress (Fig. 12).

DELAMINATION

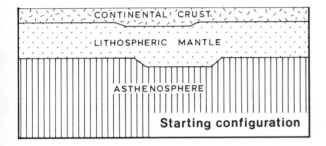

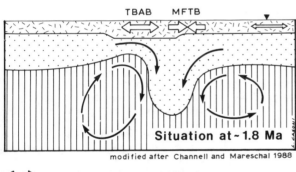

Figure 12. Cartoon simplifying the results obtained from numerical modelling of lithospheric mantle delamination. The overthickened lithosphere produced by continental collision (upper panel) has an unstable root that tends to detach and sink. This process creates a stress field (lower panel) that can account for extension in a foreland position such as that of the Strait of Sicily (black triangle). TBAB= Tyrrhenian back-arc basin; MFTB= Maghrebian fold-and-thrust belt.

One is a region of large strain rate that corresponds to the Tyrrhenian basin while in the other the strain rate is much lower. The position of this last region corresponds to the Strait of Sicily rift zone. Therefore, the two models above described, that explain the opening of the Tyrrhenian basin and the growth of the adjacent fold-and-thrust belts, can also account for the presence of extensional structures within the foreland. Tensional stresses in the foreland should be more or less parallel to compressional stresses within the belts. This kind of relationship is observed in the Strait of Sicily for both the Plio-Quaternary and the

Tortonian structures. In particular, these models may have worked in the Pliocene and in the Tortonian when significan extension in the foreland area occurred. Considering that the direction of shortening within the Maghrebian belt of Sicily rotated in time from NW-SE to N-S (Oldow et al., 1990) and that this rotation appears to be closely related to the SE-ward opening of the Tyrrhenian basin (Kastens et al., 1988), it can be inferred that the Neogene extensional structures present in the Strait of Sicily are a side-effect of the opening of the Tyrrhenian basin. As the extension within the Tyrrhenian basin propagated SE-ward in time, the western part of the fold-and-thrust belt, adjacent to the Adventure foredeep, was left almost unaffected by deformation, while contraction occurred until the Pleistocene within the Gela basin.

The above mentioned models work at the scale of the whole Central Mediterranean and thus one could expect that tensional stresses affect all of the foreland areas surrounding the Tyrrhenian basin to the east and to the south. So far, apart from the Strait of Sicily, evidence of extension occurring in these areas has not been reported. However, the presence of tensional stresses does not necessarily imply that extension will occur. The strength of the lithosphere (Kuzsnir and Park, 1987) and the presence of inherited weakness zones at the appropriate angle (Etheridge, 1987) can be major factors controlling whether or not deformation will occur. The Mesozoic faults trending NW-SE, NE-SW and N-S that are present in the Strait of Sicily (Antonelli et al., 1988) may have made this area particularly sensitive to the subsequent tectonic stresses.

Conclusions

The Strait of Sicily encompasses part of the Apennine-Maghrebian fold-and-thrust belt and its foreland and contains several basins ranging in age from Tortonian to Quaternary.

Two groups of sedimentary basins have been recognised in the upper Miocene-Quaternary sequence of the Strait of Sicily: one is Tortonian, and the other is Plio-Quaternary. These two groups show different trends and display different tectonic histories

The Gela foredeep and the rift basins in the Strait of Sicily are Plio-Quaternary basins.

Basins of Tortonian age occur in a foredeep setting (Adventure basin) and along a very narrow "rift belt" in the foreland. These basins were affected by the successive tectonic events and as a result, the foredeep basin was incorporated into the fold-and-thrust belt while the foreland basins were sructurally inverted.

The Plio-Quaternary foredeep and the rift basins have trends around WNW-ESE while data concerning the Tortonian indicate NE-SW trends for both the foredeep and rift basins.

For both the Tortonian and the Plio-Quaternary basins systems, contraction in the fold-and-thrust belt and extension in the foreland occur along the same orientation. This orientation changed from Tortonian to Plio-Quaternary rotating from NW-SE to N-S. It appears reasonable to infer that a similar tectonic regime was active during both periods but with the regional stress field rotating through time.

Processes like slab roll-back or lithospheric mantle delamination can give a sound explanation of the inferred stress field and are also supported by the extension, in a back arc position, of the Tyrrhenian basin which initiated in Tortonian time. Within this geodynamic frame, the observed rotation of the stress field can be linked to the progressive SE-wards opening of the Tyrrhenian basin.

Acknowledgements

Michael Marani contributed to improve this paper with dicussions and critical reading of the early manuscript. Luciano Casoni is gratefully acknowledged for drawing the figures. Contibution n. 908 of Istituto per la Geologia Marina - CNR.

References

Antonelli, M., Franciosi, R., Querci, A., Ronco, G.P. and Vezzani, F. (1988) 'Paleogeographic evolution and structural setting of the Northern side of the Sicily Channel', Mem. Soc. Geol. It., 41, 141-157.

Argnani, A. (1987) 'The Gela Nappe: evidence of accretionary melange in the Maghrebian foredeep of Sicily', Mem. Soc. Geol. It., 38, 419-428.

Argnani, A. (1990) 'The Strait of Sicily Rift Zone: foreland deformation related to the evolution of a back-arc basin', Journal of Geodynamics, 12, 311-331.

Argnani, A., Cornini, S., Torelli, L. and Zitellini, N. (1986) 'Neogene-Quaternary foredeep system in the Strait of Sicily', Mem. Soc. Geol. It., 36, 123-130.

Argnani, A., Cornini, S., Torelli, L. and Zitellini, N. (1987) 'Diachronous foredeep-system in the Neogene-Quaternary of the Strait of Sicily', Mem. Soc. Geol It., 38, 407-417.

Barberi, F., Civetta, L., Gasparini, P., Innocenti, F., Scandone, R. and Villari, L. (1974) 'Evolution of a section of the Africa-Europe plate boundary: paleomagnetic and volcanological evidence from Sicily', Earth Plan. Sci. Lett., 22, 123-132.

Beccaluva, L., Colantoni, P., Di Girolamo, P. and Savelli, C. (1981) 'Upper-Miocene Submarine Volcanism in the Strait of Sicily (Banco Senza Nome)', Bull. Volcanol., 44, 573-581.

Boccaletti, M., Nicolich, R. and Tortorici, L. (1984) 'The Calabrian Arc and the Ionian Sea in the dynamic evolution of the Central Mediterranean', Marine Geology, 55, 219-245.

Boccaletti, M., Cello G. and Tortorici, L. (1987) 'Transtensional tectonics in the Sicily Channel', J. Struct. Geol., 9, 869-876.

Butler, R. W. H., Grasso M. and La Manna F. (1992) 'Origin and deformation of the Neogene-Recent Maghrebian foredeep at the Gela Nappe, SE Sicily', J. Geol. Soc. London, 149, 547-556.

Calanchi, N., Colantoni, P., Rossi, P.L., Saitta, M. and Serri, G. (1989) 'The Strait of Sicily Continental Rift System: Physiography and Petrochemistry of the Submarine Volcanic Centres', Marine Geology, 87, 55-83.

Carbone S., Grasso M. and Lentini F. (1982) 'Considerazioni sull'evoluzione geodinamica della Sicilia sud-orienale dal Cretaceo al Quaternario', Mem. Soc. Geol. It., 24, 367-386.

Catalano, R. and D'Argenio, B. (1982) 'Schema geologico della Sicilia occidentale', in Catalano, R. and D'Argenio, B. (eds.), Guida alla Geologia della Sicilia occidentale, Palermo, 9-41.

Channell, J.E.T. (1986) 'Paleomagnetism and continental collision in the Alpine Belt and the formation of late-tectonic extensional basins', in Coward, M.P. and Ries, A.C. (eds.), Collision Tectonics, Geol. Soc. London, spec. Publ., 19, 261-284.

Channell, J.E.T. and Mareschal, J.C. (1988) 'Delamination and asymmetric lithospheric thinning in the development of the Tyrrhenian rift', in Coward M. P., Dietrich D. and Park R.G. (eds.), Alpine Tectonics, Geol. Soc. London, spec. Publ. 45, 285- 302.

Civetta, L., Cornette, Y., Crisci, G., Gillot, P.Y., Orsi, G. and Requiejo, C.S. (1984) 'Geology, geochronology and chemical evolution of the island of Pantelleria', Geol. Mag., 6, 541-562.

Colantoni, P. (1975) 'Note di geologia marina sul Canale di Sicilia', Giorn. Geol., 40, 181-207.

Colombi, B., Giese, P., Luongo, G., Morelli, C., Riuscetti, M., Scarascia, S., Schutte, K.G., Strowald, J. and De Visintini, G. (1973) 'Preliminary report on the seismic refraction profile Gargano-Salerno-Palermo-Pantelleria', Boll. Geofis. Teor. Appl., 15, 225-254.

Della Vedova, B. and Pellis, G. (1985) 'Age Estimates from Heat Flow and Subsidence Data in the Rifting Basins of the Southern EGT Segment', in ESF, Second EGT Workshop, The Southern Segment, Venice, 7-9 February, 235-239.

Dewey, J.F. (1980) 'Episodicity, sequency and style at convergent plate boundaries', in Strangway, D.W. (ed.), The Continental Crust and its Mineral Deposits, Spec. Pap., Geol. Assoc. Can., 20, 553-573.

Etheridge, M. A. (1987) 'On reactivation of extensional fault systems', Phil. Trans. R. Soc. London, 141, 170-194.

Finetti, I. (1984) 'Geophysical Study of the Sicily Channel rift zone', Boll. Geofis. Teor. Appl., 26, 3-28.

Finetti, I. and Morelli, C. (1972) 'Wide scale digital seismic exploration of the Mediterranean Sea', Boll. Geofis. Teor. Appl., 14, 291-342.

Grasso M., De Dominicis A. and Mazzoldi G. (1990) 'Structures and tectonic setting of the western margin of the Hyblean-Malta shelf, Central mediterranean', Annales Tectonicae, 4, 140-154.

Horvath, F. and Berckhemer, H. (1982) 'Mediterranean Backarc Basins', in Berckhemer, H. and Hsu, K.J. (eds.), Alpine-Mediterranean Geodynamics, AGU, Geodynamic Series, 7, 141-173.

Houseman, G.A., McKenzie, D.P. and Molnar, P. (1981) 'Convective instability of a thickened boundary layer and its relevance for the thermal evolution of continental convergent belts', J. Geoph. Res., 86, 6115-6132.

Hsu, K.J. (1977) 'Tectonic evolution of the Mediterranean basins', in Nairn, A.E.M., Kanes, W.S. and Stehli, F.G. (eds.), The Ocean Basins and Margins, Vol. 4A, Plenum Press, New York, 29-75.

Illies J. H. (1981) 'Graben formation - the Maltese Islands - a case history', Tectonophysics, 73, 151-168.

I.O.C., (1987) 'International bathymetric chart of the Mediterranan', Head Department of Navigation and Oceanography of the USSR.

Jongsma, D., Van Hinte, J.E. and Woodside, J.M. (1985) 'Geologic stryucture and neotectonics of the north African continental margin south of Sicily', Marine and Petroleum Geology, 2, 156-179.

Kastens, K., Mascle, J. and ODP cruise (1988) 'ODP Leg 107 in the Tyrrhenian Sea: insights into passive margin and back-arc basin evolution', Geol. Soc. Am. Bull., 100, 1140-1156.

Kincaid, C. and Olson, P. (1987) 'An experimental study of subduction and slab migration', J. Geophs. Res., 92, 13832-13849.

Kusznir, N. and Park, R.G. (1987) 'The extensional strength of the continental lithosphere: its dependence on geothermal gradient, and crustal composition and thickness', in: Coward, M.P., Dewey F.F. and Hancock, P.L. (eds.), Continental Extension Tectonics, Geol. Soc. London, spec. Publ., 28, 35-52.

Malinverno, A. and Ryan, W.B.F. (1986) 'Extension in the Tyrrhenian Sea and shortening in the Apennines as a result of arc migration driven by sinking of the lithosphere', Tectonics, 5, 227-245.

Mascle, J., Kastens, K., Auroux, C. and Leg 107 scientific party (1988) 'A land-locked back-arc basin: preliminary results from ODP Leg 107 in the Tyrrhenian Sea', Tectonophysics, 146, 149-162.

Moussat, E. (1983) 'Evolution de la mer Tyrrhenienne centrale et orientale et de ses marges septentrionales en relation avec la neotectonique dans l'Arc Calabrais', Thesis of third cycle, Villefranche-sur-Mer.

Oldow, J.S., Channell, J.E.T., Catalano R. and D'Argenio, B. (1990) 'Contemporaneous thrusting and large-scale rotation in the Western Sicilian fold and thrust belt', Tectonics, 9, 661-681.

Patacca, E., Sartori, R. and Scandone P. (1990) 'Tyrrhenian basin and Apenninic arcs: kinematics relations since late Tortonian times', Mem. Soc. Geol. It., in press.

Reuther C.D. and Eisbacher, G.H. (1985) 'Pantelleria Rift: crustal extension in a convergent intraplate setting', Geol. Rund., 74, 585-597.

Ritsema, A.R. (1979) 'Active or passive subduction of the Calabrian Arc', Geol. Mijnbouw, 58, 127-134.

Sartori R., Mascle G. and Amaudric du Chaffaut S. (1987) 'A Review of Circum-Tyrrhenian Regional Geology', in Kastens K., Mascle J. et al., Proc. ODP Init. Repts., 107, 37-63.

Scandone, P. (1979) 'Origin of the Tyrrhenian Sea and the Calabrian Arc', Boll. Soc. Geol. It., 98. 27-34.

Spence, W. (1987) 'Slab Pull and the Seismotectonics of Subducting Lithosphere', Rev. Geophysics, 25, 55-69.

Wezel, F.C. (1982) 'The structure of the Calabro-Sicilian Arc: results of a post-orogenic intra-plate deformation', in Legget, J.K. (ed.), Trench-Forearc Geology, Geol. Soc. London, spec. Publ., 10, 345-354.

Winnock, E. (1981) 'Structure du bloc Pelagien', in Wezel, F.C. (ed.), Sedimentary Basins of the Mediterranean Margins, Bologna, Tecnoprint, 445-467.

Zarudzki, E.F.K. (1972) 'The Strait of Sicily. A Geophysical Study', Rev. Geograph., Phys. et Geol. Dynam., 14, 11-28.

SEISMOLOGICAL STUDIES OF UPPER MANTLE STRUCTURE BELOW THE MEDITERRANEAN WITH A REGIONAL SEISMOGRAPH NETWORK

A. MORELLI, S. MAZZA, N. A. PINO, and E. BOSCHI
Istituto Nazionale di Geofisica
Via di Vigna Murata 605, 00143, Roma, Italy

ABSTRACT. A broadband seismograph network can provide valuable information for the study of a complex area such as the Mediterranean. Besides enabling the detailed study of source processes, the network also supplies fundamental data for analyzing lithospheric and asthenospheric structure. These are the two main goals considered when planning the MEDNET network of very broadband stations. Three dimensional tomographic images of Mediterranean upper mantle have been obtained from bulletin travel time data. However, the complexity of body wave propagation at regional distances requires the analysis of full waveforms in order to identify correctly energy which has travelled along different paths. The high quality data that are now available from broadband seismographs allow further investigations, such as those based on surface waves and broadband body waveform modelling. The geographical distribution obtained through the stations of the MEDNET and other programs makes up a dataset with a high potential which has been exploited only in part.

1. Introduction

The interest in the earth sciences is very strong in the Mediterranean as testified, for instance, by the work published in this volume. The Mediterranean area is tectonically and seismically active and is characterized by a complex geodynamic environment. Many important aspects of its evolution are not yet fully understood, and remain the subject of vigorous research by earth scientists. Seismology plays a fundamental role in Mediterranean studies for two reasons — earthquake mechanisms give information on active tectonics, and they represent the source of energy which can be used to image deep structure. These two aspects of seismological research naturally complement each other, as a reliable model for wave propagation is needed to reconstruct source mechanisms, and reliable knowledge of the source is needed to study the detailed structure of the medium. In the present paper, however, we will restrict ourselves to the discussion of the seismological contributions to the knowledge of the deep structure of the Mediterranean.

Seismological studies provide crucial information about the interior of the earth. The technique of seismic tomography, which allows us to see the three-dimensional structure, is now well known. Its importance for geodynamic investigations consists of the fact that it can provide images of the deeper parts of the large-scale structures that we see at the surface of the earth. The seismic structure of the upper mantle is now often considered as a constraint for quantitative modelling of geodynamic processes. Seismological results represent our best knowledge and in fact should be considered as constraints which kinematic models or dynamical theories should satisfy. However, the models that seismologists have produced have limitations and there are important restrictions to their use. Those limitations arise from limits and errors in the available data. These limits yield non-uniqueness in the derived models. Excessive confidence in the exact shape, location, or magnitude of a velocity anomaly, for instance, could be mistaken. This does not limit the importance of seismological models, but, rather, their application.

Just like any other experimental science, to improve models we have to improve the observations. This necessarily occurs through the installation of new, better, seismograph networks. We are currently witnessing a considerable effort by seismological institutions of European and Mediterranean regions to install new, better, instrumentation. We will discuss, in particular, the goals and potential of the MED-NET project, consisting of a network of very broadband seismographs installed in Mediterranean Countries and managed by the Istituto Nazionale di Geofisica and World Laboratory. The determination of upper mantle structure is one of the goals considered when designing the network.

2. Travel time tomography

The classical seismological approach to the study of earth structure consists of the analysis of the travel times of body waves. This approach provided us with the first information about the deep structure of the earth, such as the existence and nature of the compositional stratification in crust, mantle, and core. The nature of the dependence of the travel times of different seismic phases with distance has been known fairly well for a rather long time — the model by Jeffreys and Bullen (1940) is not very different from the most modern fits to similar data (Kennett and Engdahl, 1991; Morelli and Dziewonski, 1993). It is a common observation, however, that travel times vary with the geographical location of source and receiver — even if this variation is much smaller than that with epicentral distance. Let us for the following restrict ourselves to the first arrival, the P wave. Figure 1 shows a map of travel time deviations from a model in which seismic velocity only varies with depth, as seen from a single station. The effect of reading errors is

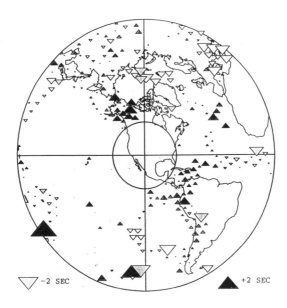

Figure 1 Average teleseismic travel time residuals for station ALQ. Each symbol represents the average residual computed for all shallow earthquakes located in the same 5° × 5° area (after Morelli and Dziewonski, 1993).

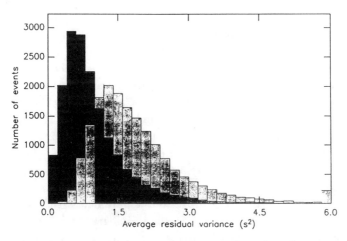

Figure 2 Histograms of frequency distribution of earthquakes according to residual travel time variance after hypocentral location. Light bars are for ISC location, dark bars show the histogram derived from locations obtained using empirical source-station corrections like those shown in Figure 1 (after Morelli and Dziewonski, 1993).

greatly reduced by averaging data from all the earthquakes located within the same same element of a 5° × 5° mesh (Morelli and Dziewonski, 1993). The presence of coherent patterns of positive (like north-western America) or negative (like the Mediterranean) residuals demonstrates the large-scale heterogeneity in deep earth structure. Another important element which we can deduce from Figure 1 is that the magnitude of the effects of heterogeneity on travel times of P waves is such that we can observe it being approximately of the same order of magnitude (say, 1 second) of the period of the incoming P wave. Therefore, we can use these observations to reconstruct lateral variations in earth structure.

The dataset used to derive Figure 1 is selected from the *Bulletins of the International Seismological Centre* (ISC). The *Bulletins* contain observations of seismic wave arrivals reported by virtually all permanent observatories distributed worldwide (Adams et al., 1982) and therefore represent a unique collection of measurements that contain information about the properties of the interior of the earth. They include all the arrival data received and compiled by ISC for individual seismic events. Some of these readings are affected by considerable error, due to a noisy station record or to misinterpretation. Because of the enormous dataflow, critical interaction between ISC and observatory analysts is limited to the mailing of a preliminary location of the event to all contributing agencies (Adams et al., 1982). This gives the analyst a chance to re-pick arrival times once their association with an event is safely determined. The resulting database is huge, but unfortunately the wealth of information it includes is difficult to retrieve completely, as some of the observations can be affected by large errors. In other words, we have a data set which is numerically very large, but also rather noisy.

Lateral variations of travel times are also important for studies of source parameters — even hypocenter location. Figure 2 shows the effect of including the empirical source-station corrections proposed by Morelli and Dziewonski (1993) in earthquake locations at teleseismic distance. The histograms show the average variance of standard ISC locations (light stipple) compared with the residual variance after inclusion of the empirical heterogeneity corrections (dark stipple). The most frequent value of average travel time variance per event is 1.3 s^2 for ISC locations, and it reduces to 0.5 s^2 applying the corrections. Therefore, lateral heterogeneities indeed modify travel times sensibly, and models of earth structure including lateral variations are important in order to improve source parameter studies.

The techniques generally designated as seismic tomography have allowed us to image the hidden structure of the earth at various depths and at scales from local to global (Iyer and Hirahara, 1993). The most important character of tomographic studies is that they permit the investigation of regions that are not directly accessible. Thus we have gained essential information about the crust, the lithosphere, and the lower mantle, which allows us the definition of local structures, the discovering of magma chambers, and the imaging of descending slabs and convective patterns.

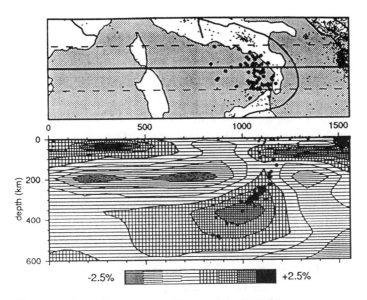

Figure 3 Cross section of a tomographic model of Mediterranean upper mantle (after Spakman, 1990). Grey scale indicates percentage variation of seismic P velocity. The slab subducted under the Eolian arc can be identified. There is a suggestion of a low-velocity layer at 200 km depth.

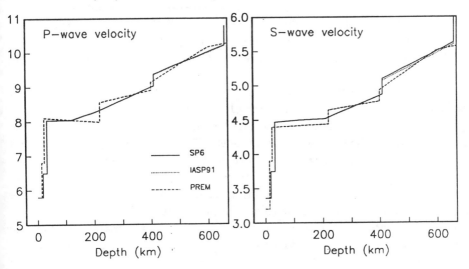

Figure 4 Upper mantle velocity profiles for model SP6, compared to *iasp91* and PREM (after Morelli and Dziewonski, 1993). Model *iasp91* differs slightly from SP6 only below 400 km.

Seismic tomography obviously is an important tool to help us to understand both the structure and tectonic processes of the Mediterranean.

Teleseismic P arrivals have been used to obtain models of velocity perturbation in the crust and upper mantle of the European and Mediterranean area (Babuska et al., 1984; Babuska and Plomerova, 1990; Amato et al., 1993). The article by Cimini and Amato in this volume presents one of these models, and discusses more extensively the results and limitations of the method. We only note here that this approach provides important and stable observations about lithospheric and asthenospheric structure beneath stations, but has limited vertical resolution. This is due to the fact that teleseismic rays, coming from events with epicentral distance between, say, 30° and 90°, cross the upper mantle with a rather steep angle. This implicit lack of resolution has only a limited effect on the application to Italy, thanks to the high density of seismograph stations. A more important limitation is that the method can only be applied to portions of the mantle beneath a seismic network, and cannot be used to derive information below the sea.

Spakman (1990) used teleseismic and regional travel times, extracted from the ISC Bulletin, to derive a three-dimensional model of the upper mantle beneath the Mediterranean. A sample cross section is reproduced in Figure 3. This model consists of blocks with a side of about 100 km inside which perturbations in slowness (inverse of seismic velocity) with respect to a laterally homogenous initial model are computed. Figure 3 shows a smoothed image of the resulting model. These images show the presence of strong and large features such as a subducting slab. By using seismic waves crossing distances less than 30°, this approach reconstructs wave velocities along paths running entirely in the upper mantle. This in principle gives good vertical resolution, and also recovers information from waves travelling in the mantle under the sea.

These images are very appealing. They have to be considered by anyone who is modelling the geodynamics of the area. However, they do have limitations. As mentioned above, any inversion of seismic data is non-unique, and therefore the same data could also be fit by a different model. This is particularly true for models of regional wave propagation, as shown in the next section. Caution should be used before drawing important conclusions if they are based only from the observation of tomographic images. To look for confirmation or modification to the travel time tomographic images, we have to consider other techniques and compare results.

An obvious limiting factor of such a strategy is given by the sparse, non-uniform distribution of earthquakes and stations. We could therefore improve the situation by an increase in the density of stations, with particular emphasis on islands or areas otherwise devoid of instrumentation. To pick travel times, relatively inexpensive short-period instruments are appropriate. However, even if it would improve the situation, this solution alone would not suffice, as it could not resolve the ambiguity connected to wave propagation at regional distance.

3. Wave propagation at regional distance

Seismic wave propagation at regional distance is significantly complex. It has almost been neglected, and has gained interest only recently. For this reason these constraints apply to travel times as well as waveforms. Complications arise from the complex upper mantle structure, which often includes discontinuities and a low-velocity layer. Seismic waves travelling to less than, say, 30° exhibit a very strong sensitivity to small changes of model parameters, which results in strong non-linearity and instability in modelling procedures.

In the following we will consider for reference model SP6 (Morelli and Dziewonski, 1993). In the depth range in which we are interested, model SP6 does not differ appreciably from model *iasp91* (Kennett and Engdahl, 1991), both of which have been derived following the guidelines set by a IASPEI Working Group for a reference model for global travel time calculations (Kennett, 1988a). Figure 4 shows P and S wave velocity in the upper mantle for SP6 as compared to *iasp91* and to PREM (Dziewonski and Anderson, 1981). The upper mantle in these models is meant to represent a global average, and is simpler than PREM, for instance. There is no low-velocity layer. Thus, this model yields seismic waveforms which are minimally complex.

Any modern upper mantle model includes discontinuities, which produce triplications in the travel time curve. For example, downgoing rays leaving a source at shallow takeoff angles travel directly to receivers, producing a prograde branch of a travel time. At steeper angles, however, rays may encounter a discontinuity at which velocity increases suddenly. The rays are reflected; the steeper the ray, the shorter the distance Δ travelled by the ray, resulting in a retrograde segment of a travel time curve. At angles steeper than the critical angle, rays penetrate the lower layer and are refracted. The travel time curve for these refracted rays is again prograde.

The result, when plotted in a travel time *vs.* distance plot, is a triplication, as those seen in Figure 5. The first two steep lines represent phases Pg and P*, which are only important at local distance. When we follow the long, almost straight, line of the Pn phase — the only arrival between 5° and 15°, corresponding to waves travelling in the lithospheric mantle — we encounter the first complication due to the discontinuity at 120 km depth . The two main triplications are due to important discontinuities at 410 and 660 km depth, respectively positioned at 22° and 28°. The location of these features of the travel time curve is of course dependent on the model used for the computations.

A seismogram analyst is generally only able to recognize one arrival — even if there are in fact more but they are not sufficiently separated. The minimum separation depends on the relative strength of incoming phases, but in seismological practice secondary arrivals are generally recognized only in particularly clear cases

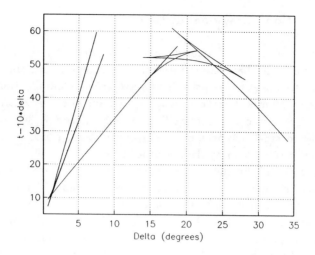

Figure 5 Travel time curve for P waves at regional distances for model SP6. For graphical purposes, travel time has been reduced by a term proportional to distance. The two steepest lines ($\Delta < 10°$) represent phases Pg and P*. Phase Pn is the first arrival between 2° and 15°. Triplications are due to mantle velocity discontinuities in the model.

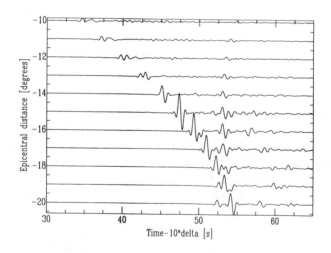

Figure 6 Broadband synthetic velocity seismograms computed for an isotropic impulsive source at distances between 10° and 20° (from top to bottom). Travel time reduction is the same as in Figure 5. Ray-theoretical arrivals on the branches of Figure 5 can be identified.

	Code	Time(s)	(min s)		dT/dD
1	Pn	239.13	3	59.13	12.5829
2	Pn	240.44	4	0.44	13.5068
3	Pn	240.48	4	0.48	13.5462
4	P	241.04	4	1.04	11.0504

Table 1: Travel times of P waves at epicentral distance of 17°. Partial derivative of travel time with respect to epicentral distance (dT/dD) is given in seconds per degree.

such as PP, PcP, PKIKP. Travel time studies are therefore based on the assumption of first arrival. An inconvenience of triplications is the presence of 'blind spots' in the model — depths at which no first arrival ray, at any distance, has reached its bottoming point. A ray bottoming at a certain depth is important to gain vertical resolution. Also, the presence of blind spots in the model is particularly bad if a smooth model — like for instance the Jeffreys and Bullen (1940) — or an otherwise inappropriate model is used to trace rays. A smooth model produces no triplications and distributes bottoming depths continuously. As a consequence, rays would be traced to grossly wrong paths.

Synthetic waveform calculations allow us to gain a general idea of the characters of seismograms at these distances. Figure 6 shows vertical traces of seismograms at epicentral distances from 10° to 20°. For the sake of simplicity, the source was isotropic and impulsive, located at a depth of 0 km. The technique used is from Kennett (1980). Comparison with Figure 5 permits us to recognize ray-theoretical travel time branches. It can be seen in the waveforms that picking arrival times may be rather difficult. Waveforms change rather quickly with distance and also pose problems for detection of polarities for computation of focal mechanisms — compare for instance waveforms at 13°, 14°, and 15°. Sometimes (e.g. at 20°) the first arrival has a smaller amplitude than the second arrival — in the presence of critical signal-to-noise ratio, the second arrival can be mistaken as the first. An example of the multiplicity of seismic rays adding up to form the wavetrain is given in Table 1, which shows ray-theoretical arrivals at 17°. The first 4 rays arrive within approximately 2 seconds, and they have rather different ray parameters (shown as dT/dD) and, therefore, sample quite different depths (Bullen and Bolt, 1985). At this distance, contributions to the wave group represented by a single travel time have sampled different portions of the model. We should bear this in mind when addressing the resolution of the results of an inversion.

The presence of these complications is demonstrated in Figure 7. This shows frequency histograms for ISC travel times at certain epicentral distances for shallow

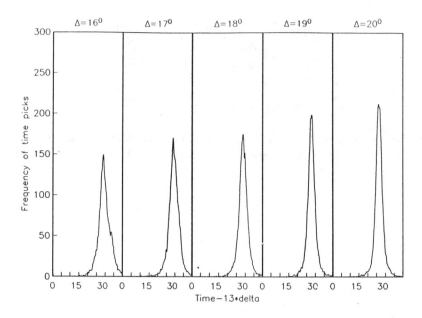

Figure 7 Smoothed histograms of frequency distribution of travel times from ISC bulletins at 5 epicentral distances. Only surface events were considered. The distributions show small peaks and broadening possibly due to time picks biased by the waveform complexity apparent in Figure 6.

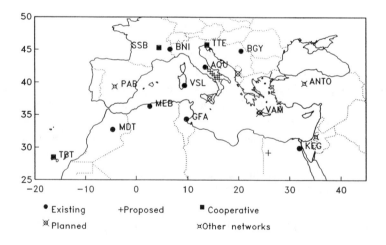

Figure 8 Map of MEDNET stations.

events. If we also compare Figures 5 and 6, we can qualitatively understand the presence of the side peak in the histogram at 16° — there is another arrival about 5 seconds after the first which apparently is often mistaken by operators contributing to the ISC. We may also notice that the second peak gets closer to the main peak as we proceed to larger distances. At 19° and 20° it can hardly be recognized as it begins to arrive before the branch with higher amplitude. This example shows that in fact energy propagation at regional distance is very complex. The complexity can confuse images based on bulletin travel time analysis. A more complete approach is based on the analysis of whole broadband waveforms.

4. The MEDiterranean NETwork

MEDNET is a network of very broadband seismograph stations installed in the Mediterranean area by the Istituto Nazionale di Geofisica (Boschi et al., 1991) within the framework of the activities of the World Laboratory. The project began in 1988 and now maintains 8 existing stations; 3 observatories are jointly maintained with other projects, 3 belong to other programs, and 3 more stations are planned within 1993 (Figure 8). Technical specifications include very broadband sensors, 24 bit analog-to-digital converters, and powerful microprocessor station controllers. Dial-up telemetry for some of the stations allows rapid transfer of data after an event of special interest. Cooperative stations are installed jointly with other institutions (TBT: IRIS/GSN, SSB: GEOSCOPE, TTE: University of Trieste). We receive continuous data from IRIS/GSN stations PAB and ANTO. Sites for planned stations have been selected, and installation will possibly follow during 1993. Proposed stations include the first 8 sites of the Italian Digital Broadband network, and one in conjunction with the new Egyptian seismograph network.

The very broadband response covers extremely-long period free oscillations of the earth and frequencies up to 10 Hz. Combined with the high dynamic range given by the digitizers, this makes MEDNET stations appropriate to record virtually all the seismic signal — only near-field, high-frequency signal needs additional and specialized seismometers. The network is therefore a tool for a variety of possible seismological studies.

The geometry of MEDNET is planned to provide a homogenous coverage of the Mediterranean. Cooperation with other countries and other instrumentation programs is an essential element to obtain this goal. Research goals include the study of earthquake sources (see the article by Giardini et al., this volume) and mantle structure.

5. Body waveform modelling

Tomographic studies require a high density of both data and stations, a situation

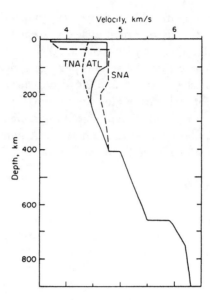

Figure 9 Comparison of three shear-wave velocity models appropriate for active tectonic (TNA), shield (SNA), and old oceanic (ATL) regions (after Helmberger, Engen and Grand, 1985)

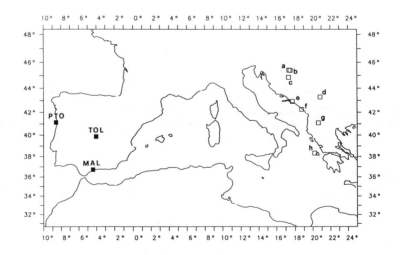

Figure 10 Map showing the WWSSN stations (■) and the events (□) used to model the western Mediterranean upper mantle structure.

which is not often obtained. On the other hand they utilize only travel times. As we previously discussed, there are strong uncertainties in phase picking, especially on a regional scale. Waveform modelling, instead, exploits all the information contained in a seismogram and can yield notable results even if only a few events and stations are available. Moreover, the same technique can be used to learn about the source, in terms of fault parameters and source-time function, once a model has been obtained for the structure.

In the last fifteen years a large number of papers describing studies of upper mantle structure by matching body waveforms from regional earthquakes with synthetic seismograms has been published (e.g. Burdick and Helmberger, 1978; Given and Helmberger, 1980; Burdick, 1981; Grand and Helmberger,1984; LeFevre and Helmberger, 1989; Zhao, Helmberger and Harkrider, 1991). Several models for P and S wave velocity have been developed for paths travelling in different regions. One of the most notable results of these studies is that upper mantle structure is strongly related to the tectonic regime. As long as paths entirely belonging to the same province are considered, the best fitting models for different regions are grouped into three categories: tectonic, shield, and old ocean. Figure 9 shows the distribution of shear velocity versus depth for a sample of each of the three groups. Tectonic models apply to tectonically active regions, like the Gulf of California spreading ridges (Walck, 1984). Shield models appear in stable continental regions, such as the Canadian shield (LeFevre and Helmberger, 1989). Old ocean models are appropriate for regions that have an old oceanic crust, like the northwestern Atlantic ocean (Zhao and Helmberger, 1992) which has a crust ranging in age from 100 to 150 Ma. Main features of these models are: a) shield models are faster than tectonic ones; b) shield models have a thick and fast lithosphere (150-180 km) above a thin and slightly pronounced low velocity layer; c) tectonic models have very thin lithosphere (few tens of kilometers) above a thick low velocity zone; d) below 200 km the velocity gradient in tectonic regions is higher than in shield ones; e) no structural difference has been found below 400 km; f) the old ocean models are in general intermediate with respect to tectonic and shield ones. When mixed paths are involved, if 1-D modeling techniques are used, the resulting structure represents an average between the different regions.

The different signatures of such models on the relative travel times and amplitudes of the triplications shown in Figure 5 can clearly be seen in the waveforms. By analyzing the data on the basis of these diagnostic patterns it is possible to preliminarily define the appropriate model. As a second step, comparison with synthetic seismograms allows the refinement of the model.

We describe now an application of this technique to some data in the western Mediterranean region. We chose earthquakes located in Greece and Yugoslavia and recorded at the WWSSN long period stations TOL (Toledo), MAL (Malaga) in Spain, and PTO (Porto) in Portugal (Figure 10). The source-station distances are

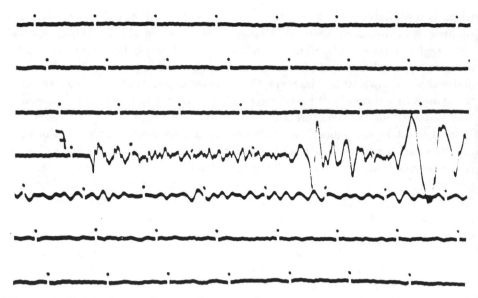

Figure 11 Original recording on the vertical component at TOL WWSSN-LP station of the Yugoslavia earthquake (5/18/80) (d in Figure 10).

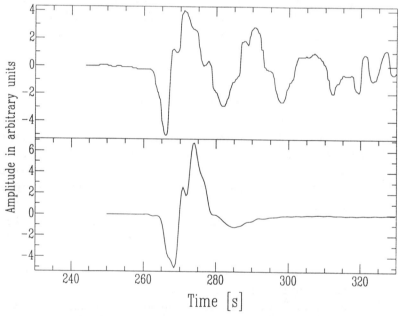

Figure 12 Data (top) and synthetic (bottom) for event d of Figure 10 at TOL. Only phases involved in the triplication of Figure 5 were modelled.

all around 19°. We used the source mechanisms obtained teleseismically from first motion polarities and waveform direct modeling (Anderson and Jackson, 1987). The original analog recordings were scanned and digitized in order to allow the application of numeric processing techniques. Figure 11 shows an original WWSSN-LP recording.

At present several models have been developed for the area investigated by using either travel time tomographic methods (Spakman, 1991) or surface wave dispersion (Panza, Mueller and Calcagnile, 1980; Snieder, 1988a). Neither approach is able to resolve fine details of the structure. The former lacks adequate data coverage to provide detailed resolution; the long wavelengths required by the second method limit the degree of resolution obtained. Modeling of body waveforms may be a powerful technique to add information and achieve better resolution.

Figure 12 shows a result of the modelling. Synthetic seismograms were computed by using Generalized Ray Theory (GRT) (Helmberger, 1983). The model obtained is very similar to an old-ocean type. This region cannot be considered laterally homogenous, hence the model should be understood as an average for the area sampled by the paths. Because of the short scale variations of the structure typical of the Mediterranean, we cannot identify paths entirely contained in the same homogeneous region. However, considering different source-station geometries allows the identification of regions with lateral extent smaller than the path length. With waveform analysis, it is possible to locate the depth at which a certain feature of the seismogram originated, a fact which sometimes allows the location of some anomaly along the path — structure near the source, in the deepest part of the ray, or near the station.

Studies of upper mantle structure were usually carried out by using long period analog data. The present availability of broadband digital recordings in the Mediterranean requires more sophisticated modeling procedures to reproduce the complexities of waveforms, since the very broadband instruments are much more sensitive to short periods, and thus to small scale structural features, than are the WWSSN-LP seismographs. As an example we present the results of modeling regional seismograms for the Turkey earthquake (03/13/1992). Figure 13 shows a comparison between the broadband waveforms and the simulations of the WWSSN-LP signals for AQU. The long period signal, related to deep reflections, allows the study of upper mantle structure. The high frequency oscillations, completely missing in the WWSSN-LP simulations, are generated by multiple and converted phases at shallow depths. By analyzing these phases it is possible to obtain a model for the crustal properties near the source and station, as well as the source depth.

Broadband signal could in fact be reconstructed from separate short and long period records (Choy and Boatwright, 1981), but this procedure cannot reach the fidelity of a true broadband instrument. Broadband modeling results are shown in Figure 14. For the computation of the synthetic seismograms a reflectivity code

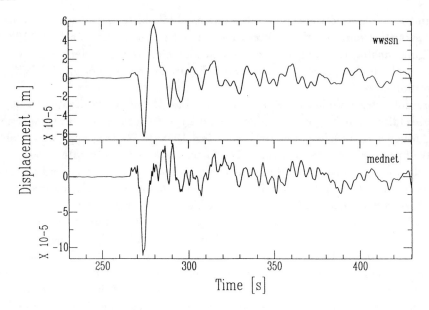

Figure 13 Comparison between the vertical component recorded by AQU broadband station and WWSSN-LP simulation for the 03/13/1992 Turkey earthquake.

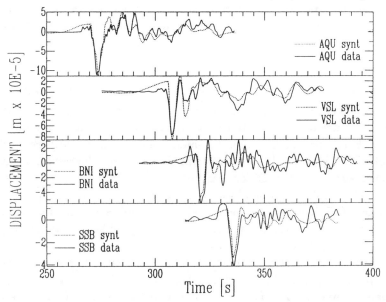

Figure 14 Data and synthetics at MEDNET stations for the same earthquake as in Figure 13.

(Kennett, 1988b) was preferred to GRT because of the automatic generation of the entire wavefield, which is particularly appropriate when many different phases are required. The strategy followed to design the final velocity model is based on the computation of a large number of synthetics. First a family of velocity profiles is generated randomly perturbing the coefficients of a reference model (SP6, Morelli and Dziewonski 1993). The synthetics are examined to extract information about the influence of the perturbations in the model on the resulting waveforms. Then, on the basis given by such fits, the final model is assembled and fine adjustments are made on it. The appropriateness of a velocity profile may be preliminarly checked by comparing some key data arrival times with theoretical values obtained with a fast and reliable method such as that described by Buland and Chapman (1983). If the agreement for such critical phases is not satisfactory, the model is discarded without computing the waveforms. Only synthetics showing the closest fits to their data are considered. Model generation could be further improved by making use of genetic algorithms (Gallagher et al., 1991; Wilson and Vasudevan, 1991), suited for nonlinear geophysical optimization problems. The strong nonlinearity affecting regional bodywave has so far prevented an effective application of inverse theory to these problems. Increasing computational resource makes Monte Carlo methods a viable direction to consider for the immediate future.

6. Surface wave modelling

Although S-wave delay times could be used for tomographic inversions in the same way as P-waves are used, two major problems obstruct this application. The number of available S-wave picks is less than the number of P arrival times and they are read with greater difficulty, as they are not first arrivals. However, the greater difficulty is derived from the presence of a low velocity layer in the upper mantle that makes the problem highly nonlinear. Chapman (1987) has shown that the tomographic inversion problem is ill-posed if a low velocity layer is present. Further complications are posed by the presence of strong lateral variations within the same layer.

An alternative approach is offered by observations coming from Love and Rayleigh waves, whose propagation is intimately influenced by S-wave velocity. Calcagnile and Panza (1980) infer strong lateral variations in the lithosphere beneath the Italian area, by the inversion of Rayleigh-wave dispersion data grouped into two main families of dispersion curves.

Even if this approach has been adopted in practice, there are some shortcomings which limit its validity in reconstructing structures whose scale is comparable to the wavelength of the surface waves. In other words, as the fundamental mode penetrates down to 200 km, with a wavelength of about 300 km, lateral heterogeneities on a scale of a few hundred kilometers can hardly be considered smooth. In such cases, the assumptions on which the ray theory is based are not strictly satisfied.

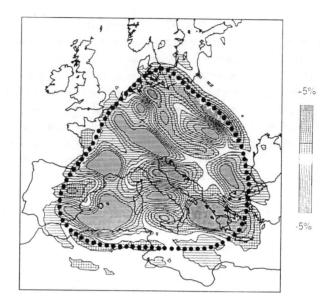

Figure 15 Relative S-velocity perturbation ($\delta\beta/\beta$) for depths between 100 and 200 km, obtained with surface wave scattering theory (after Snieder, 1988b).

Figure 16 Wavepaths of the NARS dataset used in the inversion for the S-velocity perturbation map of Figure 15 (after Snieder, 1988b).

Snieder (1988a, 1988b) shows how to take into account multipathing effects and scattering of surface waves and how they can be used to map lateral heterogeneities of S velocity. His technique consists of two main steps, a preliminary nonlinear inversion to obtain a smooth reference model and a subsequent linear inversion using linear scattering theory. The first inversion is needed to make the problem more linear, so that the second inversion is performed under the best assumptions. Figure 15 shows the model found by Snieder (1988b) for the Mediterranean–European upper mantle. Important structural features can be identified in Figure 15, showing lateral variations in the low velocity layer of the upper mantle. The high velocities under the Adriatic are related to the subduction of Africa under Europe. In continental Europe, the two major lineaments – the Rhine Graben and the Tornquist-Tesseyre line – show up respectively as slow and fast anomalies.

One of the major drawbacks of this technique is the limited availability of high-quality long-period recordings for the Mediterranean area. The model shown in Figure 15 was computed by means of the recordings of the NARS array (Dost *et al.*, 1984, Figure 16), but it is possible now to consider using a larger amount of data, as in the meantime the quantity and quality of stations in the area have been remarkably improved.

Not all earthquakes can be included in the inversion. The strongest earthquakes have to be discarded because of the complexity of their source. On the other hand, those which are too small and for which a reliable source moment tensor has not been calculated cannot be used. Even if it is possible to include in the inversion a large number of seismograms, this is not in practice the best strategy. It is preferable to limit their number, choosing events which have been recorded by many good stations and for which a reliable independent estimate of the source parameters is available. It is worthwhile mentioning that the increased number of broadband stations in the area lowers the threshold magnitude for which the moment tensors can be calculated (Giardini *et al.*, this volume). An example of the distribution of ray paths currently available is shown in Figure 17. These, and others, may now be added to the NARS data originally used. Events were recorded during the period January 1990 - July 1992 at MEDNET stations and at others made available by cooperating projects in the European area. In Figure 15 velocity anomalies are elongated in a southeast-northwest direction. This is possibly due to the prevailing orientation in this direction of the ray paths used in this study. The inclusion of paths to other stations, often intersecting the NARS dataset, will help to improve the lateral resolution of the tomographic image. Also, the availability in the current MEDNET dataset of paths to the African continent and to the Eastern Mediterranean – compare Figure 17 with Figure 16 – will allow the enlargement the model to cover the Africa-Europe boundary, and to follow it much further to the east.

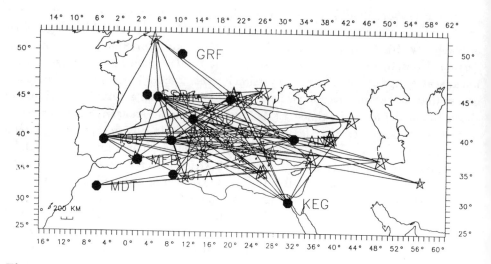

Figure 17 Wavepaths of the MEDNET dataset for events between Jan 1990 and Apr 1992, for which moment tensors are available.

7. Conclusions

Seismology represents the only means of gathering precise information on mantle structure. Several seismic techniques can be used to retrieve radial or lateral structure. Travel time tomography provides spectacular three-dimensional images of the upper mantle (Spakman, 1991), but these images may be distorted by the complications in body wave propagation due to the presence of radial discontinuities and a low velocity layer. They are of great interest, but some of their features — such as finer details of slab geometry — need confirmation before they can be used to draw geodynamic constraints. Surface wave studies are a more reliable means to map lateral heterogeneity in the asthenosphere. The long wavelengths are a classical limit to the resolution of the technique. However, inclusion of linear scattering theory in the inversion makes it possible to map sharp transitions characterized by dimensions shorter than the wavelength. Snieder (1988b) has shown how important structural elements can be identified in this way. Inclusion of more data now available enlarges the domain of the inversion, and provides a better coverage by intersecting rays. The complex propagation features of bodywaves at regional distances require the analysis of full waveforms to identify energy which travelled along different paths. Broadband waveform analysis is very sensitive to the fine structure of an earth model, and has therefore a great potential.

Methods which exploit the information contained in the full waveforms require high fidelity in recording ground motion. This is granted by modern digital broadband seismographs. The committment undertaken by several instrumentation programs (such as MEDNET, NARS, but also global networks like GEOSCOPE and IRIS/GSN) is giving us good prospects of finding more accurate models for lithospheric and asthenospheric structure of the Mediterranean region.

Acknowledgements. J. Boatwright and L.J. Sonder improved the article with their comments and suggestions.

References

Adams, R. D., A. A. Hughes, and D. M. McGregor (1982) *Analysis procedures at the International Seismological Centre*, Phys. Earth Planet. Interiors, 30, 85–93.

Anderson, H. and Jackson, J., (1987) *Active tectonics of the Adriatic region*, Geophys. J. R. Astron. Soc., 91, 937-984.

Amato A., B. Alessandrini, and G. Cimini (1993) *Teleseismic wave tomography of Italy*, in H. M. Iyer and K. Hirahara (eds.), Seismic tomography: theory and methods, Chapman and Hall (in press).

Babuska, V., and J. Plomerova (1990) *Tomographic studies of the upper mantle beneath the Italian region*, Terra Nova, 2, 569-576.

Babuska, V., J. Plomerova, and J. Sileny (1984) *Spatial variations of P residuals and deep structure of the European lithosphere*, Geophys. J. astr. Soc., 79, 363-383.

Boschi, E., Giardini, D. and Morelli, A. (1991) MEDNET: the Broadband seismic network for the Mediterranean, Il Cigno Galileo Galilei, Roma.

Buland, R. and Chapman C.H. (1983) *The computation of seismic travel times*, Bull. seism. Soc. Am., 73, 1271-1303.

Bullen, K.E., and B.A. Bolt (1985) An introduction of the theory of seismology, Cambridge University Press, 499pp.

Bulletin of the International Seismological Centre: Catalogue of Events and Associated Observations (Years 1964-1989), Vols. *1-26*, International Seismological Centre, Newbury, Berkshire, England.

Burdick, L. J. (1981) *A comparison of upper mantle structure beneath North America and Europe*, J. Geophys. Res., 86, 5926-5936.

Burdick, L. J. and Helmberger, D. V. (1978) *The upper mantle P velocity structure of the western United States*, J. Geophys. Res., 83, 1699-1712.

Calcagnile G., and G.F.Panza (1980) *The Main Characteristics of the Lithosphere-Astenosphere System in Italy and Surrounding Regions*, Pure Appl. Geophys., 119, 855-879.

Chapman, C.H. (1987) *The Radon transform and seismic tomography*, in Seismic tomography, with applications in global seismology and exploration geophysics, edited by G.Nolet, pp.99-108, Reidel, Dortrecht.

Choy, G.L., and J. Boatwright (1981) *The rupture characteristics of two deep earthquakes inferred from broadband GDSN data*, Bull. Seism. Soc. Am., 71, 691-711.

Dost, B., A. van Wettum, and G. Nolet (1984) *The NARS array*, Geol. Mijnbouw, 63, 381-386.

Dziewonski, A. M., and Anderson, D. L. (1981) *Preliminary Reference Earth Model (PREM)*, Phys. Earth Planet. Inter. 25, 297-356.

Gallagher, K., Sambridge M. and Drijkroningen G. (1991) *Genetic algorithms: an evolution from Monte Carlo methods for strongly non-linear geophysical optimization problems*, Geophysical Research Letters, 18, 2177-2180.

Given, J. W. and Helmberger, D. V. (1980) *Upper mantle structure of northwestern Eurasia*, J. Geophys. Res., 85, 7183-7194.

Grand, S. P. and Helmberger, D. V. (1984) *Upper mantle shear structure of North America*, Geophys. J. R. Astron. Soc., 76, 399-438.

Helmberger, D. V., Engen, G. and Grand, S. (1985) *Upper-mantle cross-section from California to Greenland*, J. Geophys., 58, 92-100.

Helmberger, D. V. (1983) *Theory and application of synthetic seismograms*, in Proceedings of the International School of Physics "Enrico Fermi", LXXXV Course, Earthquakes: Observation, Theory and Interpretation, pp. 174-222, eds. Kanamori, H., and Boschi, E., North Holland.

Iyer H. M., and K. Hirahara (eds.) (1993) *Seismic tomography: theory and practice*, Chapman and Hall, London.

Jeffreys, H. and K. E. Bullen, (1940) *Seismological tables*, British Association for the Advancement of Science, London.

Kennett, B. L. N. (1980) *Seismic waves in a stratified half space - II. Theoretical seismograms*, Geophys. J. R. Astron. Soc., 61, 1-10.

Kennett, B. L. N. (1988a) *Quakes to have better locations*, EOS Trans. Am. Geophys. Un., 65, 1571.

Kennett, B. L. N. (1988b) *Sistematic approximations to the seismic wavefield*, in D.J. Doornbos, Seismological Algorithms, Academic Press, San Diego, pp.237-259.

Kennett, B. L. N., and Engdahl, E. R., (1991) *Traveltimes for global earthquake location and phase identification*, Geophys. J. Int., 105, 429-465.

LeFevre, L. V. and Helmberger, D. V. (1989) *Upper mantle P velocity structure of the canadian shield*, J. Geophys. Res., 94, 17749-17765.

Morelli, A., and D. M. Dziewonski (1993) *Body wave travel times and a spherically symmetric P and S wave velocity model*, Geophys. J. Int., in press.

Panza, G. F., Mueller, S. and Calcagnile, G. (1980) *The gross feature of the lithosphere-asthenoshere system in Europe from seismic surface waves and body waves*, Pure appl. Geophys., 118, 1209-1213.

Snieder, R. (1988a) *Large-Scale Waveform Inversion of Surface Waves for Lateral Heterogeneity, 1, Theory and Numerical Examples*, J. Geophys. Res.,93, B10, 12055-12065.

Snieder, R., (1988b) *Large-Scale Waveform Inversion of Surface Waves for Lateral Heterogeneity, 2, Application to surface waves in Europe and the Mediterranean*, J. Geophys. Res.,93, B10, 12067-12080.

Spakman, W. (1990) *Tomographic images of the upper mantle below central Europe and the Mediterranean* Terra Nova, 2, 542-553.

Spakman, W. (1991) *Delay-time tomography of the upper mantle below Europe, the Mediterranean, and Asia Minor*, Geophys. J. Int., 107, 309-332.

Walck, M. C. (1984) *The P-wave upper mantle structure beneath an active spreading center: the Gulf of California*, Geophys. J. R. Astron. Soc., 76, 697-723.

Wielandt, E. and Steim, J. M. (1986) *A digital very-broad-band seismograph*, Annales Geophysicae 4B3, 227-232.

Wilson, W.G. and Vasudevan K. (1991) *Application of the genetic algorithm to residual static estimation*, Geophysical Research Letters, 18, 2181-2184.

Zhao, L-S. and Helmberger, D. V. (1992) *Upper mantle compressional velocity structure beneath the Northwest Atlantic ocean*, Submitted to J. Geophys. Res.

Zhao, L-S., Helmberger, D. V. and Harkrider, D. G. (1991) *Shear-velocity structure of the crust and upper mantle beneath the Tibetan Plateau and southeastern China*, Geophys. J. Int., 105, 713-730.

THE DETERMINATION OF EARTHQUAKE SIZE AND SOURCE GEOMETRY IN THE MEDITERRANEAN SEA

D. GIARDINI, B. PALOMBO & E. BOSCHI
Istituto Nazionale di Geofisica
Via di Vigna Murata 605, 00143, Roma, Italy

ABSTRACT. The deployment of modern digital seismic networks is changing the seismological practice of estimating the earthquake size and source geometry. Here we discuss applications and limitations of the principal parameters used in quantifying earthquakes: magnitude, fault plane solutions and moment tensors. We present a method for the rapid determination of the moment tensor using digital waveforms from a regional network, suitable for application to regional earthquakes (up to 3000 km distance) and to significant global events. The algorithms are routinely applied to significant Mediterranean and global events using digital data recorded by MEDNET; we show results for the largest global earthquakes of 1990 and for 20 significant Mediterranean earthquakes of 1990-1992, including the large shocks in Romania (May 30, 1990), Iran (June 20, 1990), Caucasus (April 29, 1991), Turkey (March 13, 1992) and Egypt (October 12, 1992). The MEDNET results are consistent with focal mechanisms and CMTs derived from global data and local networks, proving that an advanced regional seismic network is capable of providing accurate, rapid control of regional and global seismicity.

1. Introduction

The practice of seismology has been traditionally based on the exploitation of parameters extracted from analog records of seismic stations distributed worldwide. Size and source geometry of earthquakes were characterized by magnitude and fault plane mechanisms derived mostly from single-phase amplitude measures and polarities of P-arrivals; the bulk of the seismicity investigations are based on these methods of analysis.

The development of global digital seismic networks has brought forth a revolution in seismological practice and worldwide seismicity is now routinely analyzed using global digital data. For any large earthquake, we may find in the literature – in addition to standard focal mechanisms and magnitudes – moment tensor solutions derived from digital data, often complemented by waveform modelling for the determination of depth and source time function.

Moment tensor catalogues have been compiled for more than 9000 events since 1977, mainly by Harvard University (CMT; *Dziewonski et al., 1981*) and the USGS

(*Sipkin, 1986*); indeed, the CMT and USGS solutions, published routinely in the NEIC Monthly Bulletins, have become a primary database for seismological analysis (a collection of CMT solutions for Mediterranean earthquakes was derived in *Giardini et al., 1984*).

While global seismic networks (e.g. GSN, IDA, GEOSCOPE) have led a major effort to ensure the even geographical coverage needed for teleseismic studies, several regional digital seismic networks are also in preparation in all continents (e.g. MEDNET, CDSN, CNSN, MIDAS, USNSN; see *Berry, 1988* and *Boschi et al., 1991*, for reviews of regional and global digital networks).

The accurate, rapid monitoring of regional seismicity in the 100-3000 km distance range is among the outstanding seismological issues. Regional analyses have been hindered in the past by the lack of digital data of adequate dynamic range and geographical distribution, and by the inherent complication of regional broad-band waveforms, requiring the accurate calibration of the lithospheric model to ensure modelling. While for large earthquakes the consensus on source parameters obtained by different techniques, dealing with teleseismic, regional and local data, is often very satisfactorying (Figure 1), the availability of source parametrizations for mid-size regional events is still very low and the reliability of these results yet to be proven.

While the high-quality digital data collected by modern global networks have made the analysis of large earthquakes a routine practice in seismology (see the NEIC Bulletins), the methods developed for global seismicity are not suitable to study moderate regional earthquakes; indeed, only the CMT method (*Dziewonski et al., 1981*) has proven capable of analyzing global earthquakes in the 5-5.5 magnitude range. In addition, rapid moment tensor determination is routinely applied only to significant worldwide earthquakes and only by few seismological centres (Harvard, Caltech, ING), using data telemetered from a limited number of global open seismic stations. Today the installation of regional seismic networks in many areas of the world is providing digital data of unprecedented quality at regional scale and seismology is renewing efforts to study the seismic process of regional earthquakes (e.g. *Dreger and Helmberger, 1990; Karabulut et al., 1991; Ritsema et al., 1991; Nakanishi et al., 1992*).

The need for the rapid evaluation of size and geometry of the seismic source is crucial in the densely inhabited Mediterranean basin, where seismicity caused about 60.000 casualties in the last 3 years and even a shock of moderate proportions may prove disastrous (e.g. the December 13, 1990, Eastern Sicily event, with $m_b = 5.4$, killed 19 people).

Here we discuss applications and limitations of magnitude and fault plane solutions and describe waveform inversion techniques recently proposed (*Giardini, 1992; Giardini and Beranzoli, 1992; Giardini et al., 1993*) for the quantification of regional and global earthquakes using digital waveforms recorded in the Mediterranean by MEDNET, very-broad band seismographic network installed by the Istituto Nazionale di Geofisica of Roma in countries of the Mediterranean area, under the WorldLab program Plato-I (*Boschi et al., 1991*). The procedures prove to be most useful in the rapid determination of focal parameters immediately after the event, when only observations from a few worldwide open stations are available and automatic, quasi-real-time retrieval of data within a regional network allows to implement procedures for rapid moment tensor determination.

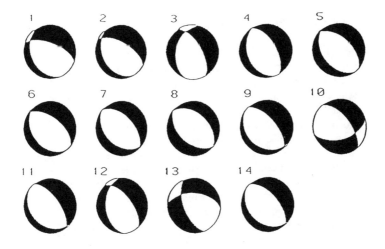

N	Fault strike	Plane dip	M_0 10^{19} N m	Reference Source	METHOD
1	290°	66°	-	Gruppo di lavoro Irpinia, 1981	Polarities
2	298°	64°	-	Gasparini et al., 1982	Polarities
3	328°	62°	-	Giardini, 1993	Polarities
4	328°	57°	3.0	Boschi et al., 1981	CMT
5	310°	54°	2.4	Giardini et al., 1984	CMT
6	305°	58°	2.4	Ekström et al, 1987	CMT
7	312°	53°	2.7	Westaway and Jackson, 1987	CMT
8	309°	45°	2.8	Kanamori and Given, 1982	Surface Waves
9	317°	63°	2.6	Nakanishi and Kanamori, 1984	Surface Waves
10	320°	60°	2.0	Deschamps and King, 1983	Surface Waves
11	317°	63°	2.6	Giardini, 1993	Surface Waves
12	322°	63°	6-10	Brustle and Müller, 1983	Love Waves
13	276°	54°	1.3	Sipkin, 1987	Body Waves
14	317°	59°	2.1	Westaway and Jackson, 1987	Body Waves

Figure 1. Summary of focal mechanisms of the November 23, 1980 Irpinia earthquake obtained from teleseismic observations. For each solution we list the fault plane, identified by independent evidence, the seismic moment, the publication reference and the data used or method of analysis. Solutions n. 1-3 were obtained by polarity data, n. 4-13 by inversion of long-period data and n. 14 by modelling body waves; n.9-10 and n. 14 also used polarities. Solution n. 2 is found in four sources in the literature. (Modified after *Giardini, 1993*).

2. The practice of seismology: magnitude

To unveil the detailed process taking place during an earthquake is a time consuming task requiring a wealth of data and sophisticated techniques, a procedure applied only to major earthquakes. The more general approach, followed for the majority of earthquakes, is to assign only one or few numbers to quantify the earthquake size and source properties.

Despite the obvious shortcoming that a single number cannot describe fault geometry, spectral scaling and source complexities, since its first introduction *(Richter, 1935)* the magnitude scale has remained the most generally used way of quantifying and cataloging earthquakes. Numerous magnitude scales have been proposed on empirical and theoretical grounds, in an effort to overcome the frequency dependence and the saturation thresholds of these measures, and much work has been devoted to comparing and equalizing different magnitude scales (Figure 2) to unify catalogues covering different time periods (*Geller and Kanamori, 1977; Abe and Kanamori, 1979; Chung and Bernreuter, 1981; Abe, 1981; Båth, 1981; Kanamori, 1983; Nuttli, 1985*). The relation of the magnitude measure to statical

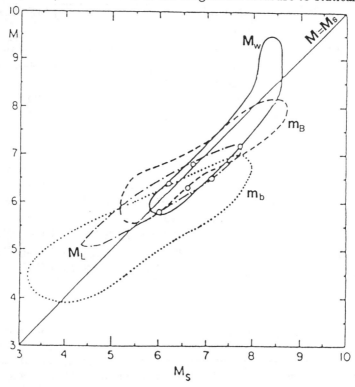

Figure 2. Regressions of magnitude scales m_B, m_b, M_W, M_D versus the magnitude M_S. (Modified after *Kanamori, 1983*).

and dynamical properties of the seismic source has been explored in detail (*Aki, 1967; Kanamori and Anderson, 1975; Geller, 1976; Vassiliou and Kanamori, 1982; Furumoto and Nakanishi, 1983*).

Magnitude scales fall roughly in the seven cathegories:

M_L (m_{bLg}): the local magnitude is measured at periods of 0.1-3 seconds on standard Wood-Anderson sensors for earthquakes up to 600 km distances.

M_S (M_{GR}, M_R, M_D, M_Z, M_V, M_{JMA}): measured from surface waves at periods of about 20 seconds, this scale became of common use in the identification of nuclear explosions with the deployment of the long-period WWSSN instruments, as a counterpart to the short-period m_b magnitude measured on short-period WWSSN sensors; it is defined only for shallow earthquakes and is routinely computed by the NEIC; it saturates at values around 8 and has been shown to display a characteristic bias related to the tectonic province of the epicenter.

m_B : the body wave magnitude (*Richter, 1935; Gutenberg, 1945*) measures various seismic phases (P, PP, S) at periods of 0.5-12 seconds and has been estimated for all major shallow and deep earthquakes since the beginning of the century.

m_b (m_{bLg}): the most common short-period, body-wave magnitude scale is m_b, measured from the largest pulse in the first five cycles of the P and P_n arrivals in the 1 second period band and routinely reported by the NEIC and ISC bulletins; due to its high-frequency character, this scale has a low saturation threshold of ~ 6.5.

M_W (M_M, M_w, M_E, M_t): the moment magnitude has been proposed to scale with the seismic moment (*Hanks and Kanamori, 1979*) and to provide a description of the source spectra at periods larger than 10 seconds, and thus to be immune from saturation.

M_C (M_d): the coda or duration magnitude measures the size of the earthquake from the duration of the signal recorded at each station; since the ground response depends strongly from the local crustal and lithospheric structure, this scale requires careful calibration and is of limited use for large earthquakes or during aftershocks sequences.

M_I (M_K): values of macroseismic intensity evaluated for historical earthquakes can be regressed to an intensity magnitude scale to construct uniform catalogues of seismicity spanning over the historical and instrumental datasets.

In recent years, the moment tensor representation of the seismic source has assumed a dominant role in seismology, owing to its capability of quantifying the size and geometry of earthquakes together with physical properties of the source. Because of its derivation from the long period portion of the seismic radiation, the seismic moment provides a frequency independent estimate of the earthquake size; it is also related in a simple way to source parameters such as fault area and average slip ($M_0 = \mu S \bar{u}$; *Aki, 1967*).

The deployment since 1977 of digital equipment and the consequent availability of high-quality digital data over a large dynamical range has made the moment tensor

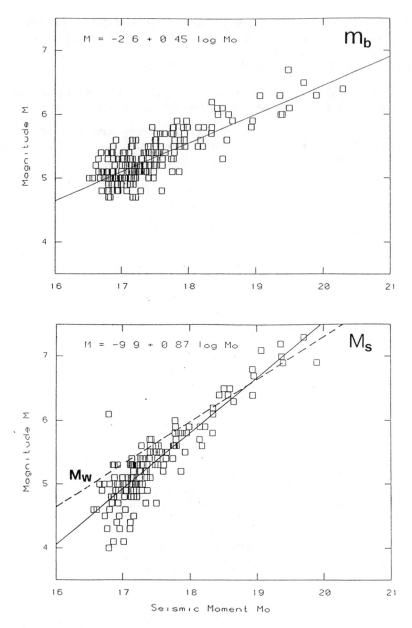

Figure 3. Regressions of magnitudes scales m_b and M_S (from the ISC Catalogue) versus the CMT seismic moments for earthquakes in the Mediterranean since 1977.

inversion a routine task; of the many methods proposed to retrieve the moment tensor from body waves, mantle waves and free oscillations, some are suitable and have been used for systematic processing of large sets of events (*Dziewonski et al., 1981; Sipkin, 1986*).

All magnitude scales are based on amplitude measures of a given phase or wave package on seismograms recorded on band-passed instruments. As such, a magnitude will suffer from several or all of the following limitations:
• the measure is nearly monochromatic and thus characterizes the source spectrum only at one frequency, relying upon the adoption of a standardized spectral model for the seismic source;
• the estimation procedure does not account for the radiation pattern of the seismic source but averages point values from a distribution of regional or global stations, which rarely can be defined as optimal;
• the calibration of the local crustal and lithospheric corrections at the station and of the instrument response should be pre-requisites for a meaningful magnitude estimation.

An estimate of the uncertainty associated with magnitude determination can be derived by looking at the spread of magnitude values reported by individual observatories, seismic networks and international organizations for significant earthquakes in the Mediterranean. For example, several estimates have been issued for the December 13, 1990, Eastern Sicily earthquake (the code in parenthesis identifies the observatory or organization reporting the parameter): seismic moments $M_0 = 4 \times 10^{24} dyne \cdot cm$ (PPT) and $M_0 = 3.3 \times 10^{24} dyne \cdot cm$ (CMT), and magnitudes $m_b = 5.5$ and $M_S = 5.3$ (NEIC), $M_D = 5.6$ (TRI), $M_D = 5.3$ (TTG), $M_D = 5.2$ (THE), $M_D = 5.1$, $M_L = 5.4$ and $M_L = 5.9$ (ING; following an initial estimate of $M_D = 4.9$); in similar way, magnitude estimates for the October 23, 1992, Morocco event ranged between 4.5 and 5.8 (as notified by EMSC) and a large earthquake in Kazhakistan (August 19, 1992) was given five magnitude values ranging between 6.0 and 7.5 (as reported by the automatic rapid warning system of the Swiss Seismic Service).

These examples indicate a wide scatter when different magnitude scales are used for the same event and even when the same type of magnitude is estimated by different organizations. If we regress the common magnitudes m_b and M_S (from the ISC catalog) against the CMT seismic moments for significant earthquakes of the Mediterranean (Figure 3), we observe scatters of $\pm 0.3 - 0.4$, which can be partly attributed to the presence of earthquakes with anomalous source spectra, but mostly to uncertainties in the estimate of magnitudes.

3. The practice of seismology: fault plane solutions

Fault plane solutions derived from polarities of first arrivals are a common tool of seismological practice, a pillar of the plate tectonics revolution (*Isacks and Molnar, 1971*). Their widespread use is tied to the physical simplicity of the source model and to the accessibility of polarity data from local networks and worldwide catalogues. High-quality, reliable fault plane solutions were obtained mostly by careful examination of long- and short-period WWSSN records, for the unambiguous identification of phase and amplitude; extensive collections of carefully compiled solutions have been assembled and constitute the backbone of global seismotectonic analyses (see *Udias et al., 1989* for the European-Mediterranean area).

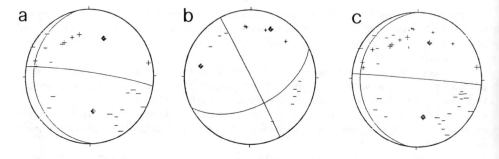

Figure 4. Fault plane solutions for the March 19, 1992, Campobasso earthquake in Central Italy, obtained using polarity data from (A) a local array of 14 stations, (B) 22 reporting stations of the Italian National Seismic Network, and (C) the combined set of 36 polarities. (*Amato, personal communication, 1992*).

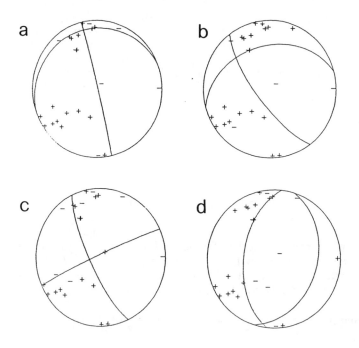

Figure 5. (A) Fault plane solution for the June 21, 1982, Tyrrhenian Sea earthquake. (B-D) Fault plane solutions obtained by inverting the sign of 16% of the polarities, chosen randomly on the focal sphere, to simulate the effects of common polarity errors.

However, it has become increasingly common to find published fault plane solutions derived using unchecked catalogue readings or produced in automatic fashion in the operation of local and national seismic networks worldwide and used in regional seismotectonic analyses (Figure 4). Although in these cases the careful screening of the data and the monitoring of the instrumentation may succeed in limiting the presence of reading errors, phase misidentifications and station reversals, the quality of fault plane solutions obtained only from polarities of the short-period P-arrivals is tied to practical and theoretical limitations of the method itself, imposing caution in the seismotectonic interpretation of the focal mechanisms (e.g. *Reasenberg and Oppenheimer, 1985*). Among the limitations are:
• the uneven density and distribution on the focal sphere of polarity readings,
• the presence of consistent percentages of erroneous readings and reversed polarities in the catalogues,
• the use of simplified structural models and ray theory algorithms in highly heterogeneous real media,
• the correspondence of the fault planes with areas on the focal sphere of minimum amplitude and least resolution of P-wave arrivals,
• the possible difference between the initial rupture and the overall fault geometry,
• the theoretical non-univocity of the fault plane solution.

The number of 'wrong' polarities expected in a focal mechanism, due to reading errors, phase misidentifications or station reversals, may be estimated by looking at events with good polarity coverages or, for a regional network, by looking at the consistency of teleseismic arrivals; in the Mediterranean, 15-25% of polarity errors are observed in fault plane mechanisms of large events with more than 100 unchecked readings. To evaluate the stability of fault plane solutions with respect to polarity errors, for test events we reverse the sign of part of the polarities (16%) chosen randomly on the focal sphere and then recompute the focal mechanisms; for events with a good coverage of polarities we observe sufficient stability of the solutions, with rotations within few degrees from the original fault geometry, while events with sparse or uneven distribution display marked variability, with focal mechanisms sometimes radically different from the original (Figure 5).

Although several ways have been proposed to make better use of polarity data, based for example on correlation methods or on the derivation of joint solutions for clusters of events (*Brillinger et al., 1980; Giardini and Velonà, 1991*), the need to constrain the source geometry for smaller events remains crucial for seismotectonic analyses in areas of low seismic activity, and one that is only beginning to be addressed by waveform inversion methods.

4. The MEDNET network

MEDNET is the very-broad band (VBB) seismographic network installed by the *Istituto Nazionale di Geofisica* of Roma in countries of the Mediterranean area, with a final goal of stations and a spacing of less than 1000 km between stations (*Boschi et al., 1991*). The project started in 1988 and will be completed within 1993, and is motivated both by research interest and by seismic hazard monitoring. To reach its goals, the network has been designed following the highest VBB technical standards: STS-1/VBB sensors, 24 bits A/D, 140 dB dynamic range, real-time telemetry.

A map of the areal distribution of the MEDNET observatories (Figure 6) illustrates how the network is now composed by 17 sites (11 MEDNET, 3 joint stations and 3 external observatories belonging to other programs).

The high standards of the VBB technology (*Wielandt and Steim, 1986*) and of the MEDNET installation procedures (*Giardini et al., 1992*) allow to record the Earth noise in the broad-band and low-frequency bands (*Mazza and Morelli, 1992*) and provide wave trains with excellent signal-to-noise ratios at regional distances for small earthquakes (Figure 7). The extended dynamic range (140 dB) of VBB instrumentation makes it possible to record on scale large local events (m=6 at less than 50km distance), while telemetering waveforms through dial-up or direct phone links allows to implement automatic procedures for the rapid evaluation of the seismic activity; several solutions presented here have been obtained within hours after the event using only telemetered records from few stations.

5. Long-period waveform modelling

The synthesis of accurate seismograms over a wide range of frequencies requires the calibration of the three-dimensional crustal and lithospheric models; these are not available but in very few spots around the world. We thus select a model-independent strategy to retrieve size and source geometry of moderate and large regional earthquakes, by modelling body and surface waves in a low-frequency band chosen as a compromise between noise performance and structural dependence of the seismic waves. In fact, the best performance of the VBB instrumentation in the long-period band is obtained in the 5-20 mHz frequency range (Figure 7); on the other hand, body- and surface-waves with frequencies below 10 mHz are insensitive to regional-scale structural heterogeneities, because of their characteristic wavelengths (e.g. Fig. 21 of *Kulhánek, 1990*) and it is thus possible to select a frequency band (5-10 mHz) where regional waves can be used to analyze moderate earthquakes using an average structural model.

Further advantages of long-period modelling are the simple parametrization of the seismic source, taken as a point source, and the availability of efficient codes for the generation of accurate synthetic seismograms, valid also in the near field.

Many techniques have been proposed for the inversion of the moment tensor from long-period digital data and some of them are suitable for routine application (e.g. *Dziewonski et al., 1981; Kanamori and Given, 1982; Romanowicz and Suarez, 1983; Sipkin, 1986*). We have developed two moment tensor inversion algorithms, correcting also for the centroid, for application to large global earthquakes and to regional events (*Giardini, 1992; Giardini et al., 1993*).

The algorithm for large global earthquakes is based on the following elements:
• we use 6 hours of three-components records, containing long-period body waves and the first 3 or 4 orbits of Rayleigh and Love waves; for a more rapid determination, we may limit the data length to include only the first surface waves packages (30-100 minutes depending on the epicentral distance);
• synthetic seismograms are generated by summation of normal modes at frequencies below 10 mHz, computed for model PREM of *Dziewonski and Anderson (1981)* using a scheme by *Woodhouse (1988)*; complete seismograms include fundamental modes and overtones to model body and surface waves; for a rapid analysis of shallow events, the summation can be restricted to fundamental modes for the synthesis of surface waves;
• inversion is performed in the frequency domain in a narrow frequency band (5-7

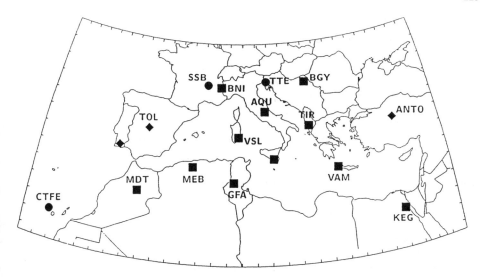

Figure 6. Geographic distribution of the MEDNET stations operating through 1990-1992 (squares represent MEDNET sites; circles are joint stations; diamonds are external installations belonging to other programs and contributing their data to MEDNET.

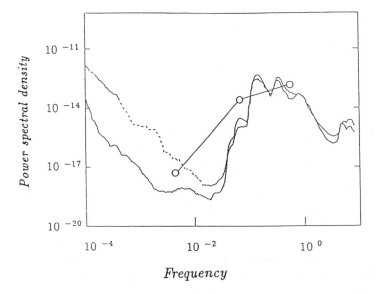

Figure 7. Experimental curves of minimum Earth noise averaged over the vertical and horizontal components of MEDNET stations (in power spectral density units; after *Mazza and Morelli, 1992*); superimposed is the acceleration expected from a $M_S = 3.6$ earthquake at 2.000 km distance.

mHz, corresponding to 140-200 seconds waves) ideal for large events;
- the moment tensor is constrained to have null volumetric component (following *Dziewonski et al., 1981*);
- a time correction is computed for each trace after the inversion for the moment tensor; an average time correction common to all stations incorporates the source half-duration and, since the network span is much smaller than the epicentral distance, also the centroid mislocation and a rough mantle correction; the use of a single phase or time correction is valid only for quasi-monochromatic waves and we select a sufficiently narrow frequency band to allow a frequency-independent term; the residual time correction account for small scale heterogeneity in the path and in the station region;
- depth is retrieved by variance minimization on different trials (*Romanowicz and Suarez, 1983*);
- we observe a larger instability associated with the $M_{r\theta}$ and $M_{r\phi}$ elements of the moment tensor for shallow earthquakes (*Giardini and Beranzoli, 1992*; also *Dziewonski et al., 1981* and *Ekström, 1989*) and we apply a weak damping scheme in the inversion to minimize the size of the two dip-slip components.

While maintaining several traits of the teleseismic algorithm, the regional algorithm has the following peculiar characteristics:
- we opt for inverting waveforms in the time domain, band-passed in the frequency band, 8-10 mHz, with the most favourable signal-to-noise ratio (see Figure 7); for noist records we choose data windows characterized by higher signal-to-noise ratio;
- we use three-component records, containing long-period body waves and the first orbit of Rayleigh and/or Love waves; we retrieve data and compute synthetics for a 20 minutes duration, sufficient to include the first surface-wave train for any regional event to a distance of 2000 km.

6. Large earthquakes of 1990

1990 was characterized by a series of large earthquakes distributed worldwide: March 3 (Fiji) and 25 (Costa Rica), May 12 (Sakhalin), 20 (Sudan) and 30 (Romania), June 20 (Iran), July 16 (Luzon) and December 30 (New Britain).

For all these events, the NEIC Monthly Bulletins report m_b values, fault plane solutions derived from polarities, CMTs and, for some of them, a long-period moment tensor solution by USGS; here we compute the moment tensor using the algorithm outlined above. Table I lists earthquake parameters used and obtained in this study: hypocentral locations from the CMTs (depths for the deep events are computed here), m_b, and seismic moments M_0 obtained in this study (in dyne·cm). Figure 8 displays the fault plane solutions on a world map; we use the best double-couple definition of *Dziewonski et al.* (*1981*). The events span the epicentral distance range between 10° and 160°; three events are deep: May 12 (Sakhalin, 590 km), May 30 (Romania, 85 km) and December 30 (New Britain, 200 km).

Figure 9 compares the fault plane solutions and seismic moments obtained here with the CMTs, the moment tensors by USGS and the fault plane solutions by NEIC. Our solutions are very stable and compare well with the CMTs, both in geometry and in size, whereas more pronounced discrepancies exist between the MEDNET and CMTs solutions and the moment tensors by USGS; of the five events analyzed with this method by USGS and reported by NEIC, the seismic moment of the March 3 event is much smaller and a different geometry is shown for the

December 30 event. Discrepancies in source geometry are observed also with respect to the NEIC fault plane solutions for the March 25 and December 30 events.

Figure 10 displays examples of waveform modelling: station VSL for the May 12, Sakhalin event and station MDT for the June 20, Iran event. We obtain a good fit between data and synthetics for the dominant, sharp surface-waves packages as well as for the body-wave trains. We note that, although only a restricted frequency window

TABLE I

date	region	lat	lon	d	m_b	M_0
3/3	Fiji	-22.04	175.16	25	6.3	3.2×10^{27}
3/25	Costa Rica	9.89	-84.89	18	6.2	7.8×10^{26}
5/12	Sakhalin	49.04	141.88	590	6.5	7.0×10^{26}
5/20	Sudan	5.32	32.29	15	6.7	5.0×10^{26}
5/30	Romania	45.87	26.67	85	6.7	2.7×10^{26}
6/20	Iran	36.96	49.41	15	6.4	1.1×10^{27}
7/17	Luzon	15.66	121.23	15	6.5	4.5×10^{27}
12/30	N. Britain	-5.09	150.98	200	6.7	1.9×10^{27}

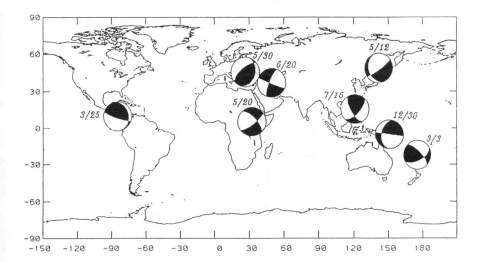

Figure 8. World map with the moment tensor solutions for the largest earthquakes of 1990, listed in Table I. For each event we show the best double-couple representation (defined following *Dziewonski et al., 1981*). (After *Giardini, 1992*).

is used in the inversion (5-7 mHz), the correspondence between data and synthetics extends to much lower frequencies for these large events (1.5-2 mHz); horizontal components are generally more noisy than vertical ones.

Slight phase discrepancies can be sometimes observed for later-orbits surface-waves trains; this small effect, induced by the mantle heterogeneity, is only partially corrected by our use of a single time term for each trace and could be further reduced by introducing a three-dimensional mantle model in the synthesis of the kernels (*Woodhouse and Dziewonski, 1984*).

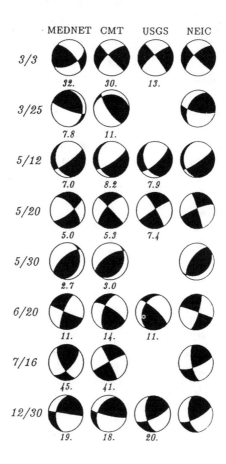

Figure 9. Comparison of fault plane solutions and moment tensors obtained with different methods for the 1990 earthquakes: the MEDNET solutions, the CMTs, the moment tensors by USGS and the focal mechanisms by NEIC. Seismic moments are in units of 10^{26} dyne·cm. (After *Giardini, 1992*).

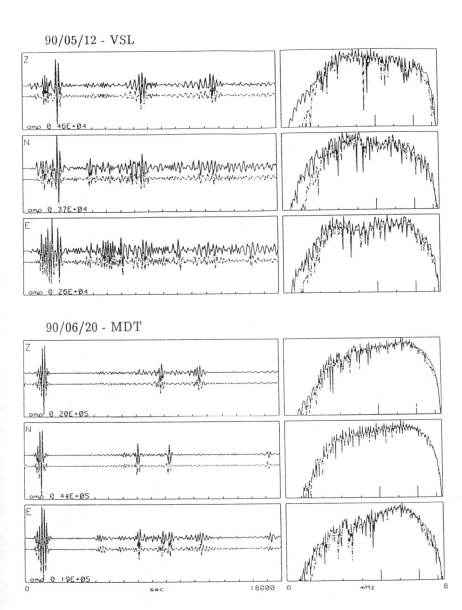

Figure 10. Waveform modelling for station VSL for the deep May 12, 1990 Sakhalin event and station MDT for the June 20, 1990, Iran event. On the right are the frequency spectra obtained from six hours of long-period seismograms, tapered at high and low frequencies; on the left are the corresponding time series. Data are indicated by continuos lines, synthetic seismograms by dashed lines. The frequency band used in the inversion is marked (5-7 mHz). (Modified after *Giardini, 1992*).

TABLE II

N	date	region	lat	lon	d	m_b	M_0
1	2/09/90	Algeria	36.75	02.15	12	5.0	1.1×10^{24}
2	5/05/90	S. Italy	40.75	15.85	26	5.3	7.3×10^{24}
3	5/30/90	Romania	45.87	26.67	90	6.7	4.0×10^{26}
4	6/16/90	Albania	39.21	20.54	34	5.5	2.7×10^{24}
5	6/20/90	Iran	36.96	49.41	10	7.7	1.3×10^{27}
6	11/11/90	Med. Sea	33.94	12.04	10	4.7	8.8×10^{23}
7	11/27/90	Yugoslavia	43.87	16.63	10	5.2	2.6×10^{24}
8	12/13/90	Sicily	37.20	15.50	10	5.4	3.7×10^{24}
9	12/21/90	Greece	40.98	22.34	18	5.8	1.7×10^{25}
10	3/19/91	Crete	34.82	26.28	18	5.4	2.0×10^{24}
11	4/10/91	Turkey	37.21	36.01	33	5.1	1.6×10^{24}
12	4/29/91	Caucasus	42.49	43.65	10	6.2	3.7×10^{26}
13	7/12/91	Romania	45.38	21.05	10	5.0	3.2×10^{24}
14	11/21/91	N.Atl.Ridge	48.76	-28.10	10	5.2	1.5×10^{24}
15	12/09/91	Azores	37.21	-24.34	10	5.2	7.3×10^{23}
16	3/13/92	Turkey	40.01	40.01	10	6.7	1.3×10^{26}
17	4/13/92	Holland	51.30	6.30	10	5.4	1.5×10^{24}
18	10/12/92	Egypt	29.90	31.00	10	5.9	9.1×10^{24}
19	10/23/92	Caucasus	42.22	45.12	12	6.7	2.8×10^{25}
20	10/23/92	Morocco	31.30	-4.40	10	5.3	2.9×10^{24}

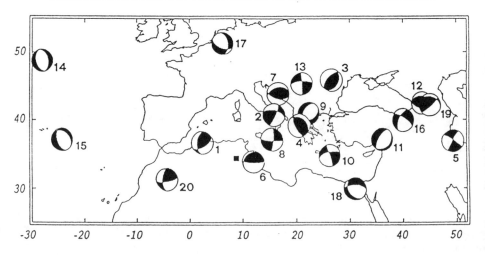

Figure 11. Moment tensor solutions of the Mediterranean earthquakes, listed in Table II. (After *Giardini et al., 1993*).

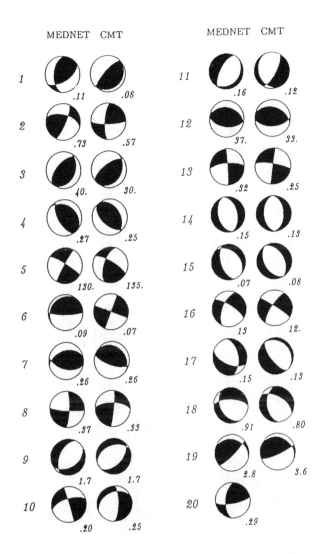

Figure 12. Comparison of the source geometries obtained here and with the CMT method for the events listed in Table II; for all solutions we list the seismic moment in units of $10^{25} dyne \cdot cm$. CMTs for events 18-19 are preliminary and unpublished (*Ekström, personal comunication, 1992*). (After *Giardini et al., 1993*).

7. Earthquakes of the Mediterranean 1990-1992

In 1990-1992 the Mediterranean was hit by disastrous earthquakes on June 20, 1990 in Iran (55.000 casualties), March 13, 1992 in Turkey (1.500 casualties) and October 12, 1992 in Egypt (550 casualties).

We analyze 20 events covering the area between Iran and the Azores; we compute the moment tensor using the algorithm outlined above and the hypocentral location and origin time broadcasted by ING and NEIC, and we compare source geometry and seismic moment with those derived by the CMT method. Table II lists significant earthquake parameters: hypocentral locations and magnitudes m_b from NEIC and ING and seismic moments M_0 obtained here (in $dyne \cdot cm$).

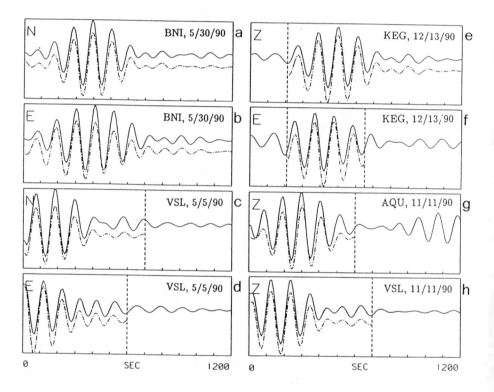

Figure 13. Examples of waveform modelling for: (a,b) the large, intermediate depth Romanian earthquake (May 30, 1990), (c,d) a close station for the May 5, 1990, Potenza event, (e,f) a distant station for the December 13, 1990, Siracusa event, and (g,h) two vertical traces for the small November 11, 1990, Southern Mediterranean earthquake. For each component we show 20 minutes of long-period seismograms, band-passed in the 8-10 mHz frequency bands (data are indicated by continuos lines, synthetic seismograms by dashed lines, displayed only within the time window selected for the inversion). (After *Giardini et al., 1993*).

Figure 11 displays the fault plane solutions on a regional map (we use the best double-couple definition of *Dziewonski et al., 1981*). In Figure 12 we compare source geometries and seismic moments obtained here and with the CMT method; the comparison is positive both in geometry and in size. Significant geometry differences are observed only for three events of small dimensions (n.1,6,11), for which only a few vertical records were available for our analysis, since the Earth is consistently noisier on horizontal components. We note, however, that for these three small events the CMT geometry produces significantly higher misfit and is thus incompatible with the MEDNET data. We do not find any event with stable, high deviation from the double couple mechanism. Figure 13 displays examples of waveform modelling.

8. Some comments on accuracy and reliability.

Having illustrated the general applications of the proposed methodologies to global and regional earthquakes, we present here the results of specific experiments designed to test the accuracy and reliability of the moment tensor solutions. Indeed, the issue of providing formal errors in moment tensor inversions has long been discussed (see *Dziewonski et al., 1981*), since standard errors are known to underestimate true uncertainties.

A first question we would like to answer is the feasibility of obtaining moment tensor solutions from single station records, to activate real-time procedures of seismic monitoring using a single VBB observatory. We show a test conducted for the December 13, 1990, Eastern Sicily, earthquake. In the immediate aftermath of the event the only available size estimates were a preliminary $M_D = 4.9$ (ING) and a reported maximum intensity I=VII-VIII; since the rapid determination procedures of EMSC and NEIC were not triggered, further magnitude estimates were provided only much later, with values ranging between 5.1 and 5.9. In addition, consistent estimates of the focal geometry were derived only later using arrivals from teleseismic stations (*Amato et al., 1991; De Rubeis et al., 1991*) and by waveform modelling (CMT). Figure 14 shows the focal mechanisms obtained by inverting the joint MEDNET data and in single station inversions; for comparison also the CMT mechanism and two fault plane solutions (S1, *Amato et al., 1991*; S2, *De Rubeis et al., 1991*) are also displayed. The agreement obtained by inversion of single station data is very encouraging, indicating that even a single VBB station permits reliable control on the seismic source of regional earthquakes (as shown for teleseismic events in *Ekström et al., 1986*). The seismic moment $M_0 = 3.7 \times 10^{24} dyne \cdot cm$ corresponds to magnitude values of $m_b = 5.5$, $M_S = 5.7$, $M_W = M_L = 5.8$ (using regression laws from *Heaton et al., 1986*).

A partial estimate of the uncertainty on the seismic moment for a well constrained moment tensor solution (the November 23, 1980, Irpinia earthquake; *Giardini, 1993*) may be estimated from a misfit curve (Figura 15), showing as error of about $\pm 0.2 dyne \cdot cm$, corresponding approximately to a ± 0.1 uncertainty in m_b, while we have previously shown (Figures 2,3) that uncertainty in magnitude scales can be estimated in $\pm 0.3 - 0.4$.

In addition, we may estimate the uncertainty associated with the single independent elements of the moment tensor; as it is common for shallow events, we observe a larger instability associated with the the $M_{r\theta}$ and $M_{r\phi}$ components of the moment tensor, which we reduce applying a weak minimization scheme in the in-

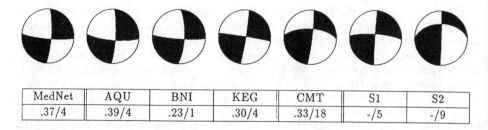

MedNet	AQU	BNI	KEG	CMT	S1	S2
.37/4	.39/4	.23/1	.30/4	.33/18	-/5	-/9

Figure 14. Focal mechanisms obtained by inverting the joint MEDNET data and in single station inversions (AQU, BNI, KEG) for the December 13, 1990, Eastern Sicily earthquake; for comparison also the CMT mechanism and two fault plane solutions (S1, *Amato et al., 1991*; S2, *De Rubeis et al., 1991*) are also displayed. For all the solutions we list the seismic moment (in units of $10^{25} dyne \cdot cm$) and the variance to the whole MEDNET dataset, defined as the ratio between the misfit and the data norm and expressed in percentiles. (Modified after *Giardini, 1991*).

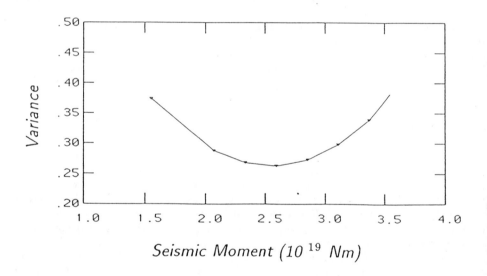

Figure 15. Normalized variance obtained by different moment tensor solutions proportional to the preferred solution for the November 23, 1980, Irpinia earthquake (solution n. 11 in Figure 1). The seismic moment is estimated to be $M_0 = 2.6 \pm 0.2 \times 10^{19} Nm$. (After *Giardini, 1993*).

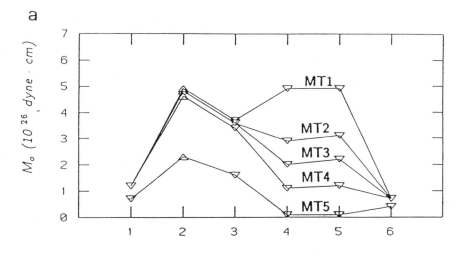

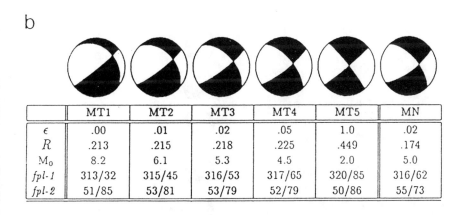

Figure 16. (A) Absolute amplitude of the six elements of the moment tensor for five inversions performed with different minimization contraints (MT1 to MT5); (B) fault mechanisms and inversion parameters for the five solutions in (A) and for a solution obtained including only MEDNET data (MN); we list the constraint ϵ, the seismic moment M_0 (in $dyne \cdot cm$ units), the normalized variance R and the fault plane geometry (strike and slip angles). (After *Giardini and Beranzoli, 1992*).

version, as shown by a test for the May 20, 1990, Sudan earthquakes (Figure 16). A non-constrained solution (MT1) has a large seismic moment of $8.2 \times 10^{26} dyne \cdot cm$, dominated by the $M_{r\theta}$ and $M_{r\phi}$ components, and reaches a good variance of R=.213. By increasing the constraint, the $M_{r\theta}$ and $M_{r\phi}$ components decrease significantly, and so does the seismic moment, while the other moment elements remain unchanged and the variance increases only slightly. It is only with a very strong contraint ($\epsilon=1.0$) that the solution (MT5) degrades considerably (R=.449), and all moment tensor elements are reduced ($M_0 = 2. \times 10^{26} dyne \cdot cm$). The geometry of the focal mechanism reflects the proportions of the moment tensor elements (Figure 16b); unconstrained solutions are dominated by the dip-slip elements $M_{r\theta}$ and $M_{r\phi}$, whereas more constrained tensors show a strike-slip geometry. The final solution MT3 is chosen on the basis of the resolution of the inversion procedure, and shows a superposition of dip-slip and strike-slip components.

The determination of the hypocentral depth is not among our goals, since we do not have sufficient data for this task; we would rather prefer the method to have limited

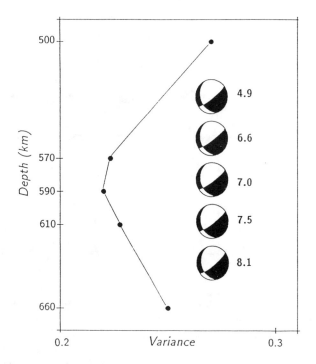

Figure 17. Variance curve obtained in the moment tensor inversion for the deep May 12, Sakhalin earthquake, perfomed at different trial depths. We display also the moment tensor solutions and the seismic moments (in units of $10^{26} dyne \cdot cm$). On the horizontal scale is the normalized variance. (After *Giardini, 1992*).

depth resolution and dependence, to ensure good stability of the results, even with wrong initial depth. To this purpose we show in Figure 17 the moment tensor and the variance obtained in inversion tests at different trial depths for the deep May 12, Sakhalin earthquake. The results are very encouraging; the variance curve constrains the depth within ±20 km, while the fault geometry remains stable over an extended depth range and the seismic moment varies by 5%.

9. Conclusions

The deployment of an advanced digital network at regional scale allows to set up reliable procedures for the rapid determination of the moment tensor for regional and global earthquakes, thus improving the seismological practice based on magnitudes and fault plane solutions.

To test the capability of an advanced digital network to provide rapid control of the regional and global seismicity, we invert waveforms recorded by MEDNET to analyze the largest earthquakes of 1990 and 20 significant earthquakes of the Mediterranean area of the 1990-1992 period, with seismic moment ranging between 7.3×10^{23} and $4.5 \times 10^{27} dyne \cdot cm$.

The estimate of earthquake size and source geometry obtained using regional waveforms are consistent with those retrieved using global data and/or local networks, indicating that an advanced regional seismic network of modern VBB standards like MEDNET allows accurate and rapid monitoring of regional and global seismicity.

ACKNOWLEDGMENTS. We are indebted to prof. A. Zichichi for his leading role in the development of MEDNET. D.G. is grateful to prof. J. H. Woodhouse for initiating him to the art of handling the Earth's normal modes and to profs. A. M. Dziewonski and J. H. Woodhouse for the experience gained on the CMT method while at Harvard University. The MEDNET Data Center is credited for collecting and promptly making available to the seismological community digital data of excellent quality; G. Ekström of Harvard University provided preliminary, unpublished CMTs; D. Riposati and G. Calcara helped in drafting the figures. We thank E. Mantovani and A. Morelli for inviting our contribution to Erice and to this volume.

References

Abe, K. (1981) *Magnitude of large shallow earthquakes from 1904 to 1980*, Phys. Earth Planet. Int. 27, 72-92.
Abe, K. and Kanamori, H. (1979) *Temporal variation of the activity of intermediate and deep focus earthquakes*, J. Geophys. Res. 84, 3589-3595.
Aki, K. (1967) *Scaling law of seismic spectrum*, J. Geophys. Res. 72, 1217-1230.
Amato A., Azzara, R., Basili, A., Chiarabba, C., Cocco, M., Di Bona, M. and Selvaggi, G. (1991) *La sequenza sismica del Dicembre 1990 nella Sicilia Orientale: analisi dei dati sismometrici*, in E. Boschi and A. Basili (eds.), Contributi allo studio del terremoto della Sicilia Orientale del 13 Dicembre 1990, Publication of the Istituto Nazionale di Geofisica 537, 57-84.
Båth, M. (1981) *Earthquake magnitude - recent research and current trends*, Earth-Sci Rev. 17, 315-398.
Berry, M. (1988) *The Federation of Digital Broadband Seismographic Networks*, in G. M. Purdy and A. M. Dziewonski (eds.), Proceedings of a Workshop on Broad-Band Downhole Seismometers in the Deep Ocean, Woods Hole Oceanographic Institution, 85-92.
Boschi, E., Giardini, D. and Morelli, A. (1991) MedNet: the Broad-Band seismic network for the Mediterranean, Il Cigno Galileo Galilei, Roma.
Boschi, E., Mulargia, F., Mantovani, E., Bonafede, M., Dziewonski, A. M. and Woodhouse, J. H. (1981) *The Irpinia earthquake of November 23, 1980*, EOS, Trans. AGU 62, 330.
Brillinger, D., Udias, A. and Bolt, B. A. (1980) *A probability model for regional focal mechanism solution*, Bull. Seism. Soc. Am. 70, 149-170.
Brustle, W. and Müller, G. (1983) *Moment and duration of shallow earthquakes from Love-wave modeling for regional distances*, Phys. Earth Planet. Int. 32, 312-324.
Chung, D. H. and Bernreuter, D. L. (1981) *Regional relationships among earthquake magnitude scales*, Rev. Geophys. Space Phys. 19, 649-663.
De Rubeis, V., Gasparini, C., Maramai, A. and Anzidei, M. (1991) Il terremoto Siciliano del 23 Dicembre 1990, in E. Boschi and A. Basili (eds.), Contributi allo studio del terremoto della Sicilia Orientale del 13 Dicembre 1990, Publication of the Istituto Nazionale di Geofisica 537, 9-44.
Deschamps, A. and King, G. C. P. (1983) *The Campania-Lucania (Southern Italy) earthquake of November 23, 1980*, Earth planet. Sci. Lett. 62, 296-304.
Dreger, D.S. and Helmberger, D. (1990) *Broadband modeling of local earthquakes*, Bull. Seism. Soc. Am. 80, 1162-1179.
Dziewonski, A. M.and Anderson, D. L. (1981) *Preliminary Reference Earth Model (PREM)*, Phys. Earth Planet. Inter. 25, 297-356.
Dziewonski, A. M., Chou, T.-A. and Woodhouse, J. H. (1981) *Determination of earthquake source parameters from waveform data for studies of global and regional seismicity*, J. Geophys. Res. 86, 2825-2852.
Ekström, G. (1989) *A very broad band inversion method for the recovery of earthquake source parameters*, Tectonophys. 166, 73-100.
Ekström, G., Dziewonski, A. M. and Steim, J. M. (1986) *Single station CMT: application to the Michoacan, Mexico, earthquake of september 19, 1985*, Geophys. Res. Lett. 13, 173-176.

Ekström, G., Dziewonski, A. M. and Woodhouse, J. H. (1987) *Centroid-moment tensor solutions for the 51 IASPEI selected earthquakes, 1980-1984*, Phys. Earth Planet. Int. 47, 62-66.
Furumoto, M. and Nakanishi, I. (1983) *Source times and scaling relations of large earthquakes*, J. Geophys. Res. 88, 2191-2198.
Gasparini, C., Iannaccone, G., Scandone, P. and Scarpa, R. (1982) *Seismotectonics of the Calabrian arc*, Tectonophysics 84, 267-286.
Geller, R. J. (1976) *Scaling relations for earthquake source parameters*, Bull. Seismol. Soc. Amer. 66, 1501-1523.
Geller, R. J. and Kanamori, H. (1977) *Magnitudes of great shallow earthquakes from 1904 to 1952*, Bull. Seismol. Soc. Amer. 67, 587-598.
Giardini, D. (1992) *Moment tensor inversion from MedNet data (1) large worldwide earthquakes of 1990*, Geoph. Res. Lett. 19, 713-716.
Giardini, D. (1993) *Teleseismic observation of the November 23, 1980, Irpinia earthquake*, Annali di Geofisica, in press.
Giardini, D. and Velonà, M. (1991) *The deep seismicity of the Tyrrhenian Sea*, Terra Nova 3, 57-64.
Giardini, D., and Beranzoli, L. (1992) *Waveform modelling of the May 20, 1990 Sudan earthquake*, Tectonophys. 209, 105-114.
Giardini, D., Dziewonski, A. M., Woodhouse, J. H. and Boschi, E. (1984) *Systematic analysis of the seismicity of the Mediterranean region using the Centroid-Moment Tensor method*, Bollettino di Geofisica Teorica e Applicata 25, 134-151.
Giardini, D., Boschi, E., Mazza, S., Morelli, A., Ben Sari, D., Najid, D., Benhallou, H., Bezzeghoud, M., Trabelsi, H., Hafaied, M., Kebeasy, R. and Ibrahim, E. (1992) *Very-Broad-Band seismology in Northern Africa under the MedNet project*, Tectonophys. 209, 17-30.
Giardini, D., Boschi, E. and Palombo, B. (1993) *Moment tensor inversion from MedNet data (2) regional earthquakes of the Mediterranean*, Geoph. Res. Lett. 20, 273-276.
Gruppo di lavoro sismometria terremoto del 23.11.1980 (1981) *Il terremoto campano del 23.11.1980: elaborazione dei dati sismometrici*, Rend. Soc. Geol. It. 4, 427-450.
Gutenberg, B. (1945) *Magnitude determination for deep-focus earthquakes*, Bull. seism. Soc. Am. 35, 117-130.
Hanks, T. C. and Kanamori, H. (1979) *A moment magnitude scale*, J. Geophys. Res. 84, 2348-2350.
Heaton, T., Tajima, F. and Mori A. W. (1986) *Estimating ground motion using recorded accelerograms*, Surveys in Geophysics 8, 25-83.
Isacks B., and Molnar P. (1971) *Distribution of stresses in the descending lithosphere from a global survey of focal-mechanism solutions of mantle earthquakes*, Rev. Geophys. Space Phys. 9, 103-174.
Kanamori, H. (1983) *Magnitude scale and quantification of earthquakes*, Tectonophys. 93, 185-199.
Kanamori, H. and Anderson, D. L. (1975) *Theoretical basis of some empirical relations in seismology*, Bull. seism. Soc. Am. 65, 1073-1095.
Kanamori, H. and Given, J. W. (1982) *Use of long-period surface waves for rapid determination of earthquake source parameters, 2. Preliminary determination of source mechanisms of large earthquakes ($M_s \geq 6.5$) in 1980*, Phys. Earth Planet. Int. 30, 260-268.

Karabulut, H., Boyd, T. M. and Sipkin, S. (1991) *Determination of earthquake source parameters from regional waveforms*, Eos Trans. AGU 72, 304.

Kulhánek, O. (1990) Anatomy of seismograms, Elsevier, Amsterdam.

Mazza, S. and Morelli, A. (1992) *Background seismic noise from MedNet very-broad band stations*, Proceedings and Activity report 1988-1990, XXII General Assembly of the European Seismological Commission, 1, 197-202.

Nakanishi, I. and Kanamori, H. (1984) *Source mechanisms of twenty-six large shallow earthquakes ($M_s \geq 6.5$) during 1980 from P-wave first motion and long-period Rayleigh wave data*, Bull. Seism. Soc. Am. 74, 805-818.

Nakanishi, I., Moriya, T. and Endo, M. (1992) *The November 13, 1990 earthquake off the coast of the Primorskij region, the Eastern Russia*, Geophys. Res. Lett. 19, 549-552.

Nuttli, O. (1985) *Average seismic source-parameter relations for plate-margin earthquakes*, Tectonophys. 118, 161-174.

Reasenberg P. A. and D. Oppenheimer (1985) *Fortran computer programs for calculating and displaying earthquake fault-plane solutions*, Open File Report 85-739, USGS.

Richter, C. F. (1935) *An instrumental earthquake magnitude scale*, Bull. seism. Soc. Am. 2, 1-32.

Ritsema, J., Kuge, J. K. and Lay, T. (1991) *Stability of time domain waveform inversion for regional earthquake source parameters*, Eos Trans. AGU 72, 304.

Romanowicz, B. and Suarez, G. (1983) *On an improved method to obtain the moment tensor and depth of earthquakes from the amplitude spectrum of Rayleigh waves*, Bull. Seismol. Soc. Am. 73, 1513-1526.

Sipkin, S. (1986) *Estimation of earthquake source parameters by the inversion of waveform data: global seismicity, 1981-1983*, Bull. Seismol. Soc. Am. 76, 1515-1541.

Sipkin, S. A. (1987) *Moment tensor solutions estimated using optimal filter theory for 51 selected earthquakes, 1980-1984*, Phys. Earth Planet. Int. 47, 67-79.

Udias, A., Buforn, E. and Ruiz de Gauna, J. (1989) Catalogue of focal mechanisms of European earthquakes, Faster, Madrid.

Vassiliou, M. S. and Kanamori, H. (1982) *The energy release in earthquakes*, Bull. seism. Soc. Am. 72, 371-387.

Wielandt, E. and Steim, J. M. (1986) *A digital very-broad-band seismograph*, Annales Geophysicae 4B3, 227-232.

Westaway, R. and Jackson, J. (1987) *The earthquake of 1980 November 23 in Campania-Basilicata (Southern Italy)*, Geophys. J. R. astr. Soc. 90, 375-443.

Woodhouse, J. H. (1988) *The calculation of eigenfrequencies and eigenfunctions of the free oscillations of the Earth and the Sun*, in D. J. Doornbos (ed.), Seismological Algorithms, Academic press, 321-370.

Woodhouse, J. H. and Dziewonski, A. D. (1984) *Mapping the upper mantle: three dimensional modeling of Earth structure by inversion of seismic waveforms*, J. Geophys. Res. 89, 5953-5986.

TECTONIC AND SEISMIC PROCESSES OF VARIOUS SPACE AND TIME SCALES IN THE GREEK AREA

G. A. PAPADOPOULOS
Department of Seismotectonics
Earthquake Planning and Protection Organization
226 Messogion Ave.
15561 Holargos - Athens
Greece

ABSTRACT. The Tertiary and active, large - scale, complex seismotectonic processes of the Greek area can be adequately interpreted by lithospheric rotations and subductions. Anomalies in the space, time and size distributions of earthquakes reflect seismotectonic complexity even in smaller scales. A new interpretation of the tomographic images is suggested to dissolve their strong contradiction with other geophysical observations. Two 2-D interrelated lithospheric models are proposed to integrate many types of observations into a unified picture of the active deformation of the Greek area. There is a general need of such multi-disciplinary approaches for understanding better geodynamic phenomena taking place in complex areas such as the Greek one.

1. INTRODUCTION

Greece is characterized by the highest seismicity in Western Eurasia. Complexity, which is the main feature of the seismotectonic processes taking place there, increases the earth scientists' interest for this region. It is , therefore, considered as a truly vast seismological laboratory.
    The review of seismic and tectonic properties of the Greek region is not an easy task because of the very long number of papers , books, and reports that have been written and the many different views that have been expressed. A review and a bibliography of the Greek seismology are given by Båth [1] for the period from 1950 up to 1982. In this paper I review the main results obtained and models developed more recently by the author and his collaborators. Some new ideas are also presented about the Greek geodynamics. Moreover, these results are discussed in connection with the most important studies that have been presented by several authors about the Tertiary tectonic evolution and active geodynamic processes.

2. LARGE - SCALE GEODYNAMIC AND TECTONIC PROCESSES

The Cenozoic geotectonic evolution of the broad Aegean area has been described from magmatic data as a discontinuous process of southward migration of an arc - trench system and associated successive subduction phases (see [2], [3], [4]). From magmatic and seismological data, Papadopoulos [4] suggested that the gradual variation in the subduction trend from the Early Miocene up to the present indicates either the counter - clockwise rotation of about 50° of the Hellenic consuming boundary or the clockwise rotation of the Aegean lithosphere or both. This interpretation is compatible with the clockwise rotation in the Aegean area postulated by paleo - magnetic measurements (e.g. [5]) , the estimate, from fault plane mechanisms, of the relative motion occurring between the Hellenic arc and the adjacent sea floor ([6]), and the idea that at least one lithospheric slab has been subducted in the Aegean area as deduced from seismotectonic , geophysical and metallogenic observations (e.g. [7], [8], [9]).

The picture of lithospheric rotations and subductions in Greece and adjacent regions is completed by the suggested thermal or gravitational spreading of the Aegean ([6], [10]) and the southwestward advance of Turkey (e.g. [11]). The large - scale Tertiary and active tectonics appears, therefore, rather complicated.

Two 2 - D interrelated lithospheric models have developed to interprete the main seismotectonic and geophysical features of the Greek and adjacent regions. The first is a sketch - map showing the tectonic deformation (Fig. 1) and the second is a cross - section (Fig. 2) which indicates the mode of shallow and deep ruptures as well as lithospheric and asthenospheric processes which may drive the properties of the area considered. An attempt has been made to integrate all the previously mentioned large - scale tectonic processes. In addition, a compilation of more than one hundred fault plane solutions ([12], [13]) have been utilized.

A stress gradient and a zonal pattern of different tectonism away from the trench axis has been postulated. Almost pure thrust faulting occurs along the Helleinic trench while the back - arc side is mainly dominated by almost N - S extension. A narrow strike - slip belt with predominantly thrust component separates these two main zones ([13], [14]). This belt seems to be a transition from the thrust - type faults to normal faults as has been recognized in other active subduction zones too ([15]). However, two other belts of strike - slip motion complicates the back arc tectonics. In the North Aegean Sea the right - lateral strike - slip motion may reflect the southwestward continuation of the North Anatolian Fault Zone.

In the South Aegean, particularly in its eastern side in the region of Amorgos, a first order stress - pattern of NW - SE to NNW - SSE extension along with second order right - lateral strike - slip movements of NE - SW direction imply that the tectonic picture of the South Aegean deviates from the simple description for pure extension in back - arc conditions ([16]).

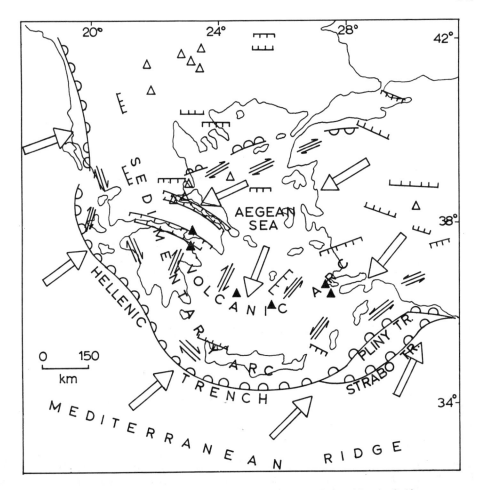

Figure 1. Sketch map of the present tectonic deformation of the Aegean area. Long curved lines = normal faults; lines with open semicircles = thrusts; double thin arrows = strike - slip motions; lonh heavy arrows = direction of relative motion of the Aegean and Mediterranean lithosphere; solid triangles = main volcanic centers; open triangles = Plio - Quaternary volcanic centers.

Convection currents in the asthenosphere of the South and North Aegean play a key role in explaining many geophysical properties. The South Aegean convection currents, which are triggered by the active lithospheric subduction in this area , may account for the active volcanism and the associated geothermal fields. In the North Aegean the existence of asthenospheric convection currents, that are presently dying - off , is also suggested. They can be explained by the paleo - subduction(s) that occurred during Tertiary times [4] and may account for the extensive Tertiary magmatism in the North Aegean area

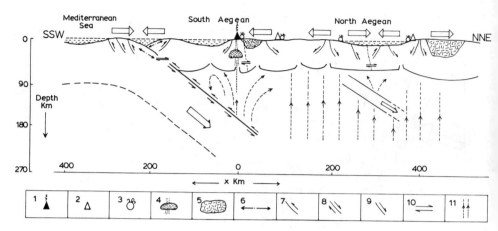

Figure 2. Lithospheric model of the Greek and surrounding regions. 1 = active volcanoes; 2 = Plio - Quaternary volcanoes; 3 = thermal springs; 4 = magmatic reservoir, 5 = plutonic intrusions; 6 = tension; 7 and 8 = reverse faults; 9 = normal faults; 10 = strike-slip motion; 11 = convection currents.

as well as for the high heat flow rate and magnetic anomalies in the North Aegean Trough.

3. CRITICAL ASPECTS OF THE TOMOGRAPHIC IMAGES

In a series of papers, Spakman and his collaborators (e.g. [17]) presented tomographic images of the Hellenic subduction zone that were interpreted as demonstrating slab penetration in the Aegean upper mantle to depths of at least 600 km, with a slab length of at least 800 km. Recently he suggested that subduction below the Aegean may occur to depths of even 800 - 1000 km, rather than 600 km (see this volume). This would extent the minimum duration time of Aegean subduction to 40 - 50 Ma.

Allowing for spatial error of 50 - 120 km, the above interpretation of tomographic images constitutes a revolution in the Greek subductology and geodynamics, thus rising strong contradiction with many geophysical and seismological observations. However, an alternative interpretation compatible with these observations could be given. The basic idea is that the tomographic images detect not only the actively subducting lithospheric slab but also remnants of slabs that were subducted during Tertiary times and that presently are dying - off in the upper mantle in depths considerably larger than the maximum depth of the active lithospheric subduction. Geometric and kinematic features of such paleo-subductions can be found in [4]. In a forthcoming paper I explain in detail the alternative interpretation of the tomographic images in the Hellenic subduction zone.

4. SPACE, TIME AND MAGNITUDE DISTRIBUTIONS OF EARTHQUAKES

Complexity in geodynamics and tectonics is associated with complexity in seismic activity which turns in anomalies in the earthquake space, time and size distributions. Figure 3 shows the spatial distribution of the relative seismic hazard in Greece and adjacent regions from a maximum likelihood estimation [18]. Only shallow seismicity is represented in this map. The seismic hazard level, considered as a function of the maximum expected earthquake magnitude and the recurrence of earthquakes of magnitude $M > 6.0$, varies from low to very high while there are large patches practically aseismic. Intermediate - depth earthquakes occur only in the South Aegean, the largest of them taking place in three separate nests situated in the western, central and eastern segments of the descending lithospheric slab [19].

Another approach of the spatial distribution of large size shocks is based on configurates of the lateral extent of their rupture zones. This concept was introduced by Sykes [20] and Kelleher [21] who concluded that rupture zones of large shallow shocks in convergent plate margins generally abut and do not overlap. This seems to be verified in the Hellenic plate margin ([22], [23]) as well as in a certain place of normal faulting in the back - arc side [24]. Recent examination of another region of normal faulting, that of Central Greece (Thessalia region), implies that the non - overlapping of large earthquake rupture zones may be a general property in back - arc conditions which is very important for earthquake prediction and hazard assessment studies. In the Thessalia region, for instance, its southern side has ruptured three times in about the last 38 years after an apparent quiescence since 1773. The rupture zones practically abut and do not overlap (Fig. 4). Its northern side, instead, did not ruptured since 1781. Should the seismicity pattern of the southern Thessalia repeat in its northern side then a migration of the activity would be expected northwards [25].

The predictive value of previous observations is being more clear when they are combined with an anomaly in the magnitude - frequency diagram which is shown in Figure 5. There is a magnitude gap of $5.8 > M > 5.2$ in the cumulative frequency, N, of events for 1964 - 1985 as integrated to 1901 - 1985. Extending N_r, the real frequency of events of $M \geqslant 6.0$ for 1901 - 1985, to smaller events, it results that the gap persists over the entire century, the magnitude error being < 0.3. Consequently, the magnitude range of about 5.8 - 7.0 is characteristic in Thessalia. Moreover, reliable b - value can be determined only for $M \geqslant 6.0$. b - values equal to either 0.88 or 0.99 have been determined depending on the data set used. Resultant mean return periods of events of $M \geqslant 5.8$ are 6.8 and 8.4 years, respectively.

Anomalies in the magnitude distribution seems to be a common feature in several segments of the Hellenic arc - trench system [26]. The main geometric expressions of them are the gap in medium size

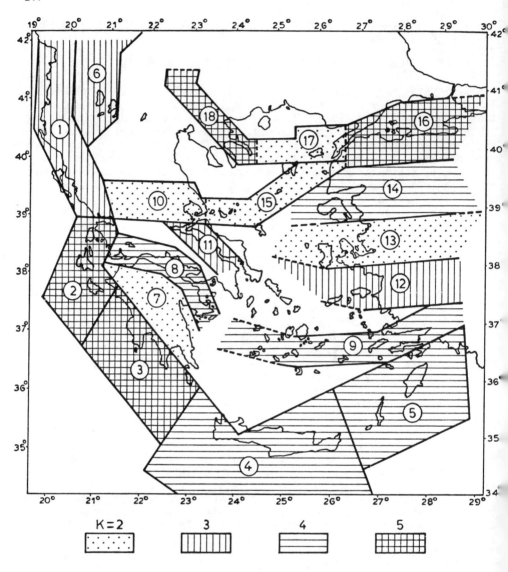

Figure 3. Distribution of the relative seismic hazard in the Greek and surrounding areas (after [18]). Low, medium, high and very high seismic hazard is signed by K = 2, 3, 4, and 5, respectively.

range and "bulge" effect in the right side of the G-R diagram, that is a bend towards the vertical of the larger magnitude frequencies. Such a non - linear structure of G-R is determined by the characteristic earthquake model and shows the general need for properly designed

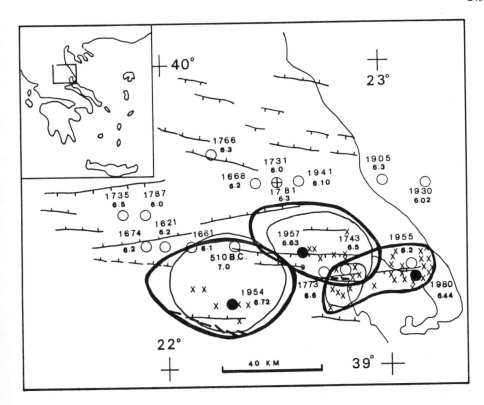

Figure 4. Main neotectonic lines (sharp lines, after [36]), surface fault breaks (heavy dashed lines) associated with the 1954 and 1980 strong shocks, and epicentres of known strong (M > 6.0) earthquakes from the ancient times up to the present in the Thessalia region. More details can be found in [25].

models of seismic hazard determination.

Statistical tests performed with some methodological innovation [27] have shown that the occurrence of mainshocks in the Aegean is stationary and random with respect to the time regardless of the segment, time interval and magnitude class considered [28]. It is suggested, however, that in the Aegean and elsewhere seismo - tectonically more inhomogeneous sources tend to generate earthquakes that are more randomly distributed in time, and that for a given seismogenic structure this may depend on the considered size scale of the structure. For example, in the Aegean back - arc side the strong mainshock activity consists of a random (M = 6.5 - 6.8) and a non-random (M = 6.9 - 7.7) component with a quasi - periodic seismic energy release, implying time clustering of the higher magnitude events [29].

The strength of the time clustering of earthquakes can be measured on the basis of the concept of the fractal geometry of nature

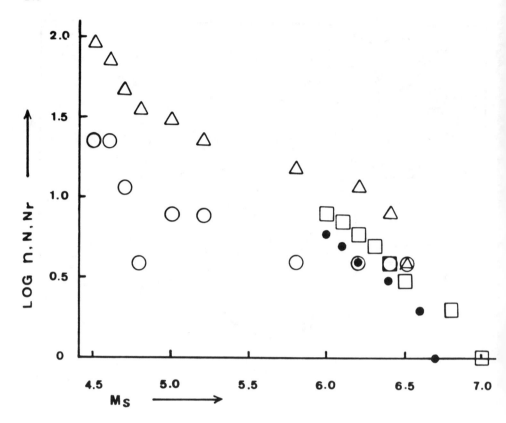

Figure 5. Magnitude - frequency diagram for the Thessalia earthquakes. For more details see the text and in [25].

as developed by Mandelbrot [30]. This approach led to the conclusion that in several segments of the Hellenic arc - trench system scale - invariant time clustering holds over very large scale lengths of time [31].

5. CONCLUDING REMARKS

Only complicated lithospheric models can account for the complex seismotectonic structure and Tertiary geodynamic evolution of the Greek and adjacent regions. Lithospheric rotations and subductions interprete adequately large - scale processes while anomalies in the space, time and magnitude distributions of earthquakes reflect seismotectonic complexity even in smaller scales. Results about the complexity in seismotectonics should influence the elaboration of proper models for seismic hazard assessment and seismicity patterns for use in earthquake prediction. For example, the design of seismic hazard approaches incorporating the concept of characteristic earthquake is an immediate nessecity. On the other hand, particular types of seismotectonic environments may result in the recognition of particular seismicity patterns useful in earthquake prediction. This has been clear in the Greek region where a variety of premonitory patterns of seismic activity were recognised such as periodicity in seismicity [29], long - term and short - term accelerating foreshock activity [23,32], synchronization in the earthquake occurrence [33], premonitory burst of seismicity [34], migration of seismic activity [19] and regularity in the earthquake occurrence [35].

6. ACKNOWLEDGEMENTS

I am grateful to Prof. E. Mantovani for his invitation to give this lecture. The part of the work referring to the Thessalia region has been supported by the European Centre on Prevention and Forecasting of Earthquakes, Council of Europe, Athens.

7. REFERENCES

[1] Bâth, M. (1983) 'The seismology of Greece', Tectonophysics 98, 165 - 208.
[2] Boccaletti, M., Manetti, P. and Peccerillo, A. (1974) 'The Balkanides as an instance of back - arc thrust belt: possible relation with the Hellenides', Geol. Soc. Am. Bull. 85, 1077 - 1084.
[3] Papadopoulos, G.A. (1982) 'Contribution to the study of the active deep tectonics of the Aegean and surrounding areas', D. Sci, Thesis, Univ. of Thessaloniki, 176pp. (in Greek with English abstr.).
[4] Papadopoulos, G.A. (1989) 'Cenozoic magmatism, deep tectonics, and crustal deformation in the Aegean Sea', in C.Kissel and C. Laj (eds.), Paleomagnetic Rotations and Continental Deformation, Kluwer Academic Publishers,Dordrecht, pp. 95 - 113.
[5] Kissel, C., Laj, C., Poisson, A. and Simeakis, K. (1989) 'A pattern of block rotations in Central Greece', in C. Kissel and C. Laj (eds.), paleomagnetic Rotations and Continental Deformation, Kluwer Academic Publishers,Dordrecht, pp. 115 - 129.
[6] Le Pichon, X. and Angelier, J. (1979) 'The Hellenic arc and trench system: a key to the neotectonic evolution of the Eastern Mediterranean area', Tectonophysics 60, 1 - 42.
[7] Mc Kenzie, D.P. (1972) 'Active tectonics of the Mediterranean region', Geophys. J.R. Astron. Soc. 30, 109 - 185.
[8] Papazachos,B.C. and Papadopoulos,G.A. (1977) ' Deep tectonics and associated ore deposits in the Aegean area', Coll. on Geol. of the Aegean Regions Proc. 3, Athens 1977, 1071 - 1081.
[9] Papadopoulos,G.A. and Andrinopoulos,A. (1984) 'Metallogenic evidence for palaeo-subduction zones in the Aegean area', Geol. Balcanica 14, 3 - 8.
[10] Mc Kenzie,D. (1978) 'Active tectonics of the Alpine - Himalayan belt: the Aegean Sea and surrounding regions', Geophys. J.R. Astr. Soc. 55, 217 - 254.
[11] Mc Kenzie,D. (1972) 'Plate tectonics of the Mediterranean region', Nature 226, 239 - 243.

[12] Kondopoulou, D.P., Papadopoulos, G.A. and Pavlides, S.B. (1985) 'A study of the deep seismotectonics in the Hellenic Arc', Boll. Geof. Teor. Appl. 27, 197 - 207.
[13] Papadopoulos, G.A., Kondopoulou,D.P., Leventakis, G.-A. and Pavlides, S.B. (1986) 'Seismotectonics of the Aegean Region', Tectonophysics 124, 67 - 84.
[14] Papadopoulos, G.A. (1985) 'Stress - field and shallow seismic activity in the Aegean region: Their tectonic implications', Quaterniones Geodaesiae 6, 65 - 74.
[15] Nakamura,K. and Uyeda,S. (1980) 'Stress gradient in arc - back arc regions and plate subduction', J. Geophys. Res. 85, 6419 - 6428.
[16] Papadopoulos, G.A. and Pavlides, S.B. (1992) 'The large 1956 earthquake in the South Aegean: Macroseismic field configuration, faulting, and neotectonics of Amorgos Island', Earth & Planet. Sci. Lett. 113, 383 - 396.
[17] Spakman, W., Wortel, M.J.R. and Vlaar,N.J. (1988) 'The Hellenic subduction zone: a tomographic image and its geodynamic interpretation', Geophys. Res. Lett. 15, 60 - 63.
[18] Papadopoulos, G.A. and Kijko,A. (1991) 'Maximum likelihood estimation of earthquake hazard parameters in the Aegean area from mixed data', Tectonophysics 185, 277 - 294.
[19] Papadopoulos, G.A. (1990) 'Forecasting large intermediate - depth earthquakes in the South Aegean', Phys. Earth Planet. Inter. 57, 192 - 198.
[20] Sykes, L.R. (1971) 'Aftershock zones of great earthquakes, seismicity gaps, and earthquake prediction for Alaska and the Aleutians', J. Geophys. Res. 76, 8021 - 8041.
[21] Kelleher, J. (1972) 'Rupture zones of large South American earthquakes and some predictions', J. Geophys. Res. 77, 2087 - 2103.
[22] Wyss, M. and Baer,M. (1981) 'Earthquake hazard in the Hellenic Arc. In: D.W. Simpson and P.G. Richards (eds.), Earthquake Prediction - An International Review, Am. Geophys. Union, Maurice Ewing Ser. 4, 153 - 172.
[23] Papadopoulos, G.A. (1988) 'Long - term accelerating foreshock activity may indicate the occurrence time of a strong shock in the Western Hellenic Arc', Tectonophysics 152, 179 - 192.
[24] Voidomatis, Ph.S., Pavlides, S.B. and Papadopoulos, G.A. (1990) 'Active deformation and seismic potential in the Serbomacedonian zone, northern Greece', Tectonophysics 179, 1 - 9.
[25] Papadopoulos, G.A. (1992) 'Rupture zones of strong earthquakes in the Thessalia region, Central Greece', XXIII General Assembly of Europ. Seism. Comm., Prague, Sept. 1992.
[26] Papadopoulos, G.A., Skafida, H.G. and Vassiliou, I.T. (1993) 'Non - linearity of the magnitude - frequency relation in the Hellenic arc - trench system and the characteristic earthquake model', J. Geophys. Res., under publication.
[27] Papadopoulos, G.A. (1992) 'On some problems about testing stochastic models of the earthquake time series', Natural Hazards, in press.

[28] Dionysiou, D.D. and Papadopoulos, G.A. (1992) 'Poissonian and negative binomial modelling of earthquake time series in the Aegean area', Phys. Earth Planet. Inter. 71, 154 - 165.
[29] Papadopoulos, G.A. and Voidomatis Ph. (1987) 'Evidence for periodic seismicity in the inner Aegean seismic zone', Pure Appl. Geophysics 125, 614 - 628.
[30] Mandelbrot, B. (1967) 'How long is the coast of Britain? Statistical self - similarity and fractional dimension', Science 156, 636 - 638.
[31] Papadopoulos, G.A. and Dedousis, V. (1992) 'Fractal approach of the temporal earthquake distribution in the Hellenic arc - trench system', Pure Appl. Geophys. 129, in press.
[32] Papadopoulos, G.A., Makropoulos, K.C. and Dedousis, V. (1991) 'Precursory variation of the foreshocks fractal dimension in time', Internat. Confer. on Earthquake Prediction: State - of - the - Art, Council of Europe, Strasbourg 15 - 18 October 1991.
[33] Papadopoulos, G.A. (1988) 'Synchronized earthquake occurrence in the Hellenic Arc and implications for earthquake prediction in the Dodecanese Islands (Greece)', Tectonophysics 145, 343 - 347.
[34] Papadopoulos, G.A. (1988) 'Premonitory burst of seismicity and its significance for predicting large Aegean earthquakes', Tectonophysics 156, 257 - 265.
[35] Papadopoulos, G.A. (1990) 'Prediction of potentially damaging earthquakes in the Cretan segment of the Hellenic Arc', Phys. Earth Planet. Inter. 59, 130 - 133.
[36] Caputo, R. (1990) 'Geological and structural study of the recent and active brittle deformation of the Neogene - Quaternary basins of Thessaly (Central Greece), Scient. Annals of the Geol. Departm., Univ. of Thessaloniki, Thessaloniki, Greece, v. 12 (in English).

A REVIEW OF THE EASTERN ALPS - NORTHERN DINARIDES SEISMOTECTONICS

D. SLEJKO

Osservatorio Geofisico Sperimentale
P. O. Box 2011
34016 Trieste
Italy

ABSTRACT. A seismotectonic model of the eastern Alps - northern Dinarides has been derived by considering all the available geological and geophysical information. The present-day seismicity, which is well documented because of the presence of a regional seismometric network, was of particular importance. The seismic area is the pre-Alpine belt which continues eastwards along the Croatian coast; it is delimited southwards by the presently active Alpine and Dinaric fronts and northwards by some significant subvertical faults. Along this belt, the most active zones are those where tectonic systems intersect.

1. Introduction

The definition of a seismotectonic model is fundamental in regional seismic hazard assessment analysis because it represents the first step in identifying the seismogenic zones. Without this piece of information, only rough methodologies, which take into account only the seismological data as they appear in the earthquake catalogues, can be applied.

One of the most seismically active regions in Italy is the Friuli sector of the eastern Alps, where even recently (May 6, 1976) a destructive earthquake occurred. Before undertaking a seismic hazard investigation within the framework of the activities of the "Gruppo Nazionale per la Difesa dai Terremoti" of the "Consiglio Nazionale delle Ricerche" (CNR), a seismotectonic study of the region between lake Garda and the border with Slovenia was made under the CNR's "Progetto Finalizzato Geodinamica". The study began by considering two separate areas, Friuli (Carulli et al. (1981), (1982), Slejko (1985)) and the Veneto piedmont belt (Panizza et al. (1981)), which were then inserted into the general seismotectonic model of northeastern Italy (Slejko et al.(1986), (1987), (1989)). Understanding the influence of external regions, such as Carinthia and Slovenia, is essential for understanding and defining the geodynamic evolution of the homogeneous seismotectonic units. Thus, also the Alps and the Dinarides were considered in the following studies (Carulli et al. (1989), (1990), Del Ben et al. (1991)). Now, a seismotectonic model is proposed for the region between lake Garda and Zagreb in Croatia, and between the Adige mouth and Lienz in the Tyrol, with less detailed analysis of the area further to the north (Figure 1).

The main difficulty faced during the study derived from the necessity to consider information (geological maps, geophysical data, etc.) related to different countries (Italy, Austria, Switzerland, Slovenia, and Croatia). This is even more relevant when considering historical seismicity, which is a key piece of information because of the very strong events in the past.

2. Geology

A sharp increase in crustal thickness (Carulli et al. (1989), (1990)), which results in a Moho depth of more than 50 km in the area included between the Venosta valley to Lienz, is observed; this discontinuity is documented towards the south and east: towards Bavaria it gradually rises up to about 30 km (Figure 2). The Adria microplate is characterized by a crustal high in the Adriatic sea - Istria region which is connected in some way to the even more pronounced high in the Verona area. Towards the Apenninic foredeep, the Moho plunges as a monocline to depths ranging from 35 to 40 km. From Istria northwards, the Moho deepens regularly and gains a depth of more than 40 km after a pronounced flexure in front of the Southern Alps (Gemona), and more than 50 km further north in the Venosta valley - Tauern mountain area. A seismic interval with velocity close to 6.7 km/s has been found in central Friuli at a depth of nearly 10 km: it is probably linked to structures present near the top of the basement. This high velocity body, at a shallower depth (about 6 km), was also found by tomographic inversion of seismological data (Bressan et al. (1992)). The Dinaric unit is characterized by a thick crust with a deep trough (thickening from 40 to 45 km) oriented NW - SE and running from the Croatian coast to Idrija. The Pannonian basin forms a crustal unit 20 - 30 km thick, which includes the inner Carpathians, the easternmost parts of the Alps, and the inner zones of the Dinarides.

The main geological elements which can be found in the study area are (Carulli et al. (1989), (1990)) as follows:
- the metamorphic basement, outcropping in the Southern Alps south of the Insubric lineament (in the Gail valley), and more widely north of the lineament itself in the Austro - Alpine domain;

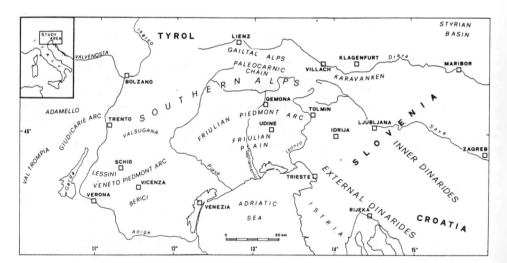

Figure 1. Index map of the study area.

- the Paleozoic units of the Austrian and Slovenian Alps with different degrees of metamorphism, and of the Hercynian non-metamorphic and archi-metamorphic sedimentary cover, which are widely present in the Paleocarnic chain;
- the Mesozoic units of the Southern Alps, of the Dinarides, the Gailtal Alps and the northern Karavanken, consisting of rigid Triassic or Cretaceous carbonate platform units detached and thrust over plastic, mainly evaporitic layers;
- the Tertiary flysch and molasse, often involved in the thrust fold system, and the Tertiary deposits of the Styrian and Pannonian basins;
- the major Quaternary deposits;
- the Tertiary peri-Adriatic intrusive masses and the Tertiary lava effusions.

The geodynamic evolution of the Adria microplate started with its detachment from the African plate, probably during the middle-upper Triassic; the microplate collided during the Paleogene against the Dinarides, and during the Neogene against the Alpine chain (Figure 3). There was great tectonic activity during the Quaternary, with the most notable areas of

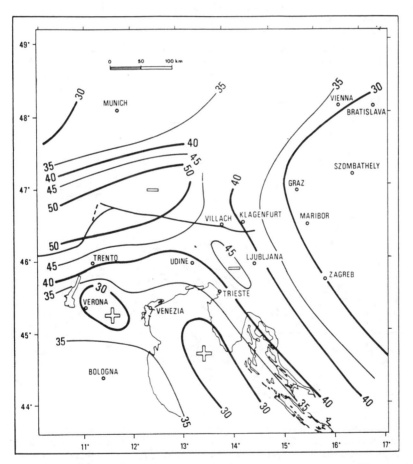

Figure 2. Map of the Moho in the eastern Alps - northern Dinarides, the Insubric lineament is also reported (from Slejko et al. (1989)).

deformation in the Veneto - Friuli pre-Alpine arc, in western Slovenia, and along the Croatian coast. At present, the Friuli pre-Alpine sector is uplifting while the plain is gently subsiding; this fact and the presence in the External Dinarides of dextral NW-SE oriented lines can be explained by a northward movement with a continuing anticlockwise rotation of the Adria plate with associated thrusting under the Southern Alps margin (Anderson and Jackson (1987)).

In the Southern Alps, the main tectonic structures (Carulli et al. (1990)) are E-W oriented thrusts, north-verging in the northern sector between the Gail line (eastern part of the Insubric lineament) and the Fella - Sava line, and south-verging in the southern (Figure 4). Minor subvertical faults, oriented N-S to NNW-SSE, can be observed in the hinge area (Tagliamento canyon, a few kilometres north of Gemona) where they act as a mechanical disengagement. The hinge area corresponds to the zone of maximum crustal shortening in the Southern Alps; moving eastwards, the major tectonic orientation changes from Alpine to Dinaric.

The Dinaric area is characterized by NW-SE oriented thrusts, and by dextral, subvertical faults with directions ranging between NW-SE and NNW-SSE, the most important of which is the Idrija line. Buried overthrusts with Dinaric direction can be found also in the northern Friuli plain.

In northern Slovenia, the main structures are E-W oriented and can be considered as the eastern continuation of the Gail line, the Fella - Sava line, and the Southern Alps thrusts. The last mentioned continue eastwards assuming gradually a NW-SE direction. In the region between Ljubljana and Zagreb the Dinaric style is complicated by ENE-WSW oriented faults, indicating tensional tectonics in the Pannonian domain.

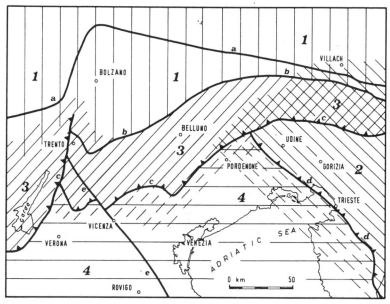

Figure 3. Map of structural units in the eastern Alps - northern Dinarides (from Slejko et al. (1986)): 1 = Alps s. s. and northern sector of the Southern Alps; 2 = External Dinarides; 3 = southern sector of the Southern Alps; 4 = Southalpine - Apenninic foreland. Tectonic limits: a = Insubric lineament, separating the Alps s. s. from the Southern Alps; b = Valsugana (westwards) and Fella-Sava (eastwards) lines; c = front of the Southern Alps; d = front of the External Dinarides; e = Schio-Vicenza line.

The Pannonian realm subsided in the Neogene and Quaternary, and has been considered a back-arc basin completely surrounded by folded arcs (Carulli et al. (1989)). It is subdivided into several sub-basins and transitional areas, the most important of which are the Inner Dinarides, the transitional zone of the eastern Styrian basin, and the Vienna basin. The tectonic activity can be attributed to the thrusting of the Adria against the European platform which induces a network of strike-slip faults separating a mosaic of blocks in the basement of the western Pannonian area (Gutdeutsch and Aric (1987)).

3. Seismicity

The study region is characterized by areas of strong seismicity and other areas practically aseismic. In general, it can be said that the pre-Alpine belt is strongly seismic while the plain and the mountain regions show low seismicity. The historical seismicity of the area is shown in Figure 5, where the earthquakes with epicentral intensity larger than, or equal to, VI Mercalli - Cancani - Sieberg (MCS) from the beginning of the Christian Age to 1984 are reported. The main concentration appears in Friuli and continues westwards along the pre-Alpine belt to lake Garda. In Slovenia the seismicity is concentrated in three parallel narrow bands: the first is elongated from Friuli to the Croatian coast, the second from Friuli to Zagreb, and the third, between the other two, encompasses the seismicity of Ljubljana. In this general scheme, the Villach earthquakes of 1348 (Io = XI MCS) and 1690 (Io = IX MCS) remain slightly apart, and could belong to an alignment of seismicity from Friuli northeastwards, which identifies the western margin of the Pannonian basin.

The major present-day seismicity (magnitude larger than 2.4), as collected by the seismographic stations of the Osservatorio Geofisico Sperimentale of Trieste (OGS), is reported in Figure 6. The main features shown by the historical seismicity map (Figure 5) are confirmed, the northeastward alignment through Villach is more clear and is seen also in the present-day seismicity map of central Europe (Gutdeutsch and Aric (1987)).

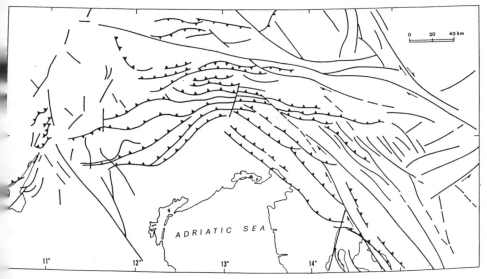

Figure 4. Tectonic map of the eastern Alps - northern Dinarides (modified from Slejko et al. (1989) and Del Ben et al. (1991)).

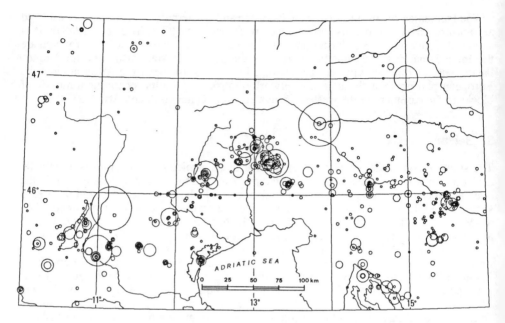

Figure 5. Epicenter map of the earthquakes with epicentral intensity larger than, or equal to, VI MCS in the eastern Alps - northern Dinarides from the beginning of the Christian age to 1984.

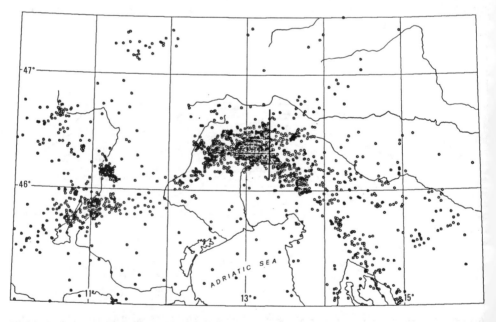

Figure 6. Epicenter map of the earthquakes with magnitude larger than 2.4 in the eastern Alps - northern Dinarides from 1977 to 1990 recorded by the OGS seismometric network.

Further information is given by the focal mechanisms of the main earthquakes of the region (Figure 7). Although the main data are for the 1976 seismic sequence in Friuli and subsequent activity in that region, the main characteristics clearly appear. The Friuli region is dominated by dip-slip mechanisms correlatable to north dipping reverse faults and to overthrusts. In the western sector of the Southern Alps, the mechanisms are of a strike-slip type with compressional axes oriented NW - SE. The compressive style remains well defined in the Alps, while towards the east some vertical movements are reported along the Sava line north of Ljubljana, and normal dip-slip as well as strike-slip motions north of Zagreb. In the External Dinarides, strike-slip and vertical patterns can be seen.

4. Seismotectonic Characteristics

A correlation can be proposed between the seismically active zones and abrupt changes in crustal thickness: from Apulia to the Dinarides, from Apulia to the Alps, from the Pannonian basin system to the Alps (Carulli et al. (1990)). The earthquakes occur mainly in the upper crust; this is consistent with the thrusting of the Adria microplate against the European platform inducing a network of strike-slip faults which created a mosaic of blocks in the basement of the western Pannonian area (Gutdeutsch and Aric (1987)). To the south, the Adria plate is thrust under the Southern Alps, forming a composite lateral slip and underthrust subparallel to the Dinarides (Carulli et al. (1990)).

Interesting additional evidence was deduced from an analysis of vertical cross-sections (Figure 8) on which geological and geophysical data and the present-day seismicity were plotted (see Slejko et al. (1989), Carulli et al. (1990), Del Ben et al. (1991)):

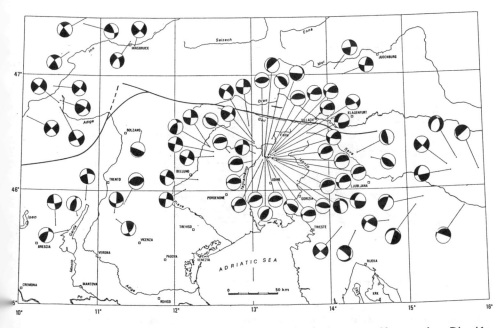

Figure 7. Map of the focal mechanisms of the major earthquakes in the eastern Alps - northern Dinarides (from Slejko et al. (1989) and Del Ben et al. (1991)).

- the hypocentres in Friuli seldom exceed 15 km depth, and mainly involve Alpine and/or Dinaric thrusts in the buried southernmost Paleozoic sequence;
- the southernmost Dinaric thrusts in the Friuli plain are not active at present; going eastwards, Dinaric faults with transpressive activity can be found and, further east, with pure transcurrent activity;
- the earthquakes in eastern Veneto have foci at greater depth, in the crystalline basement, and seem to be mainly connected with transcurrent faults reactivating old Mesozoic structures of Dinaric direction;
- around lake Garda, variable hypocentral depths and different focal mechanisms have been found, indicating activity of structures both longitudinal (with reverse dip-slip movement) and transversal (with transcurrent movement) to the Giudicarie system, in agreement with the hypothesis of an anticlockwise rotation of the Adria microplate with the pole near lake Garda (Anderson and Jackson (1987)).

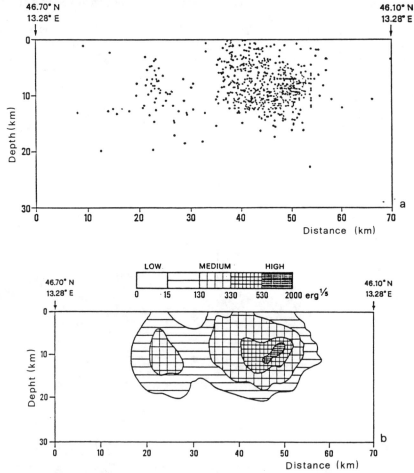

Figure 8. Vertical cross-section in Friuli (for location see Figure 6) with the seismicity 1977 - 1988 recorded by the OGS seismometric network (from Peruzza et al. (1990a), (1990b)): a) representation of the hypocentres, b) representation of the hypocentral probability.

To better quantify the evidence shown by the vertical cross-sections, a new elaboration of the present-day seismicity has recently been proposed (Peruzza et al. (1991)). By considering the uncertainties related to the hypocentral solutions, the areas with larger probability of having been earthquake foci remain identified (Figure 8b). The analysis performed for the Friuli area confirmed the evidence given by the classical cross-sections, and shows with greater detail the breaking segment under the piedmont belt which becomes quasi-horizontal at 10 km depth.

5. Conclusions

A tentative seismotectonic model for the whole area can be made by extrapolating that of Friuli, which is well constrained by the data. The movement of the Adria microplate against the European plate determines the present seismicity in Friuli and along the Croatian coast (Del Ben et al. (1991)). In Friuli, the presently observed deformation process and consequent seismicity is explained by a dominant purely compressive effect, caused by the underthrusting of the Adria microplate. In the External Dinarides, this thrusting of the Adria microplate has generated mainly transpressive deformation on the westernmost Dinaric faults. In these most active regions, a clear crustal thickening is also observed.

The main seismicity is concentrated in a narrow band between the presently active Alpine and Dinaric fronts and some subvertical faults (Slejko et al. (1989), Carulli et al. (1990)): the Giudicarie line north of lake Garda, the Valsugana line in Veneto, the Fella - Sava line in northern Friuli and the Idrija line in Slovenia, which probably act as a mechanical disengagement (Figure 9). Some nodal points, areas of high seismicity at the interference between tectonic systems, are identified as well (Slejko et al. (1989)). Two of

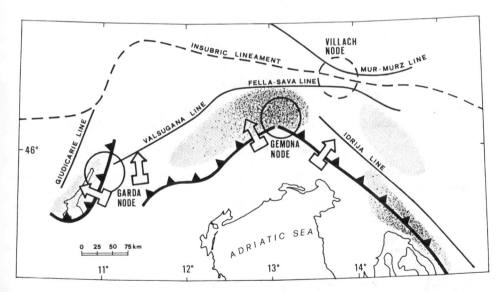

Figure 9. Seismotectonic model of the eastern Alps - northern Dinarides (redrawn from Slejko et al. (1989) and Carulli et al. (1990)): arrows indicate increases of crustal thickness, increasing dashing shows greater seismicity.

them have clear caracteristics: the Garda node, at the Giudicarie - Schio system contact (Panizza et al. (1981), Slejko e Rebez (1988)), and the Gemona one, at the Southern Alps - Dinarides contact. The third, the Villach node at the contact between the Insubric lineament and the Mur - Murz fault, shows sporadic seismicity in the Villach area, where the 1348 event was probably the greatest earthquake in the Alps. The maximum seismicity, in terms of frequency and magnitude, is in Friuli, where the maximum neotectonic activity as well as the maximum crustal shortening are observed (Carulli et al. (1981)).

References

Anderson, H. and Jackson, J. (1987) 'Active tectonics of the Adriatic region', Geophys. J. R. Astr. Soc. **91**, 937 - 983.

Bressan, G., De Franco, R. and Gentile, F. (1992) 'Seismotectonic study of the Friuli (Italy) area based on tomographic inversion and geophysical data', Tectonophysics **207**, 383 - 400.

Carulli, G.B., Giorgetti, F., Nicolich, R. and Slejko, D. (1981) 'Considerazioni per un modello sismottettonico del Friuli', Rend. Soc. Geol. It. **4**, 605 - 611.

Carulli, G.B., Giorgetti, F., Nicolich, R. and Slejko, D. (1982) 'Friuli zona sismica: sintesi di dati sismologici, strutturali e geofisici', in A. Castellarin and G.B. Vai (eds.), Guida alla geologia del Sudalpino centro-orientale, Guide Geol. reg. S.G.I., Bologna, pp. 361 - 370.

Carulli, G.B., Nicolich, R., Rebez, A. and Slejko, D. (1989) 'Some considerations on the seismotectonics of the Northern Dinarides', in E.S.C. Proceedings 21st General Assembly, Bulg. Acad. Sc., Sofia, pp. 67-75.

Carulli, G.B., Nicolich, R., Rebez, A. and Slejko, D. (1990) 'Seismotectonics of the Northwest External Dinarides', Tectonophysics **179**, 11 - 25.

Del Ben, A., Finetti, I., Rebez, A. and Slejko, D. (1991) 'Seismicity and seismotectonics at the Alps - Dinarides contact', Boll. Geof. Teor. Appl. **33**, 155 - 176.

Gutdeutsch, R. and Aric, K. (1987) 'Tectonic block models based on the seismicity in the east Alpine-Carpathian and Pannonian area' in H.W. Flugel and P. Faupl (eds.), Geodynamics of the Eastern Alps. Franz Deuticke, Vienna, pp. 309-324.

Panizza, M., Slejko, D., Bartolomei, G., Carton, A., Castaldini, D., Demartin, M., Nicolich, R., Sauro, U., Semenza, E. and Sorbini, L. (1981) 'Modello sismotettonico dell'area fra il Lago di Garda ed il Monte Grappa', Rend. Soc. Geol. It. **4**, 587-603.

Peruzza, L., Padoan, G., Rebez, A. and Slejko, D. (1990a) 'New methods in evaluating the seismicity along cross-sections' in A. Roca and D. Mayer-Rosa (eds.), ESC 22nd General Assembly: proceedings and activity report 1988 - 1990, IGOL S.A., Barcelona, pp. 673 - 677.

Peruzza, L., Rebez, A., Slejko, D. and Padoan, G. (1990b) 'Parametri per la valutazione di strutture sismogenetiche dedotti dalla sismicità regionale attuale', in Atti 9 Convegno Annuale GNGTS, ESA, Roma, pp. 3 - 6.

Peruzza, L., Rebez, A., Slejko, D. and Padoan, G. (1991) 'Weighted uncertainties used to detect seismogenic structures', Boll. Geof. Teor. Appl. **33**, 25 - 45.

Slejko, D. (1987) 'Modello sismotettonico del Friuli, scuotibilità e rischio sismico', in E. Boschi and M. Dragoni (eds.), Aree sismogenetiche e rischio sismico in Italia, Galileo, Lausanne, pp. 327 - 370.

Slejko, D. and Rebez, A. (1988) 'Caratteristiche sismotettoniche dell'area benacense', in Atti 7 Convegno GNGTS, ESA, Roma, pp. 157 - 167.

Slejko, D., Carraro, F., Carulli, G. B., Castaldini, D., Cavallin, A., Doglioni, C., Nicolich, R., Rebez, A., Semenza, E. and Zanferrari, A. (1986) 'Seismotectonic model of northeastern Italy: an approach', Geologia Applicata ed Idrogeologia **21** (part 1), 153 - 165.

Slejko, D., Carulli, G.B., Carraro, F., Castaldini, D., Cavallin, A., Doglioni, C., Iliceto, V., Nicolich, R., Rebez, A., Semenza, E., Zanferrari, A. e Zanolla, C. (1987) 'Modello sismotettonico dell'Italia Nord-Orientale', C.N.R. G.N.D.T. Rendiconto n. 1, Ricci, Trieste, 83 pp.

Slejko, D., Carulli, G.B., Nicolich, R., Rebez, A., Zanferrari, A., Cavallin, A., Doglioni, C., Carraro, F., Castaldini, D., Iliceto, V., Semenza, E. and Zanolla, C. (1989) 'Seismotectonics of the eastern Southern-Alps: a review', Boll. Geof. Teor. Appl. **31**, 109 - 136.

REGIONAL STRESSES IN THE MEDITERRANEAN REGION DERIVED FROM FOCAL MECHANISMS OF EARTHQUAKES

A. UDIAS and E. BUFORN
Department of Geophysics
Universidad Complutense
28040 Madrid, Spain

ABSTRACT. The distribution of epicenters and the focal mechanism of earthquakes are the basic tools in the determination of the regional stresses in a region. The Mediterranean basin is located at the plate boundary between the Eurasian and African plates. This boundary is of a complex nature and its dynamics is related to the opening of the Atlantic ocean and the closing of the Thetys sea. From Azores to Gibraltar the motion changes from strike-slip to reverse faulting under horizontal compression in NW-SE direction. From Gibraltar to the Caucasus, the boundary is complicated by the presence of secondary blocks, areas of extended deformation and subduction zones. The regional stresses show predominant horizontal compressions in a general NW-SE direction with some isolated areas of horizontal tensions in the Betics-Alboran sea, Apennines and north of the Hellenic arc.

1. Introduction

The Mediterranean basin is located at the boundary between the lithospheric plates of Eurasia and Africa which extends from west to east from the Azores islands to the Caucasus mountains. From Azores to Gibraltar, the boundary is relatively simple separating on both sides oceanic lithosphere. East of Gibraltar in the Mediterranean basin, the boundary is formed by the interaction of continental and oceanic lithosphere. The boundary in this region is specially complicated by the presence of small lithospheric blocks and the distribution of stresses by deformations extended over wide areas. The plate boundary itself must be interpreted in this region as an extended area that follows a complicated system of continental blocks, oceanic basins and orogenic belts, located between the stable parts of Europe and Africa. Earthquake activity in the Mediterranean region is also spread over wide areas with significant intraplate occurrence.

Many studies have been made on the seismotectonic conditions of this region derived from seismicity and focal mechanisms, among the most recent covering the whole area, are those of Constantinescu et al. (1966), McKenzie (1972), Udias (1982), Udias (1985), Jackson and McKenzie (1988) Argus et al. (1989), Westaway (1990), Udías and Buforn (1991), Müller et al. (1992). In this work we study the direction of the horizontal stresses as derived from the focal mechanisms of large shallow earthquakes. The deduction of stress directions from fault plane solutions of earthquakes is not exempt from ambiguity, as was already pointed out by McKenzie (1969). However, the stress axes derived from fault plane solutions of large earthquakes, as is our case, may serve as an indication of the direction of the regional stresses and may not deviate strongly from the expected direction. Using only large earthquakes has, also, the advantage that their solutions are well determined. The consistency in the stress directions found in this study confirms this point of view.

2. Seismicity

The distribution of epicenters with magnitude M ≥ 4 from 1940 to 1985 are shown in figure 1. Earthquakes are located in a general west-east trend from 40°W to 50°E, occupying a wide band from 30°N to 50°N. Inside this band, the trend of the epicenters changes direction several times, following orogenic belts and outlining secondary blocks. The main characteristics of the seismicity may be described as follows. At the western end of the plate boundary, earthquakes are located following the trend of the Azores islands, off shooting from the Mid-Atlantic ridge in SE direction. From 24°W, earthquakes are located along the west-east Azores-Gibraltar fault. From about 12°W epicenters are spread over a wider zone in south Iberia and northern Morocco. In northern Iberia, shocs are mainly located along the

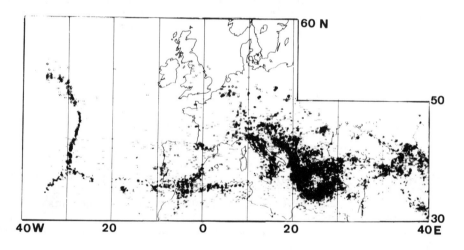

Figure 1. Epicenter distribution for 1940-1985 and M > 4 (USGS Hypocenter Data File)

Pyrenees. From northern Morocco, earthquakes continue eastward along the coast of Algeria and Tunisia. From this point , epicenters change direction to the NE in the Sicily-Calabria arc and, then, continue in NW direction along the Apennines in the Italian peninsula. In northern Italy, earthquakes form a wide arc at the Alps and continue again in SE direction along the coast of Yugoslavia and northern Greece. From this point, earthquakes form a wide arc, the Hellenic arc, with the highest intensity of seismic activity in the whole region. The arc is convex to the south and behind it, in northern Greece and western Turkey, another region of high activity is present with

E-W trend. Two lineaments of earthquakes are located in the Anatolia peninsula. One at the north in E-W direction along the north Anatolian fault and the other in SW-NE. Both merge at about 40°E where earthquakes continue W-E in the Caucasus and NW-SE along the border of the Arabian plate.

Intermediate and deep earthquakes (h > 60 km) are located at four distinct areas and related to the arc-like structures of Gibraltar, Sicily-Calabria, Hellenic and Carpathian. Deep activity associated with the Gibraltar arc is revealed mainly by the deep Spanish earthquake of 1954 at 640 km depth. Shocks of intermediate depth (60 < h < 150 km) are also present in south Spain and north Morocco. Deep earthquakes in the Sicily-Calabria arc are located at its concave side and extend to a depth of 450 km. Most are concentrated between 200 and 350 km with only a few shocks at greater depth. Deep seismic activity associated with the Carpathian arc is located in a rather small region, known as the Vrancea seismic zone. Most shocks are located at depths between 70 and 160 km. The largest concentration of intermediate and deep earthquakes is located along the Hellenic arc, which spans an area from the western coast of Greece to the southern coast of Turkey. The distribution of shocks and their depth indicate a well developed Benioff zone that dips from the convex side of the arc, reaching a maximum depth of 200 km.

3. Regional stresses

Udias and Buforn (1991) have shown the direction of stresses along the Africa-Eurasia plate boundary derived from the focal mechanisms of 83 European earthquakes with magnitudes M > 6, for the period 1935 to 1983, based on the catalogue of Udías et al. (1989). These results will be used here to study the regional stresses in the Mediterranean region. The directions of the horizontal projections of the pressure axes for shallow earthquakes (h < 60 km) are shown in figure 2, together with the trend of plate boundaries. Figure 2 shows that data are very consistent and the majority of the horizontal projections of the P axes are nearly normal to the plate boundaries. In particular, from Azores to Tunisia, P axes form an angle from 60° to 90° with the plate boundary, in a consistent NW-SE direction. They correspond to strike-slip in the central part of the Azores-Gibraltar fault and thrust mechanisms in its eastern part and from Gibraltar to Tunisia. On the north and east boundary of the Adriatic block, P axes are are normal to it in NW-SE and ENE-WSW direction. In the Hellenic arc , data are more scattered , but a direction normal to the arc is common. Behind the arc, between Greece and Turkey, P axes have E-W direction. Along the north Anatolian fault, the direction is NW-SE corresponding to strike-slip mechanisms. In the Caucasus region a nearly N-S trend is found, normal to the boundary of the Arabian plate, corresponding to thrust mechanisms.

Horizontal projections of the tension axes for shallow earthquakes are shown in figure 3. Since only those dipping less than 45° are given , the axes represented correspond to the regions under horizontal tensional regime. In the Azores, T axes are normal to the volcanic alignment in NE-SW direction. In the central part of the Azores-Gibraltar fault, horizontal T axes also in NE-SW direction correspond to the horizontal P axes in NW-SE direction resulting in strike-slip right-lateral motion along the E-W

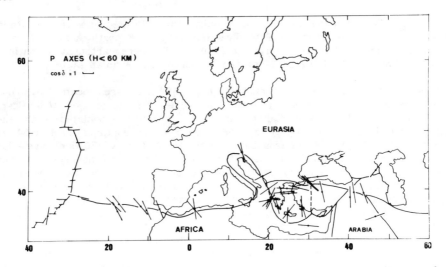

Figure 2. Horizontal projection of P axes with plunge less than 45° for shallow earthquakes (h < 60 km) and M > 6 (Udías and Buforn, 1991).

trending fault. In the Mediterranean region, horizontal T axes are present in the Apennines, Italy in NE-SW direction, normal to the trend of the mountain chain and corresponding to normal faulting. A large concentration of horizontal T axes in a general N-S direction is present behind the Hellenic arc, in Greece and western Turkey, corresponding also to normal faulting.

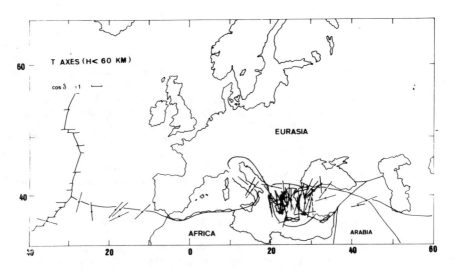

Figure 3. Horizontal projection of T axes with plunge less than 45° for shallow earthquakes (h < 60 km) and M > 6 (Udías and Buforn, 1991).

Along the north Anatolian fault, horizontal T axes in NE-SW direction correspond to the also horizontal P axes in NW-SE direction, in agreement with the strike-slip motion along this fault.

Figure 4 shows the direction of the regional stresses in the Mediterranean region derived from the data shown in figures 2 and 3, together with the trend of the plates boundaries. These boundaries, however, must be understood as a simplification of the actual situation, that involves broad areas of deformation and cannot be reduced totally to the motion of rigid blocks. In a general way, the direction of the regional stresses is a consequence of the east-west opening of the Atlantic ocean and the north-south closing of the Mediterranean sea.

Predominant horizontal stresses in the Mediterranean region are of compressive nature in a general NW-SE to N-S direction consistent with the motion about a pole of rotation of Africa with respect to Eurasia at $21°N$, $20°W$ (Argus et al., 1989). This direction of compressive stresses is also found in central and northern Europe from other type of data such as in situ stress

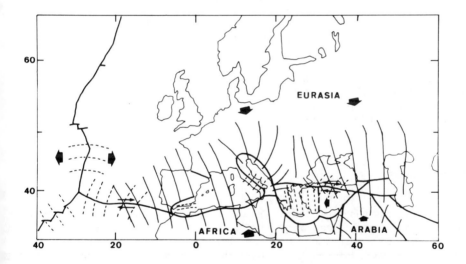

Figure 4. Direction of regional stresses in the Mediterranean region. Continuous lines, compresive; dashed lines tensional.

measurements, overcoring, hydraulic fractures, etc. (Müller et al., 1992; Grünthal and Stromeyer, 1992). Along the Sicily-Calabria and Hellenic arcs, compressive stresses are normal to their trends, varying somewhat from the general direction. On the boundary of the Adriatic block, the direction of stresses varies from that of the whole region, showing the independent motion of this block about a different pole of rotation with respect to Eurasia ($46°N$, $10°E$) (Anderson and Jackson, 1987).

Horizontal tensional stresses are present at the Mid-Atlantic and Azores ridges, oriented normal to their trend. In the central part of the Azores-Gibraltar fault, the direction of the tensional horizontal stresses is NE-SW, resulting in right-lateral strike-slip motion in east-west direction. In south Spain and the Alboran sea, there is a region of horizontal tensional stresses in east-west direction that argues in favor of the detached character of this area with westward motion (Buforn et al., 1988). As has been mentioned, the rotation of the Adriatic block about a different pole results in horizontal tensions in NE-SW direction in the Apennines with normal faulting. Horizontal tensional stresses in north-south direction are present in northern Greece and western Anatolia, as a result of the stretching of the lithosphere, due to the southward advancing of the Hellenic arc. Along the north Anatolian fault, horizontal tensions in NE-SW direction together with the also horizontal compressions in NW-SE direction result in the right-lateral strike-slip motion along this fault.

In conclusion, focal mechanism data from large earthquakes show that the predominant horizontal stresses in the Mediterranean region are compressive in NW-SE to N-S direction due to the convergent motion of the Eurasia and Africa plates. This direction of horizontal compressive stresses is also present inside the Eurasian plate. Deviations from this general direction are due to the motion of secondary independent blocks. In the east coast of the Adriatic, horizontal compressive stresses are in NE-SW to E-W direction due to the interaction of the independent Adriatic block and in northern Greece in E-W direction. There are in the Mediterranean region three isolated areas of horizontal tensional stresses, located in the Betics-Alboran in E-W direction, in the Apennines in NE-SW direction and in northern Greece behind the Hellenic arc in N-S direction. These tensional areas are also due to the interaction of independent smaller units with differencial motion with respect to the two large plates.

Acknowledgements

The authors wish to thank Dr. A. Espinosa, USGS, Golden, Col. for providing seismicity data and helpful suggestions. This work has been partially supported by the Dirección General de Investigación Científica y Tecnica, project PB-89-0097 and the European Community, project SCI-0176-C-(SMA). Publication No. 348, Departamento de Geofísica, Universidad Complutense de Madrid.

References.

Anderson, H. and Jackson, J. (1987). Active tectonics of the Adriatic region. Geophys. J. R. Astr. Soc. 91, 937-983.

Argus, D. F., R.G. Gordon, C. DeMets and S. Stein (1989). Closure of the Africa-Eurasia-North America plate motion circuit and tectonics of the Gloria fault. J. Geophys. Res. 94, 5585-5602.

Buforn, E., A. Udías and J. Mezcua (1988). Seismicity and focal mechanisms in south Spain. Bull. Seism. Soc. Am. 78, 2008-2024.

Constantinescu, L., L. Ruprechtova and D. Enescu (1966). Mediterranean Alpine earthquake mechanism and their seismotectonic implications. Geophys. J. R. Astr. Soc. 10, 347-368

Grünthal, G. and D. Stromeyer (1992). The recent stress field in central Europe: Trajectories and finite element modeling. J. Geophys. Res. 97, 11805-11820.

Jackson, J. and D. McKenzie (1988). The relationship between plate motions and seismic moment tensors and the rates of active deformations in the Meditterranean and Middle East. Geophys. J. R. Astr. Soc. 93, 45-73.

McKenzie, D. (1969). The relation between fault-plane solutions for earthquakes and the directions of the principal stresses. Bull. Seism. Soc. Am. 59, 591-601.

McKenzie, D. (1972). Active tectonics of the Mediterranean region. Geophys. J. R. Astr. Soc. 30, 109-185.

Müller, B., M.L. Zoback, K. Fuchs, L. Martin, S. Gregersen, N. Pavoni, O. Stephansson and C. Ljunggren (1992). Regional patterns of tectonic stress in Europe. J. Geophys. Res. 97, 11783-11803.

Udías, A. (1982). Seismicity and seismotectonics stress field in the Alpine-Mediterranean region. In: Alpine-Mediterranean Geodynamics (H. Berckhemer and J.K. Hsü, eds.) Geodynamics Series 7, 75-82.

Udías, A. (1985). Seismicity of the Mediterranean basin. In: Geological evolution of the Mediterranean basin (D.J.Stanley and F. C. Wezel, eds.) Springer Verlag, New York, 55-63.

Udías, A. and E. Buforn (1991). Regional stresses along the Eurasia-Africa plate boundary derived from focal mechanisms of large earthquakes. Pageoph. 136, 433-448.

Udías, A., E. Buforn and J. Ruiz de Gauna (1989). Catalogue of focal mecanisms of European earthquakes. Dept. of Geophysics, Universidad Complutense, Madrid.

Westaway, R. (1990). Present-day kinematics of the plate boundary zone between Africa and Europe, from Azores to the Aegean. Earth and Plan. Sci. Lett. 96, 393-406.

SOURCE PROCESS OF SOME LARGE EARTHQUAKES IN GREECE AND ITS TECTONIC IMPLICATION

J. DRAKOPOULOS and G.N. STAVRAKAKIS
National Observatory of Athens
Seismological Institute
P.O. Box 200 48
118 10 Athens, Greece

ABSTRACT. In the present paper the source process of some large earthquakes occurred in Greece is discussed in terms of the distribution of the asperities or barriers on the fault plane. The source process has been revealed by the inversion of teleseismic long period body waves using two different techniques. Based on the obtained source time functions an attempt is made to explain the low stress drop values for the earthquakes occurred in Greece. It seems that the stress drops computed by using Brune's model are an average estimate and do not reflect to some seismotectonic characteristics of the focal region.

1. INTRODUCTION

The Hellenic arc-trench system (fig.1) is a subduction zone of about 1000 km where the African lithosphere is subducting beneath the Aegean lithospheric plate in SW-NE direction. This motion of the Hellenic consuming boundary with respect to Africa results from three different processes: the northward motion of the African plate, the Aegean extensional spreading, and the westward motion of Turkey (McKenzie, 1972,1978; Le Pichon and Angelier, 1979). Within the extensional province, the seismicity is most intense (higher earthquake magnitudes) in the northern Aegean and coastal regions of Greece and western Turkey. Various estimates of the present rate of extension in the Aegean have been made on the basis either of kinematic arguments (McKenzie, 1978; Le Pichon and Angelier, 1979), seismic moment rates (Jackson and McKenzie, 1988a,b; Ekstrom and England, 1989; Main and Burton, 1989), or by direct measurement using satelite laser-ranging (Sellers and Cross, 1989). However, all these studies have shown that the north-south extension rate across the Aegean is considerably faster than the convergence between Africa and Europe.
Fault plane solutions of earthquakes occurred in Greece and nearby areas show the extensional stress field in Aegean and

the compressional along the Hellenic arc (McKenzie, 1972, 1978; Drakopoulos and Delibasis, 1982; among others). Recently, Papazachos et al. (1991) proposed a model for the stress pattern in the Aegean and the surrounding area using all available fault plane solutions (figure 2).

In the present study we used the results obtained by waveform modeling of some large and moderate earthquakes occurred in this area to investigate whether the partial stress drop model proposed by Brune et al. (1985) is valid and whether the low stress drop value revealed for these events is model dependent or is due to some seismotectonic characteristics of the focal region.

2. GENERAL ABOUT SOURCE PROCESS

For many years seismologists have modeled earthquake faulting by a uniform slip or a uniform stress drop over the entire fault plane. Recently, they recognized the need for an irregular slip motion (heterogeneneous slip distribution) over a heterogenous fault plane in order to explain the high-frequency radiation from large earthquakes (Das and Aki, 1977; Papageorgiou and Aki, 1983, Aki, 1984, among others).

Various terms have been proposed to describe such heterogeneity in a fault plane, but " Asperities " or "Barriers" appear to be the ones most frequently used in recent literature. Both terms refer to strong patches on the fault plane that are resistive to breaking but they are used for modeling distinctly different roles in the process of earthquake faulting. In the asperity model the stress is heterogeneous before the main shock and becomes homogeneous afterward. Thus, this model represents the main shock as a stress-smoothing process in contrast to the barrier model in which the main shock is considered as a stress-roughening process.

Recent results from paleoseismology suggest that earthquakes recurrent on a given fault may often have the same characteristic length and amount of slip. Stable asperities and barriers, which survive many earthquakes, can explain these results. Aki (1984) proposed the existence of two distinct earthquake families. One is called "asperity type" and is associated with a relatively homogeneous fault plane with stable asperities that break in a large earthquake. The other may be called the "barrier type" earthquake family and is associated with a strongly heterogeneous fault plane containing weaker barriers between strong stable ones. The term homogeneous or heterogeneous fault is related to the stress distribution on the fault plane.

By inverting the teleseismic body waves of some large and

moderate earthquakes occurred in Greece we individuate the areas with the largest seismic moment on the fault plane and we interprete them in terms of barriers and asperities.

3. SUMMARY OF THE RESULTS OF THE INVERSION OF TELESEISMIC P-WAVES OF SOME LARGE EARTHQUAKES IN GREECE (KIKUCHI AND KANAMORI'S APPROACH)

It is widely recognized that large earthquakes, consist of a sequence of smaller subevents. Kikuchi and Kanamori (1982) developed a numerical method to deconvolve body waves into a multiple shock sequence. Under the assumption that all subevents of a multiple shock have identical fault geometry and depth, the far-field source time function is obtained as a superposition of ramp functions. The height and onset time of the ramp functions are determined by matching the synthetic waveforms with the observed seismograms in the least-square sense. The individual subevents (point sources) are then identified by pairs of ramp functions or discrete trapezoidal pulses in the source time sequence. The fault plane is divided into a NxM discrete points in the strike direction with a constant spacing interval in order to compute Green's functions at all possible locations. For each subevent a set of parameters (m_i, X_i, Y_i, t_i), $i=1,2,...N$, where m_i, is the seismic moment, (X,Y) is the relative location of the ith event and t is the onset time. Details of the method are given by Kikuchi and Kanamori (1982), Kikuchi and Sudo (1985) and Stavrakakis et al.(1987).

3.1 The Thessaloniki earthquake of June 20, 1978

Figure 3 shows the source time function, the moment rate function as well as the spatial distribution of the individual point sources on the fault plane of the Thessaloniki earthquake of June 20, 1978 (Ms=6.8) obtained by inversion of teleseismic long period P-waves (Stavrakakis et al., 1987). The source time function consists of three subevents, the largest one is at the beginning of the rupture. The size of the squares is proportional to the seismic moment. The pair (5,4) indicates the relative coordinates at which the rupture started. This point was found by trial and error till the best matching of the observed and synthetic waveforms (fig.4).

3.2 The Gulf of Corinth (central Greece) earthquakes of 1981

The Kikuchi and Kanamori (1982) approach has been used (Stavrakakis et al., 1986) to invert the P-waves of the February 24, 1981 (M_s=6.7) mainshock and of the two largest aftershocks of February 25, and March 4, 1981 (M_s=6.4) (Gulf

of Corinth, Central Greece). Figures 5, 6, and 7 illustrate the obtained results. Based on these figures it appears that the rupture of the main shock can be represented by the sequential occurrence of approximately 4 significant subevents with time separations varying from 3 to 8 sec. The solid circles in figures 5,6,7 depict the nucleation points of the fault planes for the three earthquakes, respectively. These points have been obtained by the best matching criterion of the observed and synthetic waveforms (figs. 5a,6a,7a).
The shape of the source time function of the first large aftershock of Feb. 25, 1981 seems to more complex which implies that the rupture mechanism of this event was quite different from that of the main shock. Seven significant individual subevents were identified as marked in figure 6. The shape of the source time function of the second large aftershock of March 4, 1981 is also complex and six subevents can be identified (fig.7). Jackson et al. (1982) based on the fault parameters estimated from both teleseismic and field observations (fault dimensions from aftershock distribution), suggest a stress drop of 33.6, 7.0 and 4.5 bars for the three seismic events investigated in this study. Kim et al. (1984), based on body wave analysis obtained 10., 8.0 and 6.6 bars, respectively.
Bezzeghoud et al. (1986) obtained different source time functions than those obtained in this study and by Jackson et al. (1982). It might be due to the the fact that they used short period GDSN records. However, the above mentioned authors interpreted also their results in terms of a complex source model in which the rupture starts at a small asperity and then triggers the larger scale rupture of a neighbouring prestressed patch.
The striking feature of the Corinth sequence is the low stress drops of the main shock and its two largest aftershocks. The low stress drop and the long duration of the revealed source time functions suggest an overall slow energy release during the earthquake sequence.

3.3 *The Cephalonia (western Greece) earthquake of Jan.17, 1983*

The rupture evolution of the Cephalonia earthquake (Ms=7.0) of January 17, 1983 has also been studied using the iterative deconvolution method of complex P-waves (Kikuchi and Fukao, 1985). Figure 8 shows the source time function, the moment rate function, and the spatial distribution of the point sources on the fault plane (Stavrakakis et al., 1989). The shape of the source time function is also complex, reflecting the comlexity of the rupture propagation. It consists of five discrete subevents. The largest one took place during the first ten seconds of the rupture. It should be emphasized that the structural complexities near the source and along the propagation path are ignored, and the complexity of the observed waveforms is wholly attributed to the source.

4. SUMMARY OF THE RESULTS OF THE INVERSION OF TELESEIMIC BODY WAVES OF SOME LARGE AND MODERATE EARTHQUAKES IN GREECE (NABELEK'S APPROACH)

As we mentioned above, modeling of body waves either in frequency or time domain is a powerful method and provides important information on the details of the source process. The results summarized in the previous section were based on the inversion technique proposed by Kikuchi and Kanamori (1982). For analysing moderate earthquakes, Nabelek (1984) developed an inversion technique in which the double-couple earthquake source is parameterized by the strike, dip, and rake angles of one of its P-wave nodal planes, the centroid depth and the amplitudes of a specified number of overlapping isosceles triangles that represent the source time function. This method has been applied to a number of moderate earthquakes occurred in Greece (Ioannidou et al., 1989, Ioannidou, 1989; Taymaz et al., 1990; Taymaz et al., 1991), in order to obtain reliable fault plane solutions and source parameters of these events. In the following we summarize the main conclusions of the above mentioned studies.

Earthquakes occurred in the northwestern part of the Hellenic arc (Ionian Sea, fig.9) are characterized by thrust faulting. In most cases, a second point source was necessary to match observed and synthetic seismograms, implying the complexity in the source process. The shape of the source time functions (fig.9) was complex and it seems that the largest subevents occur at the beginning of the rupture. However, the second point source contributes to the last stage of the rupture.

Earthquakes occurred in the southern part of the Hellenic arc (near Crete island, fig.10) have extensively studied by Taymaz et al. (1990). These authors devided the focal mechanisms in this region into four groups: (i) normal faults with a N-S strike in the over-riding material above the subduction zone, (ii) low-angle thrust with an E-W strike at a depth of about 40 km, (iii) high-angle reverse faults with the same strike but shallower focal depths than (ii), and (iv) events within the subducting lithosphere with approximately E-W P-axes. The source time functions obtained by inversion of P- and SH-waves of the events occurred in this region are not as complicated as those of the events in Ionion Sea (Ioannidou, 1989b). A great amount of the seismic moment is released during the first five seconds of the rupture.

Earthquakes occurred in the Gulf of Corinth (central Greece, fig.11) are characterized by normal faulting. As we have mentioned above, in most cases a low stress drop has been obtained.

In north and central Aegean Sea (fig.12), the source process of the earthquakes seems to be also complex (Ioannidou, 1986, Taymaz et al., 1991). The focal mechanisms show that the faulting in the western part of the Aegean region is mostly extensional on normal faults with a NW to WNW strike and with slip vectors directed to NNW to NNE. In the central and eastern Aegean the predominant faulting is right-lateral strike-slip on faults trending NE to ENE and with slip vectors directed NE. Based on the source time functions of the earthquakes occurred in this region (Ioannidou, 1986), it seems that an amount of the seismic moment is released at the end of the rupture. However, the stress drop of these events is higher than that of the earthquakes occurred in the Hellenic arc.

5. STRESS DROP AND SOURCE PROCESS

Several attempts have been made to obtain the source parameters, such as seismic moment, source dimension, average displacement on the fault plane, stress drop for some moderate and large earthquakes occurred in Greece (Kiratzi et al., 1985; Stavrakakis and Blionas, 1990; Stavrakakis et al., 1991). The above mentioned parameters have been revealed either by waveform modeling or by computing the far-field displacement spectra of teleseismic long period body waves, making use the models proposed by Brune (1970,1971). The striking feature of the source parameters of the earthquakes in Greece is the low stress drop value (for additional to the above mentioned references: Kulhanek and Meyer, 1979; Soufleris and Stewart, 1981; Jackson et al., 1982; Kim et al., 1984). However, Madariaga's model (Madariaga, 1976) resulted in higher values (Stavrakakis et al., 1989a).

The question which has to be answered is whether the earthquakes occurred in Greece are low stress drop events or are model dependent. In the present work, we used the concept of the "partial stress drop" (Brune et al., 1985) to explain the low stress drop associated with Brune's spectral model. According to their model, partial stress drop events might occur when the stress release is not uniform over the fault plane, and performed in a series of multiple events, with parts of the fault remaining unbroken. Each subevent which occurred on the fault plane may be associated with large displacement and large stress drop (Mori and Shimazaki, 1984), but between the areas of individual events slip may be small. Therefore, the stress drop values obtained by using Brune's model may be regarded as an average estimate.

In the following we discuss a characteristic case to show that the partial stress drop model seems to be valid for some earthquakes in Greece. Following Pacheco and Nebelek (1988), the source radius for the Cephalonia earthquake of Jan.17, 1983 is estimated from the characteristic time t_c of the source time function (fig.9). This parameter is defined as the time at which half of the total moment is released. Assuming a circular rupture that propagates at a velocity $V_r = 0.75Ö$ (Ö is the shear wave velocity equal to 3.5 km/s), then the source radius is $R = V_r t_c$ and the stress drop is $0.44 M_0/R^3$. For this earthquake, the characteristic time is of about 5 sec which corresponds to a 13.2 km source radius. For a seismic moment $M_0 = 7.3 \times 10^{25}$ dyne.cm (Kiratzi and Langston, 1991) the average stress drop is 14 bars. The same value has been obtained by computing the far-field displacement spectra of this earthquake and by using Brune's model (Stavrakakis et al., 1990). On the other hand, the source time function of this earthquake indicates that the total seismic moment was released in three distinct phases corresponding probably to three asperities.

For the first asperity during which approximately 20 per cent of the total moment is released, the rise time is 1.0 sec implying a radius of 2.6 km and a stress drop of 355 bars. For the second one a stress drop of 263 bars is obtained by assuming that 50 per cent of the total moment is released. Finally, for the third asperity a stress drop of 67 bars is computed corresponding to 30 per cent of the total moment and to a rise time of 2.0 sec. Same calculations have been performed for the other earthquakes occurred in different seismotectonic regions of Greece. The results will presented in a subsequent paper due to space limitations (Stavrakakis and Drakopoulos, 1993).

6. DISCUSSION AND CONCLUSIONS

One of the most important developments in understanding of the earthquake source in the last years is the documentation of the complexity of earthquake rupture. This complexity can be due to geometrical complexities in the fault strength or heterogenities in the fault strength or tectonic stress. Whatever the cause, a common finding is that small areas of high stress drop are embedded in larger, low stress drop areas.
Considering that the shape of the source time function reflects to the complexity of the source, it is possible to explain the low stress drop values (between 1 to 30 bars) obtained for the earthquakes occurred in Greece using Brune's model. It has been shown by using the "partial stress drop event" and the complexity of the source time function that Brune's model is an average estimate of the stress drop. Of

cource, one could interprete the low stress drop values in terms of the presence of soft material near the source region, if large stress drops imply high strength, or by assuming that low stress drop earthquakes occur in zones of weakness along existing major faults in Greece. This needs further investigation and more data.

Acknowledgements

The authors would like to express their thanks to an anonymous reviewer for useful comments

7. REFERENCES

Aki, K. (1984) 'Asperities, barriers, characteristic earthquakes and strong motion prediction', J.Geophys. Res., 89, 5867-5872.

Bath, M. (1985) ' Global tectonic relations - A project for the future', Seismological Dep. Uppsala, Sweden, Rep.4-85.

Brune, J.N. (1970) ' Tectonic stress and spectra of seismic shear waves', J. Geophys. Res., 75, 4997-5009.

Brune, J.N. (1971) 'Correction', J. Geophys. Res., 76, 5002.

Brune, J.N., Fletcher, J., Vernon, F., Haar, L., Hanks, T., and Berger, J. (1985) ' Low stress-drop earthquakes in the light of new data from ANZA, California telemetered digital array', Maurice Ewing Series, Vol.6 (Am. Geophys. Un. Mon. 37), 237-245.

Das, S. and Aki,K. (1977) 'Fault plane with barriers: a versatile earthquake model', J. Geophys. Res., 82, 5613-5670.

Drakopoulos, J. and Delibasis, N. (1982) ' The focal mechanism of earthquakes in the major area of Greece for the period 1947-1981', University of Athens, Pub. No.2,

Ekstrom, and England, P.C. (1989) 'Seismic strain rates in the regions of distributed continental deformation', J. Geophys. Res., 94, 10231-10257.

Fukao, Y and Furumoto, M. (1975) 'Foreshocks and multiple shocks of large earthquakes' Phys. Earth Plan. Inter., 10, 355-368.

Ioannidou, E., Stavrakakis, G., and Drakopoulos, J.(1989) ' A source study of the Aegean earthquake of March 27, 1975, from the inversion of teleseismic body waves' Proc. of the 1st Hellenic Geophysical Congress (in press).

Ioannidou, H. (1989) ' Seismic source parameters from inversion of teleseimic body waves. A study of all moderate and large Greek earthquakes from the years 1965-1988', PhD Thesis, Athens University.

Jackson, J.A., Gagnepain, J., Houseman, G., King, G.C.P.,

Papadimitriou, P. Soufleris, C. and Vireux, J. (1982) ' Seismicity, normal faulting and the geomorphological development of the Gulf of Corinth (Greece): the Corinth earthquakes of February and March 1981', Earth Plan. Sci., Lett., 57, 377-397.

Jackson, J. A. and McKenzie, D. (1988a) ' The relationship between plate motions and seismic moment tensors, and the rates of active deformation in the Mediterranean and the Middle East', Geophys. J., 93, 45-73.

Jackson, J.A. and McKenzie, D. (1988b) ' Rates of active deformation in the Aegean Sea and surrounding area', Basin Res., 1, 121-128.

Kikuchi M. and Kanamori, H. (1982) 'Inversion of complex waves', Bull. Seism. Soc. Am., 72, 491-506.

Kikuchi, M. and Sudo, K. (1985) 'Inversion of teleseimic P-waves of the Izu-Oshima, Japan, earthquake of January 14, 1978', J., Phys. Earth, 83, 161-171.

Kim, W-Y., Kulhanek, O. and Meyer, K. (1984) ' Source processes of the 1981 Gulf of Corinth Earthquakes from body-wave analysis' Bull. Seism. Soc. Am., 74, 459-477.

Kanamori h. (1972) ' Tectonic implications of the Tonankai and the 1946 Nankaido earthquakes', Phys. Earth Plan. Inter., 5, 129-136.

Kanamori, H. and Stewart, G.S. (1978) ' Seismological aspects of the Guatemala earthquake', J. Geophys. Res., 82, 5613-5670.

Kanamori, H. and Cipar, J. (1974) ' Focal process of the great Chilean earthquake, May 22, 1960', Phys. Earth Plan. Inter., 9, 125-136.

Kiratzi, A.A., Karakaisis, G.F.,Papadimitriou, E.E>, and Papazachos, B.C. (1985) ' Seismic source-parameter relations for earthquakes in Greece', Pure and Applied Geophys.,123, 27-41.

Kiratzi, A.A and Langston C.A. (1991) ' Moment tensor inversion of the 1983 January 17 Kefallinia event of Ionian islands (Greece)', Geophys. J. Int., 105, 529-535.

Kulhanek, O., and Meyer, K. (1979) 'Source parameters of the Volvi-Langadhas earthquakes of June 20, 1978 deduced from body-wave analysis' Bull. Seism. Soc. Am., 68, 1298-1294.

Lay, T., Kanamori, H., and Ruff, L. (1982) 'The asperity model and the nature of large subduction zone earthquakes' Earth Pred. Res., 1, 3-71.

Le Pichon, X. and Angelier, J. (1979) 'The Hellenic arc and trench system: the key to the evolution of the eastern Mediterranean area' Tectonophysics, 60, 1-42.

Madariaga, R. (1976) ' Dynamics of an expanding circular fault', Bull. Seism. Soc. Am.,66, 639-666.

Main, I.,G. and Burton, P.W. (1989) ' Seismotectonics and the earthquake frequency-magnitude disrtibution in the Aegean

area', Geophys. J. R. astr. Soc., 98, 575-586.
McKenzie, D. (1972) ' Active tectonics of the Mediterranean region', Geophys. J. R. astr. Soc., 30, 109-185.
McKezie, D. (1978) 'Active tectonics of the Alpine-Himalayan belt: the Aegean Sea and surrounding regions', Geophys. J. R. astr. Soc., 55, 217-254.
Mori, J. and Shimazaki, K. (1984) ' High-stress drops of the short-period subevents from the 1968 Tokachi-Oki earthquake as observed on strong motion records', Bull. Seism. Soc. Am., 74, 1529-1544.
Nabelek, J. L. (1984) 'Determination of earthquakes source parameters from inversion of body waves' PhD Thesis, MIT, MA.
Pacheco, J. and Nabelek, J. (1988) ' Source mechanisms of three moderate California earthquakes of July 1986', Bull. Seism. Soc. Am., 78, 1907-1929.
Papageorgiou, A. and Aki, K. (1983) ' A specific barrier model for quantitative description of inhomogeneous faulting and prediction of strong ground motion. Part I: Description of the model', Bull. Seism. Soc. Am., 73, 693-722.
Papazachos, B., Kiratzi, A. and Papadimitriou, E. (1991) ' Regioanl focal mechanisms for earthquakes in the Aegean area', Pageoph, 136, 405-420.
Soufleris, C. and Stewart, G.S. (1981) ' A source study of the Thessaloniki (northern Greece) 1978 earthquake sequence' Geophys. J. R. Soc., 67, 343-358.
Stavrakakis, G., Tselentis, G-A., and Drakopoulos J.(1987) ' Iterative deconvolution of teleseismic P-waves from the Thessaloniki (n. Greece) earthquake of June 20, 1978', Pageoph, 124, 1039-1050.
Stavrakakis, G., Drakopoulos, J. and Makropoulos, K. (1986)' A rupture model for the Corinth earthquake sequence of 1981', Proc. of International Seminar on Earthquake Prognostics, 24-27 June, 1986'Berlin, 129-153.
Stavrakakis, G.N., Ioannidou, E. and Drakopoulos, J. (1989) ' A source model for the January 17, 1983 Cephalonia (W. Greece) earthquake from time domain modelling of teleseismic P-waves', Boll. di Geo. Teor. ed Apll., XXXI, N.122, 149-157.
Stavrakakis, G.N., Drakopoulos, J., Latoussakis, J., Papanastassiou, D., and Drakatos, G. (1989a) 'Spectral characteristics of the 1986 September 13 Kalamata (southern Greece) earthquake', Geophys. J. Int., 98, 149-157.
Stavrakakis, G.N. and Blionas, S.V. (1990) ' Source parametrs of some large earthquakes in eastern Mediterranean region based on an iterative maximum entropy technique', Pageoph, 132, 680-698.
Stavrakakis, G.N., Blionas, S.V., and Goutis, C.E. (1991) ' Dynamic source parameters of the 1981 Gulf of Corinth

(central Greece) earthquake sequence based on FFT and iterative maximum entropy techniques', Tectonophysics, 185, 261-275.

Taymaz, T., Jackson, J. and Westaway, R. (1990) 'Earthquake mechanisms in the Hellenic Trench near Crete', Geophys. J.Int., 102, 695-731.

Taymaz, T., Jackson, J. and McKenzie, D. (1991) 'Active tectonics of the north and central Aegean Sea', Geophys. J. Int., 106, 433-490.

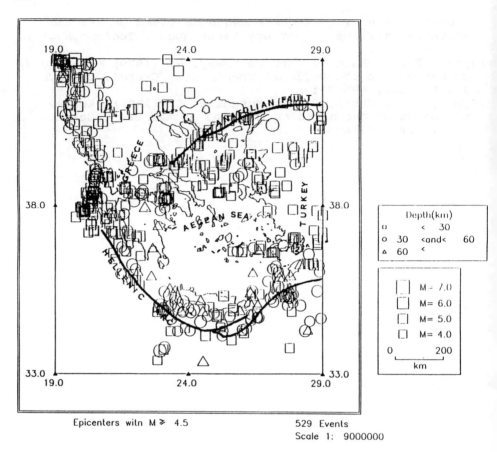

FIG.1: Seismicity map of the Aegean region. Data are taken from NOA bulletins

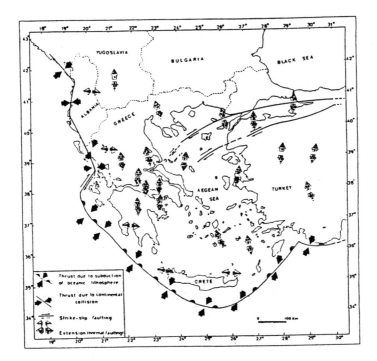

FIG.2: The main tectonic features in Greece and the nearby area (Papazachos et al., 1991)

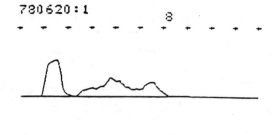

FIG.3: Far-field displacement source-time function(upper) obtained after 25 iterations and moment rate function (lower).

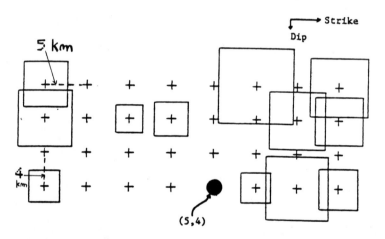

Fig.3: (A) Source time function, (B) moment rate function and (C) Spatial distribution of the areas of the largest seismic moments on the fault plane of the Thessaloniki earthquake of June 20, 1978 (Stavrakakis et al., 1987). The soloid circle depicts the rupture start point.

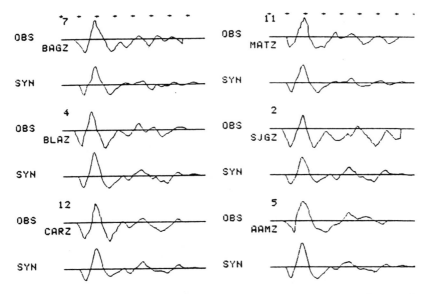

Fig.4: Observed and synthetic seismograms for the Thessaloniki earthquake of June 20, 1978 (Stavrakakis et al., 1987)

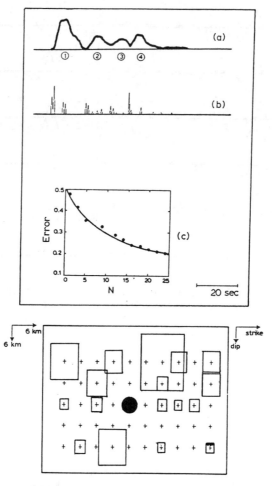

Fig.5: (a) Source time function, (b) moment rate function, (c) normalized approximation error versus number of iterations and spatial distribution of the areas of the largest seismic moment for the main shock of Feb. 24, 1981 (Stavrakakis et al., 1986). The solid circle depicts the rupture start point.

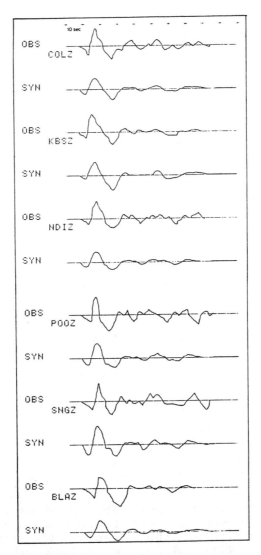

Fig.5a: Observed and synthetic waveforms for the main shock of the Feb.24, 1981.

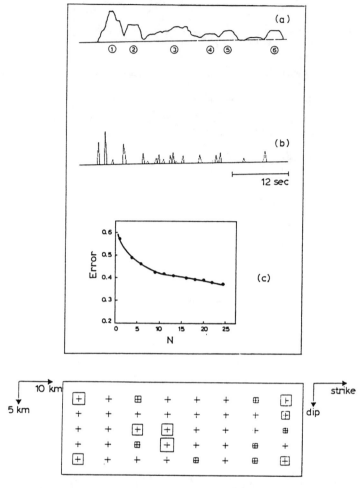

Fig.6: (a) Source time function, (b) moment rate function, (c) normalized approximation error vs number of iterations and spatial distribution of the areas of the largest seismic moment for the first large aftershock of Feb.25, 1981 (Stavrakakis et al., 1986). The solid circle depicts the rupture start point.

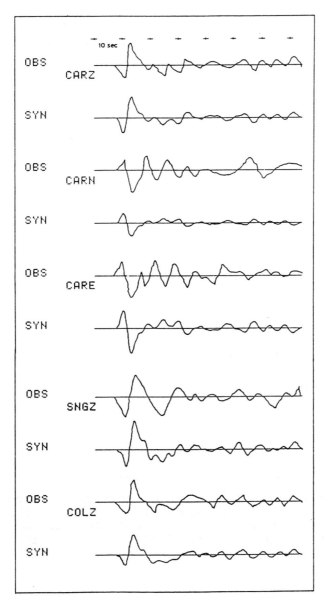

Fig.6a: Observed and synthetic waveforms for the first large aftershock of Feb.25, 1981.

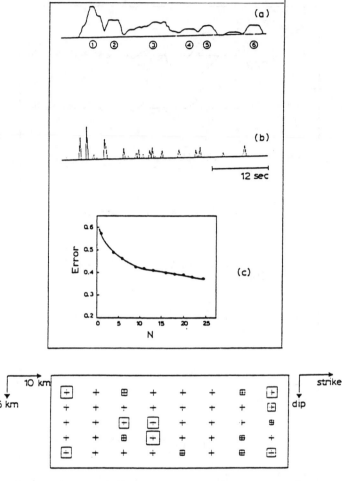

Fig.7: (Source time function, (b) moment rate function, (c) normalized approximation error vs number of iterations and spatial distribution of the second large aftershock of March 4, 1981 (Stavrakakis et al., 1986). The solid circle depicts the rupture start point.

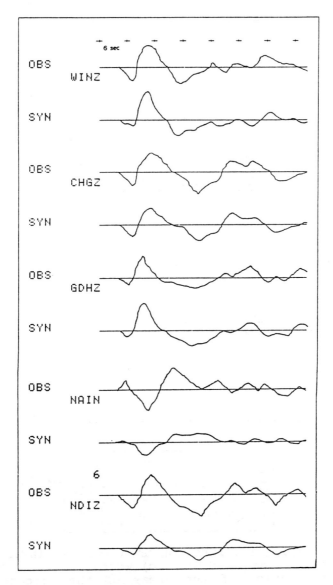

Fig.7a: Observed and synthetic waveforms for the second large aftershock of March 4, 1981.

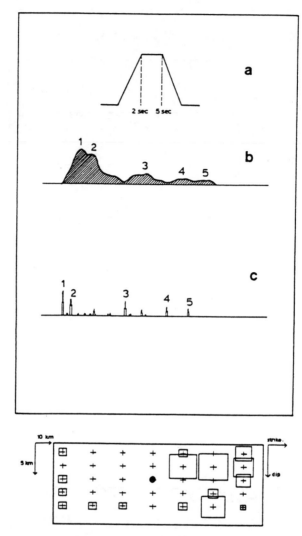

Fig.8: (a) Source time function used for the inversion, (b) inferred source time function, (c) moment rate function, and spatial distribution of the areas of the largest seismic moment for the Cephalonia island earthquake of Jan.17, 1983 (Stavrakakis et al., 1989)

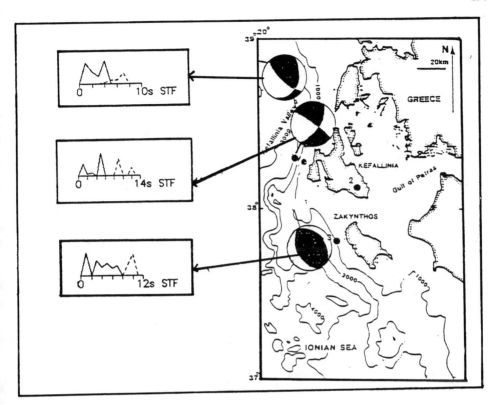

Fig.9: Fault plane solutions and source time functions of some large and moderate earthquakes occurred in Ionion Sea (after Ioannidou, 1989).

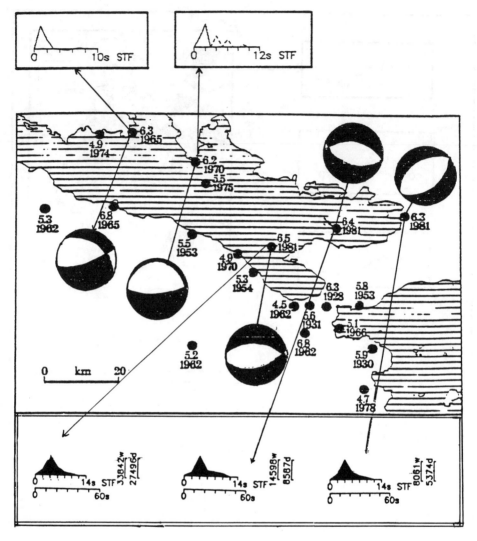

Fig.10: Fault plane solutions and source time functions of some large and moderate earthquakes occurred in the Gulf of Corinth (Central Greece) (After Ioannidou, 1989).

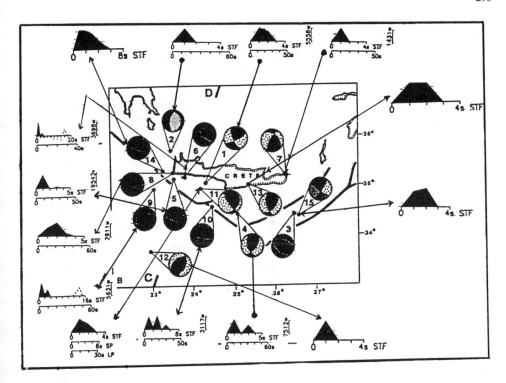

Fig. 11: Fault plane solutions and source time functions of some large and moderate earthquakes occurred in the southern part of the Hellenic arc (After Taymaz et al., 1990).

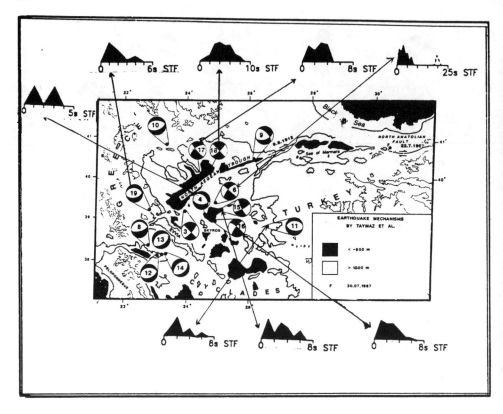

Fig.12: Fault plane solutions and source time functions of some large and moderate earthquakes occurred in north and central Aegean (after Taymaz et al., 1991).

RECENT SEISMIC ACTIVITY AND EARTHQUAKE OCCURRENCE ALONG THE APENNINES

M. Cocco, G. Selvaggi, M. Di Bona and A. Basili
Istituto Nazionale di Geofisica
Via di Vigna Murata 605, 00143, Roma, Italy

ABSTRACT. In this paper we analyze the seismicity along the Apennines in order to relate the seismicity pattern to the seismogenic environment and to the active tectonic processes. The analysis of crustal and subcrustal seismicity show that the northern and southern Apennines are characterized by two distinct patterns of seismicity. These two domains are separated by two important lithological discontinuities (the *Ancona - Anzio* and the *Ortona - Roccamonfina* lines). The seismicity along the Apennines is mainly concentrated in a narrow belt running along the chain, with an evident geometrical offset which corresponds to the *Ancona-Anzio* line. We discuss these observations considering both the seismological evidence and the present tectonic regime in which the earthquakes occur. We focus on the southern Apennines, where the largest earthquakes ($6 \leq M \leq 7.0$) have occurred. We analyze the main features of some recent seismic sequences in order to discuss the seismicity patterns in terms of the hypothesis of segmentation of the southern Apennine seismogenic belt. This work aims to be a preliminary contribution to the important goals of understanding the spatial and temporal patterns of seismicity and of identifying active faults.

1. Introduction

The heterogeneity and complexity of individual fault zones are related to the epicentral and hypocentral patterns which develop over time and during aftershock sequences. In this work we analyze the seismic activity in southern Italy, detected in the past years by permanent and local networks in order to relate the seismicity patterns to the seismogenic structures of the Apennines. The analysis of seismicity patterns provides important information about the earthquake preparatory processes and the occurrence of large earthquakes.

The Apennines are characterized by an extensional tectonic setting in which several moderate-sized ($5 < M < 7$) earthquakes have occurred during the last two decades (Figure 1). The strongest event that occurred in the region is the 1980 Irpinia earthquake ($M_S = 6.9$). The faulting process of the 1980 earthquake is well known, thanks to the analysis of the observed seismic radiation, the evidence for surface rupture, and the crustal deformation pattern (*Westaway and Jackson, 1987; Bernard and Zollo, 1989; Pantosti and Valensise, 1990; Amato et al. 1992*). Both rheological heterogeneities and geometrical irregularities affected the faulting process producing an evident fault fragmentation (*Pantosti and Valensise, 1990; Cocco and Pacor, 1992*).

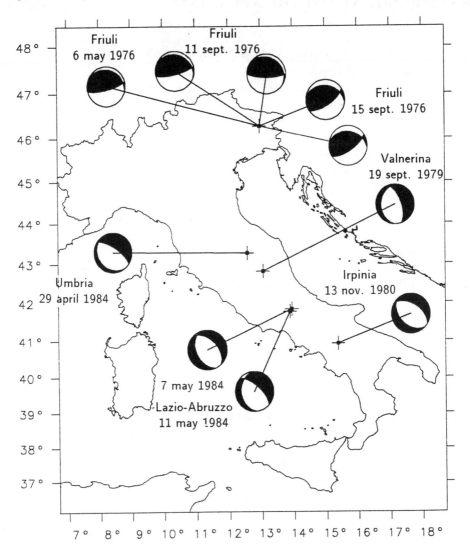

Fig. 1 Fault plane solutions of recent strong earthquakes in Italy. The figure points out the different tectonic setting of the north-eastern Alps with respect to the Apennines. Focal mechanisms of the Friuli earthquakes are taken from *Cipar (1981)* and *Anderson (1985)*; those of the Apenninic earthquakes from *Giardini et al. (1984)* and from *NEIS* bulletins.

Unfortunately, the 1980 Irpinia earthquake is the only Italian seismic event for which the geometry of the activated faults and their position are well known. More work is needed to obtain a reliable map of active faults in Italy and to investigate the interactions between the different fault segments of the main seismogenic areas. With this in mind, we focus in this paper on the southern Apennines. The existence of a seismogenic belt along the southern Apennines was originally proposed by *Omori* (1909) by analyzing the alignment of isoseismals of the strongest earthquakes (Figure 2). Because of the large number of strong earthquakes that occurred in the past along the southern Apennines, the geophysical data available for this area, and the accurate knowledge of the rupture process of the recent 1980 Irpinia earthquake, the Southern Apennine seismogenic belt represents the only favourable opportunity to shed light on fault segmentation in Italy.

Recently, *Pantosti and Valensise* (1989) proposed the existence of a segmented fault zone striking along the Apenninic chain. Each segment can be fragmented in several faults that can rupture together during the major earthquakes, as clearly shown by the 1980 Irpinia earthquake. Figure 3 shows the Mercalli intensity X isoseismal curves of the largest historical earthquakes occurred in the central and southern Apennines together with the fault segments proposed by *Pantosti and Valensise (1989)* and the distribution of seismicity during the past 12 years. The highest rate of seismicity during this period occurs in the zone struck by the 1980 Irpinia earthquake (*Amato and Cocco, 1992*) whose isoseismals are very similar to those of the 1694 event (drawn in the Figure). This Figure shows that seismicity is distributed along the seismogenic belt and that is clustered in particular areas.

This work aims to point out the important contribution of seismicity to the tectonic setting of southern Apennines. We first analyze the seismicity along the Apenninic chain, considering the present tectonics and the main features of the seismogenic areas. Afterwards, we analyze the details of the recent seismic sequences for which accurate microseismic data are available in order to explain the patterns of seismicity in terms of the fault segmentation hypothesis.

2. Analysis of the seismicity along the Apennines

We re-analyze the distribution of crustal and subcrustal seismicity that occurs along the Apennines in order to associate the seismic data with the recent tectonic processes that the Apennines are following.

The data set is a collection of more than 8,000 earthquakes detected in the past 6 years (1986-1992), for which more than 7 P arrivals are available. The data set includes arrival times at the 74 seismic stations, mostly vertical, of the *ING* national network. It also includes arrivals at stations of other regional networks in order to improve the azimuthal coverage. We have re–located the earthquakes using the program Hypoinverse (*Klein, 1989*) with a three-layer laterally homogeneous velocity model (*Selvaggi and Amato, 1992*). The use of different velocity models and V_P/V_S values led to changes in the focal coordinates within the estimated errors, thus ensuring the stability of the hypocentral

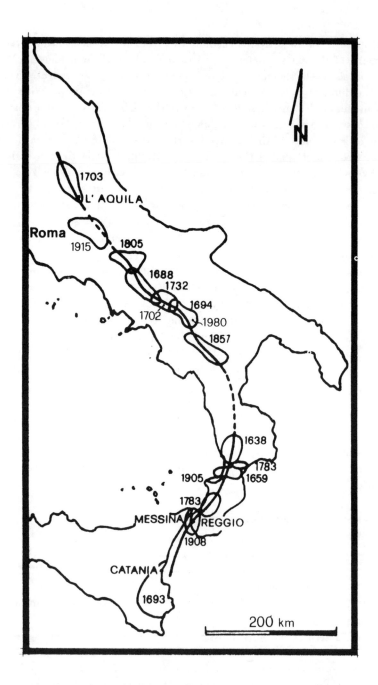

Fig. 2 Alignment of isoseismals of the larger earthquakes along the Apennines, re-drawn from *Omori, (1909)*. Also included are the isoseismals of the 1915 Avezzano earthquake, and the zone struck by the 1980 Irpinia earthquake.

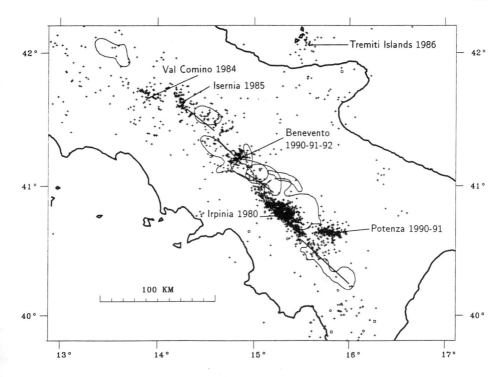

Fig. 3 Distribution of seismicity from 1980 to 1992 detected in the past 12 years by permanent and local networks. This time window has been selected for the quality of the data and in order to include the Irpinia 1980 seismic sequence. This figure also shows the isoseismal curves of Mercalli intensity X for the largest historical earthquakes, and the fault segments proposed by *Pantosti and Valensise (1989)* belonging to the Southern Apennine seismogenic belt.

solutions. Hypocentral locations were selected considering both the vertical and horizontal formal errors, the RMS, and other criteria based on the convergence of the solutions. As the S phases were mostly picked from the vertical components, we also located the events using P phases only, producing negligible changes in the calculated hypocenters. Due to the particular shape of the Italian peninsula and hence to the geometry of the network, the error ellipses of the earthquake locations are slightly elongated towards the E–W direction.

Figure 4 shows a map of the re-located seismicity from 1986 to 1992. This time window has been selected because after 1986 the network configuration allows reliable earthquake locations. The data plotted in Figure 4a represents a subset of more than 4,000 earthquakes whose horizontal errors are less than 8 km and that satisfy the imposed criteria on RMS errors and solution convergence. Figure 5 shows a simplified structural sketch of the italian region. Three main seismotectonic domains can be delineated from SW to NE across the Italian peninsula, based on the crustal seismicity distribution: (a) the Tyrrhenian belt,

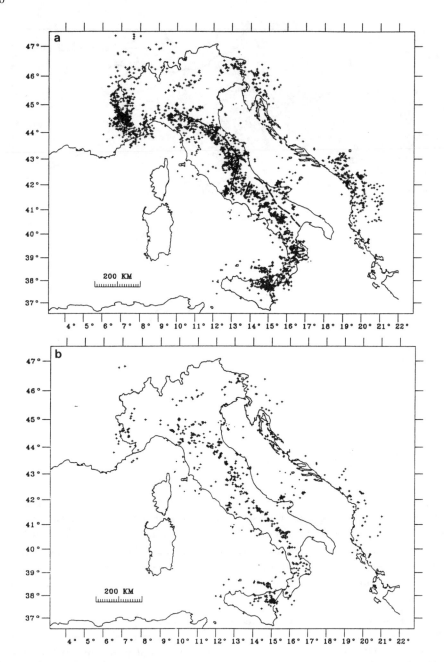

Fig. 4 a) Distribution of instrumental seismicity from 1986 to 1992 detected by local and national networks. Only hypocenters with horizontal and vertical formal errors less than 8 km are plotted. The compressive margin of the Apeninnes (dashed line), the *Ancona - Anzio* and the *Ortona - Roccamonfina* lines (solid lines) have been drawn in the figure b) Distribution of the epicenters of earthquakes with magnitude larger than 3.0 (Erh and Erz less than 3.0 km).

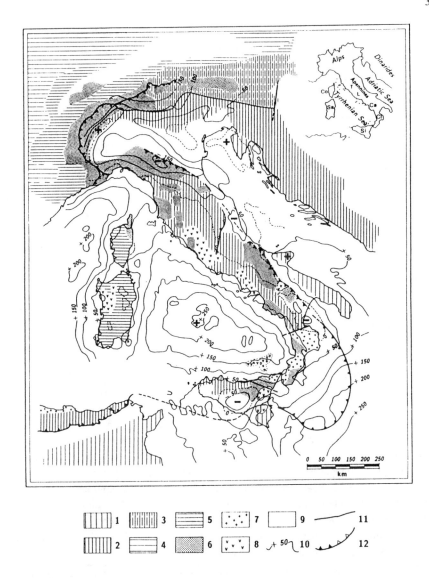

Fig. 5 Structural sketch of the Italian region [simplified and redrawn after *Ogniben et al.*, 1975; *Funiciello et al.*, 1980; *Malinverno* and *Ryan*, 1986; *Amato et al.*, 1993]. 1 = African foreland; 2 = deformed African continental margin; 3 = Austroalpine nappes; 4 = European foreland; 5 = deformed European continental margin; 6 = oceanic remnants (ophiolites, oceanic sediments - Ligurides, Sicilides, etc. -); 7 = "internal massifs" (metamorphosed Alpine and pre-Alpine complexes); 8 = volcanic complexes (includes both the Tertiary volcanism of Sardinia and the (Plio-)Quaternary volcanism of Tuscany, Latium, Campania, and Sicily); 9 = terrigenous sediments of the late- and post-orogenic phases; 10 = Bouguer gravity anomalies; 11 = important tectonic lines with uncertain meaning (i.e. Insubric Line in the Alps); 12 = main thrust front system of Alps and Apennines (solid triangles) and boundary (*Ancona-Anzio* line) between northern and central Apennines (open triangles).

with shallow ($\leq 7km$) small magnitude earthquakes, mostly concentrated in the vicinity of the Quaternary volcanoes and the geothermal areas; (b) the Apenninic chain, characterized by stronger and deeper events (≤ 20 km); and (c) the Adriatic foredeep, where the density of seismicity is generally less.

The seismicity along the Apennines is mainly concentrated in a narrow belt that follows the highest elevations. The distribution of epicenters along the Apennines shows an evident offset in the central section of the chain that corresponds with a well known lithological structural discontinuity, the *Ancona-Anzio* line (Figure 4a). The offset is roughly 50 km in a direction perpendicular to the trend of the Apennines, and is a factor of 10 larger than the earthquake location errors. Thus, even if the error for Apenninic events is greatest in the EW direction, the offset of seismicity is not an artifact of the location procedure. Figure 4b shows the distribution of the earthquakes with magnitude larger than 3.0. Even with fewer events, the offset between the northern and southern Apennine seismic belts is still evident. This offset separates two domains each of which shows a peculiar seismicity pattern. The northern domain coincides with the northern Apennines and is characterized by a diffuse seismicity, rarely exceeding magnitude 5, with a high rate of earthquake occurrence. The crustal seismicity is concentrated along the Apenninic arc and bends following the curvature of the axis of the chain (Figure 4a). The southern domain coincides with the portion of the Apenninic chain where the largest earthquakes ($M \approx 7$) have occurred during historical times; the crustal seismicity is mainly concentrated along a narrow ($\sim 30-35$ km) and straight belt. Moreover, intermediate-depth earthquakes occur only in the northern domain: *Selvaggi and Amato (1992)*, found 40 anomalously deep earthquakes ranging from 30 to 90 km beneath the northern Apennines located slightly to the southwest of the axis of the chain. This subcrustal seismicity deepens westward or southwestward, approximately delineating a $35° - 45°$ dipping wedge from the Adriatic to the Tyrrhenian Sea (*Selvaggi and Amato, 1992*). The depth distribution of these earthquakes may suggest that the Adriatic lithosphere is presently subducting beneath the northern Apenninic chain, even if it can not be clearly related to a well – defined dipping Wadati–Benioff zone. The presence of a slab in the northern Apennines has been already suggested both by teleseismic tomography (*Amato et al., 1993*) and by other geophysical data (*Panza et al., 1990*). No intermediate-depth and deep earthquakes occur along the southern Apennines (South of the *Ancona-Anzio* line); the deep seismicity is concentrated in the Calabrian arc and in the southern Tyrrhenian sea.

All the observations concerning crustal and subcrustal seismicity suggest that northern and southern Apennines exhibit different seismotectonic behaviors. Our interpretation is that the *Ancona-Anzio* discontinuity may represent the northern end of the Apenninic segmented seismogenic belt. *Westaway (1992)* investigating the tectonic deformation rates in and around the Apennines found that northern Apennines are tectonically distinct, at present, from the areas farther south. He concluded that the northern Apennines show different rates of deformation from the southern section of the chain.

From a geological point of view the *Ancona - Anzio* structural line marks the border between the pelagic facies of the northern Apennines and the platform units of the central and southern Apennines (Figure 5). South of the *Ancona - Anzio* line there is another structural discontinuity named the *Ortona - Roccamonfina* line. Both these two lithological discontinuities run from the Tyrrhenian belt to the Adriatic foredeep and they presumably delimit a transition zone between the pelagic facies and the platform units (Figure 5).

The geometrical offset shown by the distribution of seismicity (Figure 4) corresponds to the increase of the curvature of the northern compressive margin proposed by different geodynamical models of the Apennines (*Elter et al., 1975; Patacca and Scandone 1989*). This offset demonstrates that the spatial pattern of both microseismicity and moderate-sized earthquakes reflects the geometry of the major lithological discontinuities (such as the *Ancona - Anzio* or the *Ortona - Roccamonfina* lines). Moreover, the analysis of the seismicity in the Southern Apennines shows that the regional pattern is mostly associated with the largest crustal earthquakes that occur along a segmented seismogenic belt.

3. Analysis of recent seismic sequences

In this section we analyze the main features of recent seismicity in order to investigate whether the seismicity patterns are related to the deformation processes along the seismogenic belt. We accept the hypothesis that the seismogenic belt is segmented into different fault systems each of which may be further fragmented into single fault elements. Paleoseismic studies (*Pantosti et al. 1993*) did not support the hypothesis of a single segmented fault striking along the Apenninic chain; rather they suggested the existence of a segmented belt.

The analysis of regional seismicity along the southern Apennines shows the highest rate in particular areas (see Figure 3): the Irpinia area struck by the 1980 ($M_S = 6.9$) earthquake, the Valcomino area where two earthquakes of magnitudes 5.2 and 5.5 have occurred in 1984 (see Figure 1), the Isernia area where several swarms occurred in 1985 and 1986, and finally the Potenza and Benevento areas where the seismic sequences discussed in this paper have occurred. Furthermore, Figure 3 depicts several areas where large historical events occurred and only few earthquakes have been recorded in the past decade. Even if the highest seismicity rates occur in the areas where local networks were operating (Potenza, Benevento and Irpinia), the data set plotted in Figures 3 and 4 is complete for magnitude larger than 3.0. This implies that the lack of seismicity observed in the areas struck by the 1857, 1805 and 1732 earthquakes is not an effect of the detection threshold of the *ING* national network.

Because the geometry and the position of the Irpinia fault system is well known, we decided to examine the seismicity patterns in two adjacent areas: the Potenza and Benevento areas which lie close to the southern and to the northern limit of the Irpinia fault, respectively (see Figure 6). High quality microseismic data are available for both these areas, thanks to the presence of local permanent and portable networks. We present the main features of the seismicity pattern in these two areas in this section, leaving the dis-

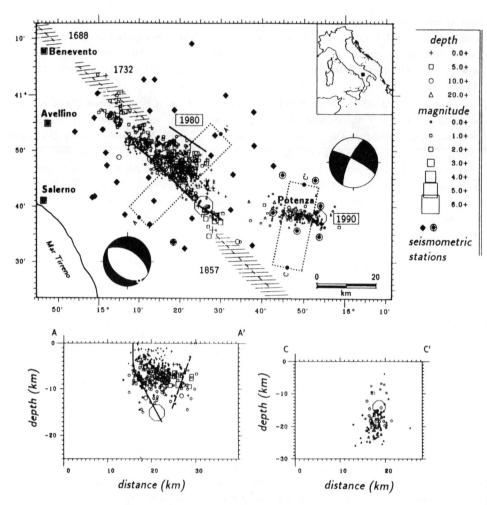

Fig. 6 Epicentral distribution of the earthquakes of the 1980 Irpinia and 1990 Potenza sequences. The focal mechanisms of the two mainshock are shown. Bold diamonds and circled stars indicate the seismic stations of the temporary networks. Cross-sections A-A' and C-C' show the hypocentral depths of the 1980 and 1990 earthquakes, respectively. The fault fragments that ruptured during the 1980 event are drawn in the map, and their geometry at depth is shown in cross-section A-A'. Shadowed areas indicate the zones struck by the largest historical events.

cussion of the relation between local seismic activity and regional seismicity to the last section.

The Potenza sequences of 1990 and 1991. Numerous moderate magnitude seismic sequences ($M_L \leq 5.0$) are clustered at the south–east end of the Irpinia normal fault, in the Potenza region (*Baratta, 1901*). In particular, in the past decade two earthquakes with magnitudes greater than 4.0 occurred in this area, on February 2, 1983 and on July 23, 1986; in addition a small magnitude seismic sequence ($M \leq 4.0$) occurred in 1988. The two most recent seismic sequences occurred in 1990 and 1991; the availability of a local digital network allowed the collection of accurate seismic data. Figure 6 shows the location of the May 5, 1990 ($M_L=5.2$) earthquake and the distribution of aftershocks. This figure also shows the distribution of epicenters of the 1980 Irpinia sequence, the surface expression of the four fault fragments that ruptured during the 1980 event, and the largest historical earthquakes that occurred in the adjacent zones. The focal mechanism (CMT) of the 1990 mainshock shows an almost pure strike slip solution with a N46E T axis and a N138E P axis (Figure 6). The T axis lies in the same direction as the 1980 earthquake (Figure 6) and parallel to the general extensional direction of central and southern Apennines (Figure 1).

The distribution of the Potenza aftershocks depicts an overall seismicity pattern comprised by an east–west striking zone of epicenters which extend for about 15 km, consistent with the $\sim$ east-west nodal plane of the mainshock focal mechanism (Figure 6). The hypocentral depths (see cross section C-C' in Figure 6) are mainly concentrated between 15 and 25 km. Figure 6 compares the distribution of hypocenters of the 1980 Irpinia and the 1990 Potenza sequences. Noteworthy is the observation that the aftershock depths of the Irpinia sequence are generally concentrated within the first 15 km of the upper crust (*Deschamps and King, 1984; Amato et al., 1992*), while those of the 1990 Potenza sequence are deeper, with only a few occurring above 10 km depth.

On May 26, 1991, a $M_L = 4.7$ earthquake occurred within a few kilometers of the 1990 mainshock (Figure 7). We installed 6 three–component digital stations in the epicentral area that allowed us to accurately locate more than 40 aftershocks with magnitudes ranging between 2.0 and 3.0 during 20 days. The fault plane solution for the 1991 event shows an orientation of the P and T axes similar to those of the 1990 event. Figure 7 shows the epicentral map of the best located earthquakes (errors less than 2.0 km) of the 1990 and 1991 sequences. The epicenters of the earthquakes of the 1991 sequence are located slightly westward of those of the 1990 sequence. Hypocentral depths are mostly between 10 and 20 km: the 1991 hypocenters are shallower (see the cross-section A-A' in Figure 7) than those of the 1990 seismic sequence and have similar depths to those of the 1980 Irpinia sequence (Figure 6). The overall pattern of hypocenters delineates an almost vertically dipping plane (see cross-section B-B' in Figure 7).

These two sequences occurred close to the southern termination of the Irpinia fault, but the earthquakes of 1991 sequence occurred closer to the southern termination of the 1980 segment than did the events of the 1990 sequence (Figure 7). South of the 1980 segment occurred the large 1857 *Neapolitan* earthquake. The position and the geometry of the faults that ruptured during the 1857 earthquake are unfortunately unknown. This circumstance limits the possibility of interpreting the seismicity pattern in the region near

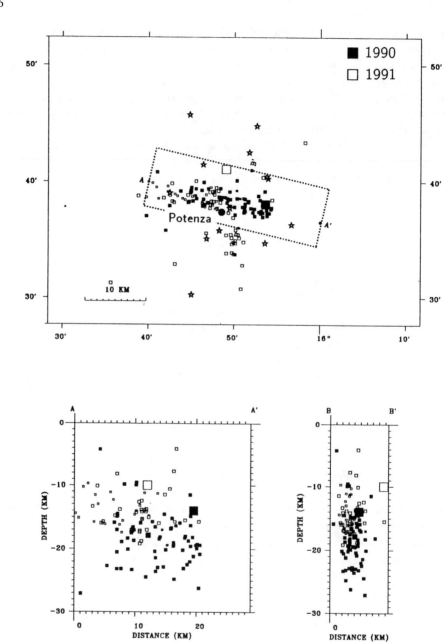

Fig. 7 Epicentral map of the best located earthquakes (errors less than 2.0 km) of the Potenza 1990 (solid squares) and 1991 sequences (open squares). The location error of the 1991 mainshock is larger than 2.0 km. Stars indicate the seismic stations. Cross-section A-A' shows that the seismicity during the 1991 sequence is slightly shallower and occurred to the west of the 1990 seismic events. B-B' shows that the hypocenters delineate a vertical dipping plane.

Potenza in terms of seismicity associated with the interaction between the 1980 and the 1857 fault zones.

The seismic activity in the area of Potenza shows one of the highest rates of occurrence in Italy during the last 8 years. The analysis of recent seismicity shows that earthquakes ($2.0 \leq M \leq 5.2$) occur along a E-W striking, vertical dipping zone. We cannot conclude, with the presennt level of knowledge, if the deformation, evident in the seismicity, is related to the loading of the 1857 fault segment or if it is an independent local process.

The Benevento sequence of 1992. The northern termination of the 1980 Irpinia fault lies in a zone where large historical earthquakes occurred in 1732, 1702 and 1688 (Figures 2 and 3). The Irpinia area was also struck by a large earthquake in 1694. It is interesting to observe that within 44 years (1688-1732) four large earthquakes occurred between the Irpinia and Benevento areas.

Figure 8 shows a map of seismicity detected by permanent and local networks in the northern section of the southern Apennine seismogenic belt. The Mercalli intensity X isoseismal curves of the 1688, 1702 and 1732 earthquakes, the focal mechanism and the epicenter of the ($M_L = 6.1$, Mercalli intensity IX) 1962 earthquake (*Westaway, 1987*) are also plotted. The 1962 earthquake occurred between the areas struck by the 1688 and 1732 events and within the area struck by the 1702 earthquake.

Even if the historical data suggest a significant stress release in this zone, the analysis of recent seismicity reveals only a low rate of seismicity with a swarm-type pattern. Moreover, the recent seismicity is mostly clustered near the town of Benevento (Figure 8), where two swarms occurred in 1990 (*Iannaccone et al.*, 1990) and in 1992 (*Cocco et al., 1993*), in which the largest magnitudes rarely exceed $M_L = 3.0$ ($M \leq 3.5$). In 1991, the ING deployed a permanent microseismic network between Benevento and Isernia, close to the area struck by the 1688 earthquake. The goal was the monitoring of an area that we consider to be a candidate for a large earthquake in the future: that is a seismic *gap* whithin the southern Apennine seismogenic belt.

The microseismic network recorded only 200 microearthquakes ($M \leq 3.0$) within one year. However, during March 1992 a seismic swarm occurred close to Benevento (Figure 8-a) and almost 300 events were detected within one week with the strongest earthquakes having magnitudes close to 3.2. The cross-section A-A' in Figure 8-b shows that the earthquakes in 1991 and 1992 mostly occur in two well defined zones: the 1991 events occurred in a vertical dipping plane N-W of the town of Benevento, while the 1992 sequence occurred in a well defined 45° S-W dipping plane. The seismicity is mostly clustered in the middle of the 1688 area to the north of the 1702 fault zone. Note that the 1962 earthquake occurred between the 1688 and the 1732 zones and in the middle of the 1702 area.

Despite of the high seismogenic potential, the area of Benevento is characterized by a low rate of seismicity and by a low energy release. In conclusion, we observe that the pattern of seismicity, in the past 10 years, in the area of Benevento is different from that of the Potenza area, even if both these areas have a high seismogenic potential and may be considered as candidates for future earthquakes.

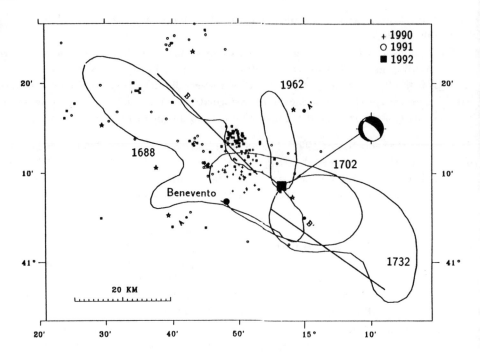

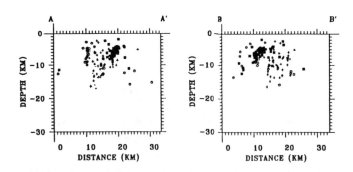

Fig. 8 Map of seismicity of the Isernia and Benevento areas as detected from permanent and temporary local networks between 1990 and 1992. Crosses, open circles and solid squares are referred respectively to 1990, 1991 and 1992 sequences. The best located events (Erh and Erz less than 2.0 and 3.0 km, respectively) are shown. Figure (8-a) shows the map of instrumental seismicity since 1990, the isoseismals of Mercalli intensity X of the 1688, 1702 and 1732 events, the isoseismals of Mercalli intensity IX of the 1962 earthquake with its epicenter and focal mechanism (taken from *Westaway, 1987*), and the fault segments of the historical earthquakes (proposed by *Pantosti and Valensise, 1989*). In (8-b) is plotted the SW-NE cross section of the hypocenters of the earthquakes located within 10 km of A-A'. In (8-c) is shown the NW-SE cross section of the hypocenters of the earthquakes located within 10 km of from B-B'.

4. Discussions and conclusions

Our understanding of the geological and seismological evolution of the Apennines is far from being accomplished. Different geodynamical models have been proposed to explain the orogenesis of the Apenninic chain (*Malinverno and Ryan, 1986; Lavecchia, 1988; Patacca and Scandone, 1989; Mantovani et al., 1990; Doglioni, 1991*, among many others). Some of these studies use seismic data to confirm their theses, even if it is difficult to assess how seismicity data may support the conclusions of geodynamical studies because the seismic energy release is only a small fraction of the energy involved in geological deformation processes. However, the spatial distribution of seismicity does reflect some important features of the present active tectonics of the Apennines. The most important seismological observations involving relevant geodynamical inferences are: (i) the deep seismicity of southern Tyrrhenian sea that demonstrates the presence of a subducting slab (*Anderson and Jackson, 1987; Giardini and Velonà, 1990*); (ii) the subcrustal seismicity of the northern Apeninnes that suggests that subduction may be still active, even if at slow rate (*Selvaggi and Amato, 1992*). In this paper we add a new result, that is, the geometrical offset in the distribution of epicenters in the central Apennines that separates two domains each of which shows a peculiar seismicity pattern. This observation is discussed in terms of the present tectonic environment rather than of active geodynamic processes.

The distribution of epicenters of the re-located earthquakes follows the geometry of the chain and it is mainly concentrated in a narrow belt showing a geometrical offset in the central section corresponding to the increase of the curvature of the northern Apennines compressive margin. Our conclusion is that the offset is due to the presence of important structural discontinuities, such as the *Ancona-Anzio* line. However, the seismological data do not demonstrate that these structures are seismogenic. As the focal mechanisms of the recent largest earthquakes both in central and southern Italy show an extensional faulting regime, we suggest that the lithological discontinuities do not affect the direction of regional stress field. Our interpretation is that the *Ancona-Anzio* line represents the northern limit of the southern Apennine seismogenic belt and that it separates two different seismotectonic domains which exhibit different seismicity patterns. The northern Apennine domain is characterized by a diffuse seismicity concentrated along the arc, rarely exceeding magnitude 5, with a high rate of earthquake occurrence. The southern Apennine domain is characterized by the largest coseismic energy release along the chain with 9 earthquakes with magnitude $M \geq 6.5$ in the past 300 years. The crustal seismicity recorded in the last decade is mostly concentrated along a narrow and straight zone indicating the elongation of the seismogenic belt.

We have focused our attention on the southern Apennines. The analysis of regional seismicity along the southern Apennines shows that only few earthquakes have been recorded in the past decade within the areas struck by large historical events (such as the 1857, 1805 and 1732 earthquakes), with the exception of the Irpinia zone where the 1980 earthquake occurred. On the contrary, several sequences, characterized by different seismicity patterns, occurred in the adjacent areas. We have analyzed the main features of two seismic sequences for which accurate microseismic data are available in order to discuss the main features of seismicity in areas between adjacent segments of the seismogenic belt. The two seismic sequences occurred in two areas with a high seismogenic potential and close to the zone recently struck by the 1980 Irpinia earthquake (namely Potenza and Benevento). The area of Potenza is characterized by a high rate of seismicity with numerous moderate-sized earthquakes; two mainshocks of magnitude close to 5 occurred in 1990 and 1991. The epicenters of the 1991 Potenza sequence are closer to the southern termination of the 1980 fault zone than do those of 1990 sequence. Moreover, the hypocentral depths of the earthquakes of 1991 are similar to those of the 1980 Irpinia sequence ($3-15$ km) while the hypocentral depths of 1990 Potenza events are deeper ($15-25$ km). The understanding of the unusual depths of the 1990 earthquakes remains an open question that has to be investigated in the future. The Benevento zone is, at present time, characterized by a lower rate of seismicity with small magnitudes ($M \leq 3.5$). The seismic activity shows a swarm type behavior with shallow earthquakes mostly concentrated in small volumes of the crust. In conclusion, we observe a significant regional variations of the seismicity patterns among the areas having one of the highest seismogenic potential in Italy. The heterogeneity and/or the complexity of individual fault zones can be responsible for the observed variations. The results obtained in this study stress the importance of a continuous monitoring of seismogenic areas by means of local networks.

Acknowledgments

We thank Alessandro Amato, Daniela Pantosti, Gianluca Valensise, for the helpful discussions. We wish to thank Enzo Boschi, Enzo Mantovani and Andrea Morelli to invite us to partecipate at the workshop *Recent Evolution and seismicity of the Mediterranean region*. We would also thank Domenico Giardini and Leslie Sonder for reviewing this paper. We thank Daniela Riposati for drawing many of the figures.

References

Amato A., C. Chiarabba, L. Malagnini and G. Selvaggi, (1992). Three-Dimensional P-Velocity Structure in the Region of the M_S 6.9 Irpinia, Italy, Normal Faulting Earthquake, *Physics of the Earth and Planetary Interior*, **75**, 111-119.

Amato, A., B. Alessandrini, and G.B. Cimini, (1993). P-wave Tomography of Italy, in Seismic Tomography, H. M. Iyer and K. Hirahara eds., Chapman and Hall, in press.

Amato, A., and M. Cocco (1992). Monitoring Seismogenic Areas with Local Networks, a Basic Tool for Earthquake Prediction: Seismicity Patterns and Faulting Mechanisms in Italy. *International Workshop on Earthquake Prediction, Erice, Italy*.

Anderson H.J. (1985). Seismotectonics of Western Mediterranean, Ph. D. Thesis, University of Cambridge, U.K.

Anderson, H., and J. Jackson (1987). The Deep Seismicity of the Tyrrhenian Sea, *Geophys. J. R. Astr. Soc.*, *91*, 613-637.

Baratta M. (1901). I Terremoti d'Italia, *Forni edit.*, 950 pp. (in italian).

Bernard, P., and A. Zollo (1989). The Irpinia (Italy) 1980 Earthquake: Detailed Analysis of a Complex Normal Fault, *J. Geophys. Res.*, **94**, 1631-1648.

Cipar J. (1981). Broadband Time Domain Modeling of Earthquakes from Friuli, Italy, *Bull. Seism. Soc. Am.*, **71**, 1219–1231

Cocco, M., and F. Pacor (1992) The Rupture Process of the 1980, Irpinia, Italy Earthquke from the Inversion of Strong Motion Waveforms, *Tectonophysics*, in press.

Cocco, M., R. Di Maro, P. Federici, and A. Marchetti (1993). Analisi della Sismicità dell'Area del Sannio-Matese negli Anni 1991-1992. *Proc. 9° Annual Meeting Gruppo Nazionale Geofisica della Terra Solida*, Rome, in press (in Italian).

Deschamps A., and G.C.P. King (1984). Aftershocks of the Campania - Lucania Earthquake of 23 November 1980, *Bull. Seism. Soc. Am.*, **74**, 2483-2517.

Doglioni C. (1991). A Proposal for the Kinematic Modelling of W-dipping Subductions. Possible Applications to the Tyrrhenian-Apennines System. *Terra Nova*, **3**, 423-434.

Elter P., G. Giglia, M. Tongiorgi, and L. Trevisan (1975). Tensional and Compressional Areas in the Recent Evolution of the Northern Appenines, *Boll. Geof. Teor. e Appl.*, **XVII**, 65.

Giardini, D., A.M. Dziewonski, J.H. Woodhouse, and E. Boschi (1984). Systematic Analysis of the Mediterranean Region Using the Centroid-moment Tensor Method, in A. Brambati and D. Slejko: *The OGS Silver Anniversary Volume*, 121-142, Trieste.

Giardini, D., and M. Velonà (1991). The Deep Seismicity of the Tyrrhenian Sea, *Terra Nova*, **3**, 57-64.

Klein F. W. (1989). HYPOINVERSE, a Program for Vax Computers to Solve for Earthquake Location and Magnitude, United States Department of the Interior Geological Survey, *Open file report* 89-314, 6/89 version.

Lavecchia, G. (1988). The Tyrrhenian-Apennines System: Structural Setting and Seismotectogenesis, *Tectonophysics*, **147**, 263-296.

Iannaccone G., R. Romeo, G. Tranfaglia, L. Errico, E. Lentini, P. Bernard, A. Deschamps, and G. Patau (1990). Analisi della Sequenza Sismica di Benevento (Aprile-Maggio 1990), *proc. 9° annual meeting Gruppo Nazionale Geofisica della Terra Solida*.

Malinverno, A., and W.B.F. Ryan (1986). Extension in the Tyrrhenian Sea and Shortening in the Apennines as Result of Arc Migration Driven by Sinking of the Lithosphere, *Tectonics*, 5, 227-245.

Mantovani, E., D. Babucci, D. Albarello,and M. Mucciarelli (1990). Deformation Pattern in the Central Mediterranean and Behavior of the African-Adriatic Promotory, *Tectonophysics*, 63-79.

Omori, F. (1909). Preliminary Report on the Messina-Reggio Earthquake of December 28, 1908, Bull. Imperial Earth. Invest. Comm., 3-2,37-46.

Pantosti, D., and G. Valensise, (1989). Riconoscere il "Terremoto Caratteristico": il Caso dell'Appennino Meridionale, in "I Terremoti Prima del Mille: Storia Archeologia Sismologia", E. Guidoboni eds., Società Geofisica Ambiente, Bologna, Italy, pp. 536-553.

Pantosti, D., and G. Valensise (1990). Faulting Mechanism and Complexity of the 23 November, 1980, Campania-Lucania Earthquake Inferred form Surface Observations, *J. Geophys. Res.*, 95, 15319-15341.

Pantosti, D., D.P. Schwartz e G. Valensise (1993). Paleoseismology Along the 1980 Surface Rupture of the Irpinia Fault: Implications for Earthquakes Recurrence in the Southern Apennines, Italy. *J. Geophys. Res.*, in press.

Panza, G.F., A.G. Prozorov, and P. Suhadolc (1990). Is There a Correlation Between Lithosphere Structure and the Statistical Properties of Seismicity?, *Terra Nova*, 2, 585-595.

Patacca, E., and P. Scandone (1989). Post-Tortonian Mountain Building in the Apennines. The Role of the Passive Sinking of a Relic Lithospheric Slab, in *The Lithosphere in Italy*, Accademia Nazionale dei Lincei (eds A. Boriani *et al.*), Rome, pp. 157-176.

Selvaggi G., and A. Amato (1992). Subcrustal Earthquakes in the Northern Apennines (Italy): Evidence for a Still Active Subduction?, *Geophysical Research Letters*, 19, 21, 2127–2130.

Westaway R.W.C. (1987). The Campania, Southern Italy, Earthquakes of 1962 August 21, *Geophys. J. R. astr. Soc.*, 88, 15-24.

Westaway R.W.C., adn J. Jackson (1987). The Earthquake of 1980 November 23 in Campania-Basilicata (Southern Italy), *Geophys. J. R. astr. Soc.*, 90, 375-443.

Westaway R.W.C. (1992). Seismic Moment Summation for Historical Earthquakes in Italy: Tectonic Implications, *J. Geophys. Res.*, 97, B11, 15,437-15,464.

P-WAVE TELESEISMIC TOMOGRAPHY: CONTRIBUTION TO THE DELINEATION OF THE UPPER MANTLE STRUCTURE OF ITALY

G. B. Cimini and A. Amato
Istituto Nazionale di Geofisica
Via di Vigna Murata 605, 00143, Roma, Italy

ABSTRACT. In recent years seismic tomography has emerged as a powerful tool to study the crustal and upper mantle structure and to constrain kinematic models of tectonic evolution in regions of plate margins. We applied a well established and robust inversion technique ("ACH" technique) to compute a three-dimensional model of the compressional wave velocity in the lithosphere-asthenosphere system beneath Italy. We used a high quality dataset of about 4700 selected arrival times of teleseismic events digitally recorded at the National Seismic Network. The results show a substantial improvement, in terms of variance reduction, model resolution, and standard errors, with respect to similar inversions (same tecnique, same starting model, same station and event distribution) performed with data derived directly from bulletins.

Our interpretation of the computed velocity model, mainly based on the observation of regions of positive velocity anomalies (up to about 5%), substantiates the hypothesis that a complex subduction system has developed beneath the Italian peninsula. The most prominent high velocity regions are the Alps, the northern Apennines, and the southern Tyrrhenian Sea. Low velocity regions are found in the upper mantle of the Adriatic microplate and beneath Sicily. Some of the high velocity regions are also characterized by intermediate and deep seismicity, whereas some others are apparently aseismic (like the Alps), mainly depending on the age of subduction. In the southern Tyrrhenian sea, where the subduction is still active, a high velocity zone as deep as about 500 km is associated with the well known seismicity related to the Wadati-Benioff zone. In the northwestern portion of the Apennines, a strong high velocity anomaly is observed in the upper 200-250 km. We found that well located subcrustal earthquakes as deep as 90 km occur in the same region, gently dipping to the southwest. We propose that they delineate the upper portion of the Adriatic lithosphere sinking beneath the Tuscany-Corsica block, whose deepest part is depicted by the high velocity anomaly. The high velocity becomes weaker and deeper moving southeastward along the Apenninic chain, suggesting that the subducted slab is not continuous at depth. The complex geometry of the subducted lithosphere may be due to the stretching of the downgoing slab in response to arc migration, and/or to an irregular geometry of the two converging plates.

1. Introduction

Seismic tomography is one of the most powerful techniques to investigate the Earth's interior, particularly in regions of plate boundaries, where deep tectonic processes such as subduction, continental collision, and volcanism generate strong velocity anomalies in the crust and the upper mantle (Michaelson and Weaver, 1986; Spakman, 1990; Harris et al., 1991; Benz and Zandt, 1992).

In our work of the last few years, we tried to obtain more and more reliable three-dimensional P-velocity models of the upper mantle beneath the Italian region (Figure 1), following two main requirements, namely (a) a high-quality dataset, and (b) a robust inversion technique (Alessandrini et al., 1989; Cimini et al., 1990; Amato et al., 1991a, 1991b, 1993). The data that we used derive from selected teleseisms digitally recorded by the National Network of the Istituto Nazionale di Geofisica (Figure 2) in the last four years. The technique that we adopted is perhaps the most widespread inversion technique, called ACH technique after the name of the authors who originally developed it (Aki, Christoffersson, and Husebye, 1977).

We believe that any geodynamic model of this portion of the Mediterranean has to (critically) take into account the results of tomographic studies (see also Spakman, 1990), since they represent one of the more reliable geophysical information, together with earthquake distribution, that help to constrain the presence and geometry of subducted slabs, deep lithospheric roots, and so on.

In next section, we briefly describe the basic concepts, the validity, and the limitations of the ACH technique. Then, we summarize the results of the application on different ensembles of teleseismic events, demonstrating the reasons why we use (a few) digital data instead of (a lot of) bulletin data (par. 3). Finally (par. 4), we discuss the main features of the upper mantle structure of Italy as delineated by our tomographic images, trying to give a geodynamic interpretation, also based on comparisons with other geological and geophysical data.

2. The ACH method

In order to image the upper mantle structure of the Italian region, we applied a modified version of the "ACH" inversion procedure (Aki et al., 1977; Evans, 1986) to teleseismic events recorded by the stations of National Seismic Network (RSNC) of the Istituto Nazionale di Geofisica (Figure 2) between 1988 and 1991. A selected dataset of the best recorded events is shown in Figure 3 (see next section).

The original and still dominant use of this technique is to invert teleseismic travel time relative residuals (station residual minus network mean residual for the given event) to estimate lateral velocity variations beneath the receiver array. For this goal, the studied area is divided into a three-dimensional grid of rectangular blocks, while the Earth outside the "target volume" is considered homogeneous. An initial P wave velocity is assigned to each layer, and the teleseismic phases are considered as plane waves incident on the bottom of the model. Disregarding refraction due to anomalous structure, the ACH inversion method results in a linearization of the travel time integral to the form

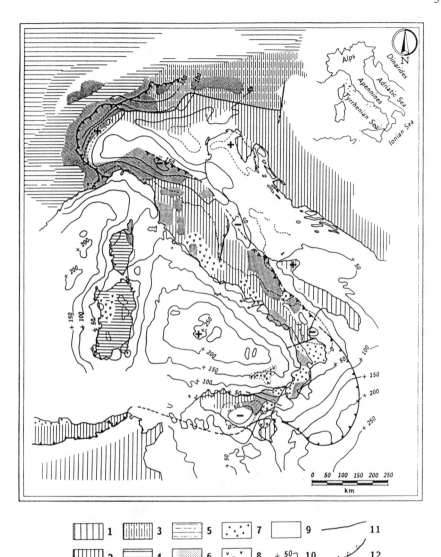

Fig. 1 Structural sketch of the Italian region (modified from *Amato et al.*, 1993). 1 = African foreland; 2 = deformed African continental margin; 3 = Austroalpine nappes; 4 = European foreland; 5 = deformed European continental margin; 6 = oceanic remnants (ophiolites, oceanic sediments - Ligurides, Sicilides, etc.); 7 = "internal massifs" (metamorphosed Alpine and pre-Alpine complexes); 8 = volcanic complexes (includes both the Tertiary volcanism of Sardinia and the (Plio-)Quaternary volcanism of Tuscany, Latium, Campania, and Sicily); 9 = terrigenous sediments of the late- and post-orogenic phases; 10 = Bouguer gravity anomalies; 11 = important tectonic lines with uncertain meaning (i.e. Insubric Line in the Alps); 12 = main thrust front of the Alps and the Apennines (solid triangles) and tectonic boundary

Fig. 2 Stations of the National Seismic Network of the Istituto Nazionale di Geofisica (ING) used in this study.

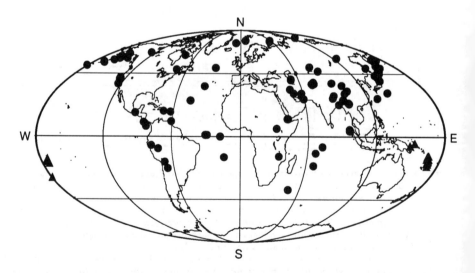

Fig. 3 Geographical distribution of the P (solid circles) and the PKP (solid triangles) events analyzed in this study.

$$\mathbf{Gm} = \mathbf{d} \qquad (1)$$

where $\mathbf{G}$ is the matrix containing the unperturbed travel times of the different ray segments in each block, $\mathbf{m}$ is the vector of the unknown velocity perturbations, and $\mathbf{d}$ is the residuals vector. An estimate $\hat{\mathbf{m}}$ of the $\mathbf{m}$ vector, can be obtained by damped least squares (Aki et al., 1977)

$$\hat{\mathbf{m}} = (\mathbf{G}^T\mathbf{G} + \theta^2 \mathbf{I})^{-1} \mathbf{G}^T \mathbf{d} \qquad (2)$$

where the term $\theta^2\,\mathbf{I}$ controls the trade-off between the residual data variance after inversion and the squared length of the solution. The damping parameter θ^2 must be carefully selected, balancing overdamping of real anomalies versus overfitting of the noise in the data (Harris et al., 1991).

A common method to asses the reliability of the computed velocity perturbations, is the analysis of the resolution and covariance matrix. The resolution matrix $\mathbf{R}$ shows how well the "true" Earth model $\mathbf{m}$ is mapped by the inversion result $\hat{\mathbf{m}}$, according to the relationship

$$\hat{\mathbf{m}} = \mathbf{Rm}. \qquad (3)$$

Given a block n of the modeled volume, the elements of the n-th column of the (symmetric) matrix $\mathbf{R}$ are representative of the contribute of all blocks in the model. Ideally, a perfectly resolved block would have a diagonal value equal to 1.0, with all zeroes on the off-diagonal elements. In our inversions, the real data show values around 0.7-0.9 for well resolved blocks. The largest values around the diagonal reveal the directions along which the anomaly for that block is smeared out into adjacent blocks. As an example, we show in Figure 4 the block resolution for an anomalous high velocity block in the northern Apennines. The diagonal value of the $\mathbf{R}$ matrix for that block is 0.74, the velocity perturbation relative to the starting model is +5.2%, and the standard error is ±1.1%. The map and the vertical section crossing the block show that it is well resolved with respect to the blocks of the same layer (off-diagonal elements are lower than 0.1, see Figure 4a), whereas the anomaly is slightly smeared in the vertical direction (off-diagonal elements are as high as 0.2, see Figure 4b).

The covariance matrix can be expressed against the remaining data variance σ_r^2 by means of the formula

$$\mathbf{C} = \sigma_r^2 (\mathbf{G}^T\mathbf{G} + \theta^2 \mathbf{I})^{-2} \mathbf{G}^T \mathbf{G} \qquad (4)$$

in which $(\mathbf{G}^T\mathbf{G} + \theta^2 \mathbf{I})^{-1} \mathbf{G}^T \mathbf{G}$ represents the explicit formulation for the resolution matrix.

The square root of the diagonal elements of this *a posteriori* covariance matrix (Evans and Achauer, 1993), are the standards errors, or uncertainties, associated with the model velocity anomalies. The standard errors observed for the above mentioned selected dataset

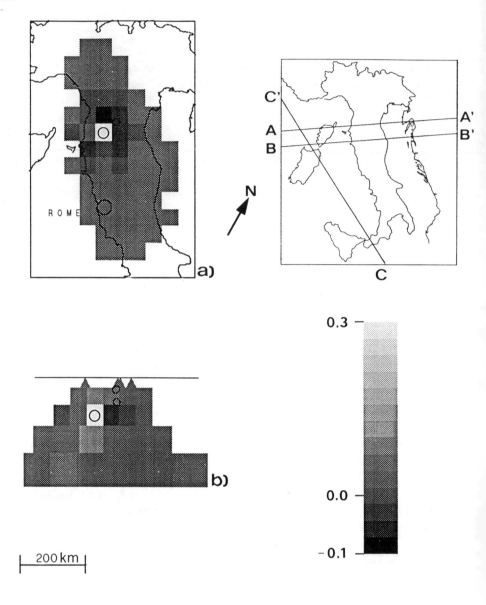

Fig. 4 a) Map view of the column of the resolution matrix for a representative "fast" block in the third layer (80 – 140 km) of a northern Apennines velocity model. b) Vertical section, crossing the block along the AA' direction. The block has a resolution of 0.74, the surrounding blocks (which ideally should have zeroes) have values lower than 0.1 in the same layer (a), and slightly higher values in the lower and the upper layers (b), witnessing a smearing of the velocity anomaly in the vertical direction. This is due to the steep incidence angles of teleseismic rays.

of teleseismic events and for an "equivalent" bulletin dataset (see Table 1) are on average of about $\pm 1.0\%$ and $\pm 1.5\%$, respectively.

However, although a careful analysis of the resolution and covariance matrix allows to evaluate how well a model is resolved, there are some general limitations of the ACH technique, as well as of other inversion techniques, that have to be taken into account if one wants to interpret objectively the tomographic results. As pointed out by Evans and Achauer (1993), the most severe limitations are related to the geometry of the raypaths, since teleseismic rays have steep incidence angles. This leads to well imaging objects elongated on the vertical direction, and to fuzzy pictures of "flat" objects. Other numerical tests show that if a small anomaly is coincident with a vertical boundary of the model, it cannot be recovered in the tomographic images, although the resolution matrix is fairly good. In this case, we avoid this parametrization artifact using an "offset and average" inversion scheme, in which the final velocity model is the average over n^2 inversions performed shifting the block grid along the x and y directions by $1/n$ of the bock size (Evans, 1986), with n generally equal to 3.

3. Data and Results

In the application of the ACH technique, both accurate model parametrization (number and size of blocks against station spacing and travel times spent in each layer, starting velocity model, optimal damping, etc.) and homogeneity of the seismic ray distribution are required (Evans and Achauer, 1993).

To provide adequate sampling of the target volume, a broad distribution of both ray backazimuth and slowness is needed. We used P events, at epicentral distances restricted to $\Delta \geq 25°$ to reduce regional effects on teleseismic propagation, and PKP events ($\Delta \geq 110°$) (Figure 3). According to the Herrin's tables (Herrin, 1968), the PKP travel time residuals have been computed relative to PKP_{DF} branch (PKIKP phases) and PKP_{AB} branch.

Also the quality of data is important in teleseismic tomography, because of the weights that large errors have in least square (L_2 norm) fitting methods (Menke, 1989). In order to achieve a high precision in the computation of the arrival times, we picked about 4700 digitized seismograms of 134 well recorded teleseisms (hereinafter called "PICK" dataset), shown in Figure 3.

The principal advantage of using these re-picked data, with respect of using arrival times derived from bulletins (the "BULL" and "BULK" datasets; see Table 1), is the high accuracy in the computation of residuals. A direct analysis of the teleseismic waveforms avoids phase mispick problems (as for example early arrivals due to diffractions around an anomalous body, bad interpretations of PKP branches), or cycle skipping caused by reverse polarities or emergent arrivals (Amato et al., 1993; Evans and Achauer, 1993). As an example, we show in Figure 5a the alignment obtained by our picking procedure (see Alessandrini et al., 1989) for a Malawi earthquake (03/10/1989, $m_b=6.2$). For comparison, we show in Figure 5b the seismograms of the same event as aligned considering the bulletin

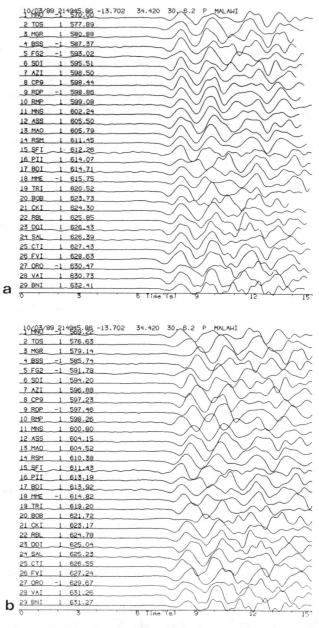

Fig. 5 a) Alignment of 29 digital seismograms of a P teleseism (Malawi earthquake, 10 March 1989, m_b 6.2) as obtained after polarity analysis, amplitude scaling and band-pass filtering using an interactive graphic procedure. b) The same waveforms aligned considering bulletin arrival times.

arrival times.

In Table 1 we show, in terms of data variance, the results of the ACH inversions that we have performed with the same starting model on the above mentioned three dataset (PICK is the repicked dataset, BULL are the bulletin data for the same events, BULK is a larger bulletin dataset containing about eight times the residuals of PICK and BULL).

It is worth noting the substantial improvement for both initial (σ_d^2) and remaining data variance (σ_r^2) after inversion, when the "PICK" dataset is used. The variance improvement $(\Delta\sigma^2)$ using the PICK data is as high as 72%, whereas that observed for the BULL and BULK data are 54% and 38%, respectively (Table 1). As regards the choice of the damping parameter (equation 2), we show in Figure 6a,b the trade-off between the unmodelled data variance (residual variance) and the model complexity (model length). For both the PICK and the BULL dataset we selected a damping value of $200 s^2$, trying to find the best compromise between the effect of the noise in the data and the fit of the model (see Figure 6).

Table 1 - Inversion results for the three datasets (see text).

Dataset	Eqs.	Obs.	σ_d^2 (s^2)	σ_r^2 (s^2)	$\Delta\sigma^2$ (%)	$\theta^2(s^2)$
PICK	134	4687	0.85	0.24	72	200
BULL	131	4993	1.05	0.48	54	200
BULK	863	28320	1.02	0.63	38	200

Figures 7 and 8 show the results of the inversions with the PICK and the BULL dataset (same parametrization, same data distribution). It can be seen that, although the main anomalies are detected by both inversions, the BULL model appears to be more "fuzzy" than the PICK model. Also based on the parameters reported in Table 1, we believe that the PICK model better depicts the velocity structure of the upper mantle of Italy.

Finally, in order to assess the stability of the tomographic results in an area of particular interest, we computed an inversion of the "PICK" data, using only the northern Apennines stations (17) and a five-layer model reaching a depth of 320 km. The results of this "regional" inversion (see Figure 9 for two vertical sections) are quite similar to those obtained with the Italian dataset, ensuring that the computed anomalies are meaningful (see Amato et al., 1991b for more details).

4. Discussion

The main features detected by our inversions in the upper mantle of Italy are shown in Figures 7 through 10. As already discussed in previous works (Cimini et al., 1990; Amato et al., 1991a; Amato et al., 1993), the most relevant anomalies are high velocities beneath

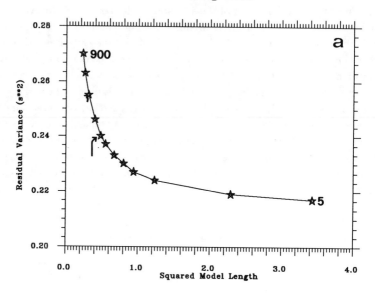

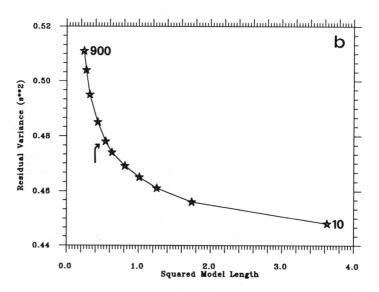

Fig. 6 Trade-off curves computed to select the optimal damping parameter for the "PICK" dataset (a) and the "BULL" dataset (b), respectively. Note for the bulletin data (Figure 6b) the greater values of the remaining data variance. The arrows indicate the selected damping values (200 s^2 for both the datasets).

the Alps, the northern Apenninic arc, and the Tyrrhenian, and low velocities beneath the Adriatic and Sicily. Our interpretation is mainly based on the assumption that highs and lows in the upper mantle velocity structure are due to lithospheric slabs (or "roots") and "soft" asthenospheric layers, respectively.

In the Alps, we detect a region of high velocity from the Moho to about 270 km (Figure 7). This anomaly becomes broader with increasing depth. It appears to be shallower in the eastern Alps, whereas it is continuous beneath the western Alps up to 250-300 km, where it joins the high velocity anomaly of the Apennines, described further on. As proposed by Panza and Mueller (1979), Babuška and Plomerová (1990), and Spakman (1990), we interpret the Alpine anomaly with the existence of deep lithospheric roots, left from the continental collision which followed the Tethyan subduction. As already proposed by Spakman (1990), the European lithosphere seems to underthrust the Adriatic plate.

In southern Italy, we find a region of high velocity in the southern Tyrrhenian (Figure 7 and 10), approximately coincident with the slab delineated by the intermediate and deep seismicity (McKenzie, 1972; Anderson and Jackson, 1987). High velocities and deep seismicity do not correspond exactly (see Figure 7), due either to low resolution in the tomographic images (caused by the absence of seismic stations on the sea side), or to a systematic mislocation of deep earthquakes (routinely located with homogeneous velocity structures), or both.

From our computed models and the seismicity distribution, it seems that the slab dip changes from nearly vertical in the upper 150 km to approximately $45°$ at greater depth (see also Anderson and Jackson, 1987). The bending corresponds with a rarefaction of seismicity (around 150 km). This observation led some authors to propose that the slab is detached between 100 and 200 km. Another zone with almost no earthquakes is observed at about 350 km (Figure 10). Also in this case, we may hypothesize that the slab is not continuous, although the two earthquake clusters (between 150 and 300 km and between 350 and 500 km) are aligned (see Figure 10) and no bending is observed between the two. Nonetheless, it is worth noting that most of the deepest earthquakes are located close to the coast of Campania and southern Latium, as for instance the well known Gaeta (Latium) earthquake of 27/12/1978 at about 400 km depth. The depth extent of the slab is probably more than 540 km (the base of our model, see Figure 10), as we detect a velocity contrast in the lowest layer of the inversion model, that is likely due to an anomaly located beneath the base of the model.

In the northern Apennines, we found original and perhaps more interesting results. The building of this mountain range is far more controversial than the Alps or the Tyrrhenian arc, where more geophysical information has been collected in the last twenty years. Since early 70's, the models proposed for the growth of the Apennines can be schematically grouped in two main classes (for a more detailed overview see Serri et al., 1991), depending on what was supposed to be the "engine" of the orogenic process. The first class invokes plate tectonics models, or derivations of these, to explain the genesis of the system Tyrrhenian-Apennines-Adriatic (Boccaletti et al., 1971; Scandone, 1979; Reutter, 1980; Malinverno and Ryan, 1986; Mantovani et al., 1992, among many others). The second

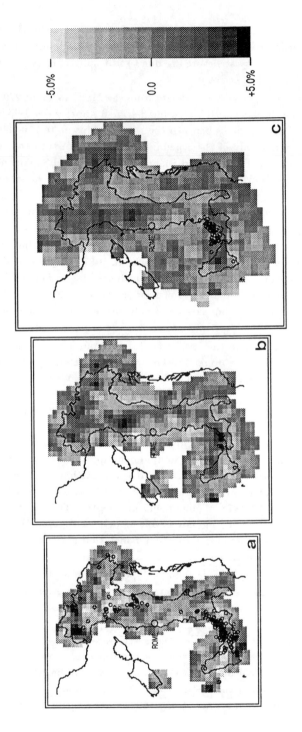

Fig. 7 Velocity anomalies in the upper mantle of Italy for three layers of a six-layer model. a) 35-100 km; b) 100-180 km; c) 180-280 km. The images are relative to the "PICK" dataset. The starting velocities are 8.05 km/s, 8.20 km/s, and 8.40 km/s for a), b), c), respectively. The empty circles are the earthquakes located within each layer.

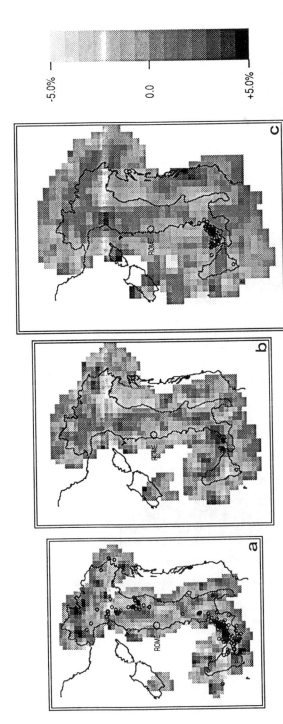

Fig. 8 Same as Figure 7. The images are relative to the "BULL" dataset.

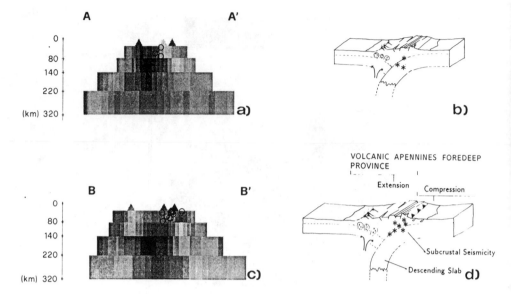

Fig. 9 Vertical sections (a,c) for a five-layers velocity model of the northern Apennines. The traces of the sections are shown in Figure 4. The scale of the velocity perturbations is the same as in Figure 7. b) and d) are simple interpretative schemes of a) and c), showing the hypothesized geometry of the Adriatic subducted lithosphere, constrained by the subcrustal seismicity (empty circles) and by the high velocity anomaly (dark grey).

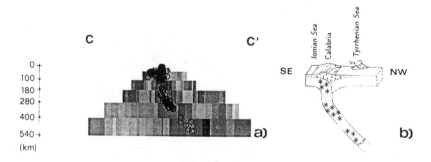

Fig. 10 a) Vertical section, through the Calabrian arc and the southern Tyrrhenian (see Figure 4 for the trace), of the velocity model shown in Figure 7. b) A simple interpretative scheme of the slab delineated by the computed high velocity anomalies and by the intermediate and deep seismicity.

group hypothesizes that vertical movements are the main cause of the evolution of the Apennines (Elter et al., 1975; Wezel, 1985; Locardi, 1985; Lavecchia, 1988). According to the first group of models, the Tyrrhenian is considered as a back-arc basin, whereas the "verticalistic" hypothesis depicts it as a rifting feature, whose opening is related to an asthenospheric plume. Also, geochemical and petrological studies performed on the perityrrhenian volcanoes of Tuscany, Latium, and Campania (see Figure 1), could not definetely solve this problem, interpreting the potassic volcanism of the so-called Roman Magmatic Province as a product of either continental rifting or subduction processes (see Serri et al., 1991, and references therein). In our opinion, more evidence substantiate the subduction hypothesis, at least as far as the northern Apennines are concerned. However, some points are still a matter of discussion, like: when the westward "Adriatic" subduction started, if it is still active, whether it has been preceded by subduction of the Corsica lithosphere towards the east, whether the lithospheric slab is oceanic or continental, whether the slab is detached or not.

Our tomography results show a clear, strong high velocity anomaly in the concave side of the northern Apenninic arc (Figures 7 and 9), at depth between 80 and 140 km in the northern sector, and up to 220 km depth moving to the southeast. This observation, together with that of recently reported subcrustal earthquakes in this area (Amato et al., 1991; Selvaggi and Amato, 1992) gives evidence in favour of a subduction mechanism to explain the genesis of this arc.

The tomographic images reveal that the depth of the maximum high velocity anomaly becomes deeper moving southeastward (Figures 7 and 9). In the northernmost section AA' (Figure 9a) the anomaly is in layer 3 (80-140 km), whereas in section BB' (Figure 9c) is located in layer 4 (140-220 km), and it is even deeper beneath central Apennines (Amato et al., 1993). In the interpretative sections of Figure 9b,d, we schematically depict these two situations, that correspond with the two tomographic cross sections of Figure 9a,b. In our interpretation, the Adriatic slab is delineated in its upper part by the seismicity distribution, while its deeper portion is delineated by the high velocity anomaly. As mentioned earlier, the shape of this anomaly cannot be perfectly resolved by our inversions, particularly as far as the dip is concerned. The apparent vertical elongation of the anomaly (Figure 9) may be an artifact of the ray geometry, thus the curved shape of the Adriatic lithospheric slab depicted in Figure 9b,d is speculative.

Beneath 400 km, the Apenninic high velocity anomaly is located approximately beneath the coast of Campania, and corresponds with the deepest seismic zone of the Tyrrhenian slab (Amato et al., 1993). This pattern of southeast deepening high velocities is coherent with an oblique subduction of the Adriatic lithosphere beneath the Corsica-Sardinia-Tuscany block, along the Apennines. If we assume that the Adriatic subduction started approximately 15-25 Ma, the convergence rate can be estimated to about 1 cm/year across northern Apennines, and up to 3-4 cm/year in the southern Apennines, considering a slab length of 150-250 km in the northern Apennines, and of at least 600-700 km for the Tyrrhenian slab.

However, the onset of the subduction process is uncertain. The values suggested above

(15-25 Ma, i.e. between the end of the Oligocene and the Lower-Middle Miocene) are based upon estimates proposed in the literature (see e.g. Reutter et al., 1980; Malinverno and Ryan, 1986), and on the first documented occurrence of subduction-related volcanism, that of Sisco (Corsica), dated at 13.5-15.0 Ma (Serri et al., 1991). According to these authors, this represents a minimum age for the beginning of subduction, because the intrusive rocks of Sisco are probably related to the first post-collisional phase between Adriatic and Tuscany blocks. The estimates of average convergence rates reported above are in agreement with the geological estimates (Patacca and Scandone, 1989).

Moreover, it must be noticed that the deepening of the high velocity anomaly along the strike of the Apennines, shown in the two sections of Figure 9 and observed throughout the Apenninic chain (Amato et al., 1993), is not uniform, showing a step in central Italy. The step coincides approximately with the boundary between the northern Apenninic arc and the southern Apennines (see Figure 1). This evidence, together with the observation that no subcrustal earthquakes are located in the southern Apennines (see Selvaggi and Amato, 1992; Cocco et al., this volume), suggests that the northern and the southern Apennines underwent different geodynamic evolutions, not only at the surface, but also at lithospheric depths.

The "cut" of the subducted lithosphere beneath central Apennines may be due to the increasing curvature of the Apenninic arc with time, that may have stretched the subducting lithosphere (see Malinverno and Ryan, 1986; Doglioni, 1991).

5. Conclusive remarks

We have shown how seismic tomography provides original information on the Earth's structure, particularly in regions of young tectonic evolution. We believe that any geodynamic model proposed for our region should take into account the indications provided by these studies, as well as those derived from seismological observations. Both these data represent pictures of the present tectonic setting, that tell us what is the shape, the extent and the depth of the upper mantle structures, and how they are deforming. We have also given some suggestions on how to evaluate the validity of the computed tomographic images. In particular, we warned about the inadequacy of the ACH tecnique in imaging horizontal anomalies and its capability in resolving vertical or oblique bodies.

The availability of high quality data of teleseisms digitally recorded by the National Seismic Network of the Istituto Nazionale di Geofisica, allowed us to obtain reliable pictures of the lithosphere-asthenosphere structure of Italy. The technique adopted to calculate travel time residuals, and hence the tomographic images, has proven more effective, though more time consuming, than simply using data from bulletins.

We imaged high velocities beneath the Alps and the southern Tyrrhenian, the former interpreted as old, aseismic lithospheric rots, and the latter as the image of the slab also depicted by the intermediate and deep seismicity. In addition, we obtained original results for the region of the northern Apennines, where a strong high velocity anomaly in the upper 200 km, compared with the seismicity distribution beneath the crust, provides a strong constraint for the existence of subducting Adriatic lithosphere beneath the Tuscany-Corsica block. The depth extent of the Apenninic subduction increases towards the southeast, suggesting an oblique convergence between the two plates. It seems as well that this process has not been continuous along the Apennines, enhancing the differences in tectonic evolution underwent by the northern arc and the southern Apennines, where the slab is imaged at depth of 400-500 km.

Acknowledgements

We thank E. Mantovani and A. Morelli for inviting us at the Erice workshop. We also thank E. Boschi, R. Funiciello and A. Rovelli for their encouragement and suggestions, and J.R. Evans and H.M. Iyer for inspiration and helpful comments. D. Riposati helped us in drafting figures.

References

Aki, K., A. Christofferson e E.S. Husebye (1977) Determination of the three-dimensional seismic structure of the lithosphere, *J. Geophys. Res.*, **82**, 277-296.

Alessandrini, B., A. Amato and G.B. Cimini (1989) Distribuzione dei residui telesismici nella regione italiana dall'analisi di forme d'onda digitali, *Proc. 8th Congress of the Gruppo Nazionale Geofisica della Terra Solida*, Rome, 347-360, (in Italian).

Amato, A., B. Alessandrini, G.B. Cimini et al. (1991a) Three-dimensional P-velocity structure and seismicity in the upper mantle of Italy, *EOS, Trans. Am. Geophys. Union*, **72**, 44, 349, (abstract).

Amato, A., G.B. Cimini, and B. Alessandrini (1991b) Struttura del sistema litosfera-astenosfera nell'Appennino settentrionale da dati di tomografia sismica. *Stud. Geol. Cam.*, Special Volume **1991/1**, 83-90, (in Italian).

Amato, A., B. Alessandrini and G.B. Cimini (1993) Teleseismic wave tomography of Italy, in *Seismic Tomography: Theory and practice*, H.M. Iyer and K. Hirahara eds, Chapman and Hall, London, 361-397.

Anderson, H., and J. Jackson (1987) The deep seismicity of the Tyrrhenian sea, *Geophys. J. R. astr. Soc.*, **91**, 613-637.

Babuška, V., and J. Plomerová (1990) Tomographic studies of the upper mantle beneath Italian region, *Terra Nova*, **2**, 569-576.

Benz, H.M., G. Zandt and D.H. Oppenheimer (1992) Lithospheric structure of northern California from teleseismic images of the upper mantle, *J. Geophys. Res.*, **97**, 4791-4807.

Boccaletti, M., P. Elter and G. Guazzone (1971) Plate tectonic models for the development of the western Alps, and northern Apennines, *Nature*, **234**, 108-110.

Cimini, G.B., B. Alessandrini, and A. Amato (1990) Tomografia telesismica della regione italiana, *Proc. 9th Congress of the Gruppo Nazionale Geofisica della Terra Solida*, Rome, 321-336, (in Italian).

Doglioni, C. (1991) A proposal for the kinematic modelling of W-dipping subductions - possible applications to the Tyrrhenian-Apennines system, *Terra Nova*, **3**, 423-434.

Elter, P., G. Giglia, M. Tongiorgi, and L. Trevisan (1975) Tensional and compressional areas in the recent (Tortonian to present) evolution of north Apennines, *Boll. Geof. teor. Appl.*, **17**, 3-18.

Evans, J.R. (1986) Teleseismic travel time residual analysis system: a user's manual, U.S. Geological Survey, Menlo Park, California.

Evans, J.R., and U. Achauer (1993) Teleseismic velocity tomography using the ACH method: theory and application to continental-scale studies, in *Seismic Tomography: Theory and practice*, H.M. Iyer and K. Hirahara eds, Chapman and Hall, London, in press.

Harris, R.A., H.M. Iyer, and P.B. Dawson (1991) Imaging the Juan de Fuca plate beneath southern Oregon using teleseismic P wave residuals, *J. Geophys. Res.*, **96**, 19,879-19,889.

Herrin, E. (1968) Introduction to "1968 seismological tables for P phases", *Bull. Seism. Soc. Am.*, **58**, 1193-1242.

Lavecchia, G. (1988) The Tyrrhenian-Apennines system: structural setting and seismotectogenesis, *Tectonophisics*, **147**, 263-296.

Locardi, E. (1985) Neogene and Quaternary mediterranean volcanism: the Tyrrhenian example, in: *Geological evolution of the Mediterranean basin*, D.J. Stanley and F.C. Wezel eds, New York, Springer Verlag, 273-291.

Malinverno, A., and W.B.F. Ryan (1986) Extension in the Tyrrhenian sea and shortening in the Apennines as result of arc migration driven by sinking of the lithosphere, *Tectonics*, **5**, 227-245.

Mantovani, E., D. Albarello, D. Babbucci, and C. Tamburelli (1992) Recent geodynamic evolution of the central Mediterranean region - Tortonian to present, *Siena, Italy*, 88pp.

McKenzie, D.P. (1972) Active tectonics of the Mediterranean region, *Geophys. J. R. Astron. Soc.*, **30**, 109-185.

Menke, W. (1989) Geophysical Data Analysis: Discrete Inverse Theory, *International*

Geophysics Series, **45**, Academic Press, Inc., 285 pp.

Michaelson, C.A., and C.S. Weaver (1986) Upper mantle structure from teleseismic P arrivals in Washington and northern Oregon, *J. Geophys. Res.*, **91**, 2077-2094.

Panza, G., and S. Mueller (1979) The plate boundary between Eurasia and Africa in the Alpine area, *Mem. Sci. Geol.*, **33**, 43-50.

Patacca, E., and P. Scandone (1989) Post-Tortonian mountain building in the Apennines, the role of the passive sinking of a relic lithospheric slab. In: the Lithosphere in Italy - Advances in Earth Science Research. Atti dei Convegni Lincei, **80**, 157-176.

Reutter, K.J. (1981) A trench-forearc model for the northern Apennines, in *Sedimentary basins of Mediterranean margins*, F.C. Wezel editor, Tecnoprint, Bologna, 433-443.

Scandone, P. (1979) Origin of Tyrrhenian sea and Calabrian arc, *Boll. Soc. Geol. Ital.*, **98**, 27-34.

Selvaggi G. and A. Amato (1992) Intermediate - Depth Earthquakes in northern Apennines (Italy): Evidence for a still active subduction? *Geophysical Research Letters*, **19**, 2127-2130.

Serri, G., F. Innocenti, P. Manetti, S. Tonarini, and G. Ferrara (1991) Il magmatismo Neogenico-Quaternario dell'area Tosco-Laziale-Umbra: implicazioni sui modelli di e- voluzione geodinamica dell'Appennino settentrionale. *Stud. Geol. Cam.*, Special Volume **1991/1**, 429-463, (in Italian).

Spakman, W. (1990) Tomographic images of the upper mantle below central Europe and the Mediterranean, *Terra Nova*, **2**, 542-552.

Wezel, F.C. (1985) Structural features and basin tectonics of the Tyrrhenian sea, in: *Geological evolution of the Mediterranean basin*, D.J. Stanley and F.C. Wezel eds, New York, Springer Verlag, 153-194.

GEOLOGICAL AND SEISMOLOGICAL EVIDENCE OF STRIKE-SLIP DISPLACEMENT ALONG THE E-W ADRIATIC-CENTRAL APENNINE BELT

P.FAVALI[1], R.FUNICIELLO[1,2], F.SALVINI[3,4]
[1] *Istituto Nazionale di Geofisica, Roma*
[2] *Dip. Scienze della Terra, 3ª Università Roma*
[3] *Dip. Scienze della Terra, Università di Pisa*
[4] *present address: Istituto Nazionale di Geofisica, Roma*

ABSTRACT. New geological, geophysical and seismological data produced in the last decade allow an improved definition of the kinematics of the Tyrrhenian Sea-central Apennines-Adriatic Sea system. The integration of these different types of dataset outlines a complex E-W deformational belt running from the Adriatic Sea, at Gargano latitudes, to the Latium-Abruzzi Platform domain (central Apennines). Its possible prosecution in the Tyrrhenian Sea is discussed. This belt has its own seismotectonic characteristics, and divides the Apennines into two zones with different seismotectonic behaviour. All these data were integrated into a multidisciplinary model that results in the introduction of a strike-slip displacement sometimes in Pliocene to Recent times along the central Apennines-Adriatic Sea E-W belt. This relative motion can take for the strike-slip block tectonics observed in the central Apennines as well as for the uplift of Gargano and Tremiti Islands areas, including their seismicity. The displacement is related to a differing behaviour of the chain-foreland systems across this bend and can be justified either by assuming a partially active sinking of the southern Adriatic lithosphere or a slower sinking processes in the northern Adriatic one.

1. Introduction

In the sixties, the global plate tectonic model underwent general acceptance among the earth scientist community. Although this theory successfully explained the gross geodynamic framework of the earth surface, it was only a ten years later that the fit of complex areas like the Mediterranean was attempted.

McKenzie (1972) proposed the first model mostly based on seismological evidence. The Author interpreted the Adriatic Sea area as a northward prominence of the African plate. The boundaries were outlined by the geographic distribution of the seismicity. Such a model, still valid in its general premises, was affected by strong disomogeneities in the distribution of data, as the Author himself underlined.

Subsequentely, Channell and Horváth (1976), Channell et al. (1979), and D'Argenio and Horváth (1984) gave to the Adriatic prominence a geological characterization, by proposing the existence of Adria, a regional scale promontory that controlled the evolution of the area since Upper Triassic times.

Among more recent models, Mantovani et al. (1985) relate the opening of the Tyrrhenian basin and coeval development of the Apenninic collisional chain to the SW subduction of the Adriatic lithosphere.

Malinverno and Ryan (1986), Royden et al. (1987), Moretti and Royden (1988), Patacca and Scandone (1989), and Patacca at al. (1990) indicate a sinking mechanism of such subduction. According to this modelling, the Tyrrhenian Sea is interpreted as a back-arc basin. The kernel of this modelling provides a satisfactory unification of the causes that lead to the coeval developing of basins and chains. More recently, Doglioni (1990) explains the abrupt westward sinking of the Adriatic plate with the relative westward wandering of the lithosphere.

All these models clearly identify a different kinematic behaviour in the Apennines-Adriatic Sea interaction and placed a generic separation line across the central Apennines between a northern sector and a southern one. Some of them, tentatively interpreted the regional geological discontinuities and identified this sector as the surficial effects of such separation (namely the Ancona-Anzio line or the Ortona-Roccamonfina one).

In this context, we consider the geological and seismological data recently collected in the central Apennines and in the facing Adriatic Sea areas to provide a more detailed picture of the tectonic evolution of this E-W belt in Pliocene to Recent times.

One of the major difficulties - we are aware - lies in the different time span sampled by the different datasets. Seismological data give a precise picture of the stress release, but limited to the tectonic activity of a short period. The geological data, that we considered, *viceversa* have seldom precise temporal identification, yet they give an integrated picture of the Plio-Pleistocenic tectonic activity. These differencies often led scientists from the two fields to interpret the same regions with slightly different models.

We made two reasonable assumptions to make possible the useful integration of these data. From one side, we consider tectonic processes as rather discontinuos. This can justify both partial discrepancies between the two sets of data and within the geological evidence itselves, in that it can also justify the coexistence of deformations ascribed to different concomitant tectonic processes. From the other side, we consider that the variations in time and space of geodynamic processes develop at low rate. This is justified by the tectonic rate on our planet that is limited to few centimetres per year, and the huge volumes of masses involved in plate tectonic processes with their inertia. According to these assumptions we combined the two sets of data and prepared the proposed model.

2. Geological Evidence

2.1 Eastern Tyrrhenian Margin of Central Apennines

The Eastern Tyrrhenian Margin tectonics (Fig.1) is the westward continuation of the offshore extensional N-S ans NW-SE tectonic trends (Bigi et al., 1989). It is characterized by a Recent volcanic activity (starting 0.8 Ma) that developed further North with respect

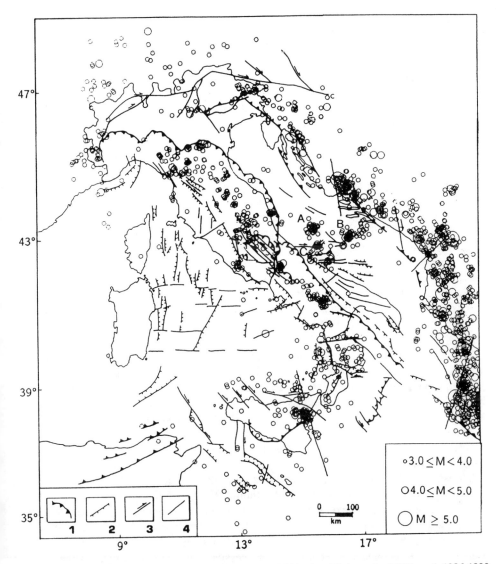

Figure 1. Plio-Pleistocenic geodynamic framework (modified after Bigi et al., 1989) and 1986-1990 seismicity (ING, 1990) of the Italian peninsula and surrounding areas with magnitude >= 3.0. A: 1986-87, B: 1988 and C: 1989-90 indicate the Adriatic seismic sequences; 1= thrust; 2= normal fault; 3= strike-slip fault; 4= sub-vertical fault.

to the coeval one in the Tyrrhenian Sea (Bigi et al., 1989). This activity depicts an important NW-SE regional positive thermal anomaly (Mongelli et al., 1989), responsible for a weakening in the rheology of the area, that separates the Apenninic chain from the Tyrrhenian basins.

The presence of N-S right-lateral strike-slip faults with evidence of Pleistocenic activity further characterizes this belt. The best known is the Sabina Fault Zone (Alfonsi et al., 1991a; 1991b), but indications of strike-slip faulting are also to the south, near Formia and in the Recent Roccamonfina volcanic area (Cerisola and Montone, 1991). The correspondance between periods of activity of the volcanic centers and of activity of these faults is still under study, but coeval ages of activity support the hypothesis that their activity can explain the presence of rising of sub-crustal magma with a relatively low crustal contamination (Alfonsi et al., 1991a; Turi et al., 1991).

2.2 APENNINES

The Apennines (Fig.1) are a rather atypical Alpidic thrust chain developed in two main tectonic phases (Patacca and Scandone, 1989). The first one was active in Early Cenozoic times, and it is responsible for the closing of the pre-existing Upper Triassic-Cretaceous oceanic basin, the Tethys, and for the overthrust of oceanic units (ophiolites, abyssal plain sediments) over a sedimentary wedge. The second event started in Upper Tortonian times and consists in the partial SW subduction of the Adriatic microplate beneath the Apennines, originating the present-day basin-chain-foreland setting (Patacca et al., 1990) with a series of E and NE verging thrusts. The Apenninic thrust chain is characterized by major changes in tectonic trends in its central sector, between the N 41°-43° parallels. The northern sector shows a NNW-SSE tectonic trend that seems to rotate to N 10° W and N-S to the South.

Across the N-S Olevano-Antrodoco line we assist to an abrupt change in this orientations, since structures trend NW-SE, leaving no space for the hypothesis of a simple progressive rotation. East of this line, in the Latium-Abruzzi Platform (LAP) domain, NW-SE trending structures dominate. They originated initially as ramp-flat thrusts, successively reactivated as regional left-lateral strike-slip faults (Salvini, 1992). This trend abruptly ends to the East along an other N-S complex line, the Ortona-Roccamonfina. Southeast of this line the structures reflect mainly ramp-flat thrusting with a NE vergence.

The structural data collected in the area indicate a relative chronology of eastward shortening (N 70° E) younger than NE (about N 35°-40° E) one (Salvini and Tozzi, 1986). Re-activation of preexisting NW-SE planes as strike-slip or oblique faults with left-lateral component east of the Olevano-Antrodoco is therefore supported starting in lower Pliocenic time (Bigi et al., 1989). Within the LAP domain, a series of N-S oriented small thrusts are the secondary effects of the successive strike-slip tectonics (Montone and Salvini, 1991). In this fashion, the LAP domain frames in a block-rotation tectonic environment (Salvini, 1992), as a series of NW-SE elongated blocks within two complex N-S right-lateral strike-slip belts. The presence of a series of coeval small basins and ridges in the LAP results from the interactions among the various blocks (Salvini, 1992). This sector of the chain shows a maximum of elevation up to 2.7-3.0 km that is significantly higher than the surrounding parts of the Apennines, where the maximum elevations remain below 2.2 km, and testifies the intense recent uplift of the LAP domain.

2.3 APULIA AND ADRIATIC FORELAND

This sector of the central Mediterranean is generally considered almost undeformed (Fig.1). This is confirmed both by surface geology and by offshore geophysics, that showed the presence of regional NW-SE extensional faults with limited displacement (Bigi et al., 1989). An exception to this picture is localized along the Gargano-Tremiti Islands belt. In the Gargano strike-slip tectonics is present, being the Mattinata fault one of the most striking structure associated. This fault trends E-W and shows evidences of left-lateral strike-slip faulting such as pull-apart and restraining bends (Funiciello et al., 1988; 1991).

Associated with this feature are a number of second order strike-slip faults oriented NW-SE and N-S. NW-SE extensional faults are also present, and relate to the regional extension. In the Tremiti Islands evidence of strike-slip activity was found, with the islands themselves interpreted as a series of uplifted ridges along it (Montone and Funiciello, 1989). The Gargano Promontory shows a structural uplift of almost 1 km, that can be related to the compression it experienced as the result of the activity of E-W left-lateral strike-slip faults in its southern part together with the presence of a regional E-W to ENE-WSW right lateral strike-slip faulting zone in correspondence of the Tremiti Islands to the N. The offshore continuation of the Tremiti and Mattinata fault systems is testified by some papers based on seismic profile interpretation (Finetti, 1982; De' Dominicis and Mazzoldi, 1987; Finetti et al., 1987; Argnani et al., 1992).

Similarly to the Tyrrhenian Sea the morphology of the Adriatic basin shows a dual pattern. North of 42° N it is a shallow-water basin with a maximum depth of about 200 m, whereas to the south there is an abrupt deepening to depths greater than 1200 m (Giorgetti and Mosetti, 1969).The age of these deformation is still questionable, even if considerations about their morphology and the presence of seismic activity in the area suggest that these structures could well be still active.

3. Seismological and Geophysical evidence

The completion of the Italian Telemetered Seismic Network has allowed a better detection of the seismic activity including the surrounding regions. In particular for the first time a seismicity within the Adriatic basin was clearly observed. The analysis of this seismicity has proved the existence of Recent activity along the transversal seismic belt cutting the Adriatic basin at the Gargano latitudes (Console et al., 1989; 1992). Three remarkable seismic sequences were recorded between 1986 and 1990 and were located in this offshore area demostrating that the Adriatic area is not aseismic inside. Fig.1 shows the distribution of the recent seismicity (5 years) and the tectonic features that have shown Plio-Pleistocenic activity. The choosen magnitude threshold is 3.0, the seismic dataset over this value is considered complete.

The three seismic sequences are indicated in Fig.1 with letters (sequence A in 1986-87 with main shock m_b= 4.2, B in 1988 with main m_b = 5.3, and C in 1989-90 with main m_b= 4.7).

On the other hand, a detailed study of the historical sources points out that the Gargano zone has always presented a strong inland and offshore seismicity, probably mislocated. The list of the 24 stronger earthquakes (I_o >= VIII M.C.S.) occurred in the Gargano area from the year 1000 is reported in Table 1. In this table the magnitudes of the old events are obviously derived from the intensities.

TABLE 1. Earthquakes with I_o >= VIII occurred from the year 1000 in the Gargano area.
I_o - epicentral intensity; M - magnitude. (Postpischl, 1985b; ING, 1990)

Year	m	d	Epic. coord.	I_o	M	Epic. area
1087	9	-	41.000 17.000	IX	5.6	Apulia
1223	4	-	41.630 15.890	X	6.1	Gargano
1267	4	10	41.166 17.000	VIII	5.1	Adriatic Sea
1414	-	-	41.880 16.180	IX	5.6	Gargano
1560	5	11	41.300 16.500	IX	5.6	Gargano
1627	7	30	41.800 15.280	XI	6.4	Gargano
1627	7	30	41.800 15.280	IX	5.8	Gargano
1627	8	7	41.800 15.280	X	6.1	Gargano
1627	9	6	41.680 15.380	IX	5.8	Gargano
1646	5	31	41.920 16.000	IX	5.6	Gargano
1689	9	21	41.200 16.500	VIII	5.3	Adriatic Sea
1731	3	20	41.460 15.550	VIII	5.3	Foggia
1731	10	17	41.333 16.500	VIII	5.1	Adriatic Sea
1841	2	21	41.667 15.583	VIII	5.1	Gargano
1875	12	6	41.700 15.700	VIII	5.1	Gargano
1876	5	29	41.700 15.700	VIII	5.1	Gargano
1893	8	10	41.700 16.200	IX	5.9	Gargano
1937	7	17	41.700 15.167	VIII	5.1	Gargano
1941	8	20	41.700 15.400	VIII	5.1	Gargano
1947	2	26	41.000 17.500	VIII	5.1	Adriatic Sea
1948	8	18	41.500 16.600	VIII	5.1	Gargano
1955	2	9	41.800 15.883	VIII	4.8	Gargano
1975	6	19	41.650 15.730	VIII	5.1	Gargano
1988	4	26	42.298 16.593	VIII	5.3	Adriatic Sea

A good example of mislocation is the earthquake occurred on July 30th, 1627 (XI M.C.S.) that was among the most destructive ones. This event was located inland, if epicentral location was based on the damage levels (Postpischl, 1985a), while the effects in the Tremiti Islands and the tsunami waves both along the northern and southern Gargano coasts could suggest an offshore epicentral area and the strongest macroseismic effects would be due to local geological conditions. The Adriatic seismicity coincide with a deformational belt in which complex regional fault systems exist. These systems have been just described in the previous paragraph, but it is important again to underline that they are present only in this portion of the Adriatic basin. At roughly the same latitudes, a strong positive Bouguer gravity anomaly (+110 mGal) corresponds to the Gargano

Promontory (Finetti and Morelli, 1973). The magnetic basement pattern, derived from the AGIP aeromagnetic survey (Cassano et al., 1986), shows an uplift of the top of the basement in a longitudinal strip from the Adriatic Sea to the Tyrrhenian Sea, changing the average depth values of 12-14 km to 8-10 km. The depth of the lithosphere/asthenosphere boundary also has a very strong gradient at the same general location (Suhadolc and Panza, 1989). A moderate lithospheric significance was given to this seismic belt interpreted to dissect the Adriatic microplate in two minor blocks (Favali et al., 1990).

The Adriatic belt has a continuation towards the West in the uplifted and deformed central Apennines just described in the geological section. The seismicity in this area is caracterized by peaks of high energy release at shallow normal depth (e.g., L'Aquila 1703 earthquake, $M = 6.5$; Avezzano 1915 earthquake, $M = 6.9$), and with long return time of thousands of years (Ward and Valensise, 1989). In particular the Avezzano earthquake had a NW-SE left-lateral strike-slip focal mechanism with a slight normal component (Basili and Valensise, 1991), coherently with the Recent geological evolution.

Altogether, the seismotectonic belt divides the Apennines in two seismotectonic provinces with differing dynamic behaviour. If we sample seismic data from two sectors about 2° of latitude wide above and below the central belt, obviously excluding the already mentioned Gargano-Adriatic seismicity, it is possible to outline different seismic characteristics between the northern and southern provinces.

The northern province is characterized by frequent activity with low energy level occurring within the crust (<= 20 km), except for slightly deeper earthquakes that sporadically take place (depths ranging in about 50-90 km) (Selvaggi and Amato, 1992). In this zone about 135 events in 100 years (1890-1990) with magnitudes ranging in only one degree (4.5-5.5) took place (ING, 1990). Nevertheless, analysing the historical seismic catalogues, we found only 7 earthquakes with intensity greater than *IX* M.C.S. (Postpischl, 1985b; ING, 1990) and maximum intensity that reached *X* M.C.S.

The southern province is characterized by higher energy activity at normal depth (<= 20 km) along the chain. About 75 events in 100 years (1890-1990) within two degree range of magnitude (4.5-6.5) occurred (ING, 1990), the difference with the magnitude range of the northern province is notable. In the historical catalogues seismic events with intensity greater than *IX* M.C.S. are 16 in the Istituto Nazionale di Geofisica catalogue (ING, 1990) and 18 in the Italian Geodynamic Project catalogue (Postpischl, 1985b). Furthermore, in many cases the eathquakes reached intensities of *XI* M.C.S. The stronger seismicity is confined in a strip along the chain and, taking into account the depth range, below the sedimentary covers of the Adriatic lithosphere that many Authors consider sinking beneath the Apennines (e.g., Malinverno and Ryan, 1986; Royden et al., 1987; Moretti and Royden, 1988).

The Tyrrhenian Sea, if the shallow seismicity linked to the Eolian volcanic Islands (Barberi et al., 1973) is excluded, has deep seismicity which occurs to the west of peri-Tyrrhenian volcanic belt into a roughly triangular area between Gulf of Gaeta to the North and Sicilian coasts to the South, and in a 200-500 km depth range (e.g., Caputo et al., 1970; 1972; Anderson and Jackson, 1987; Giardini and Velonà, 1991). The stronger positive Bouguer gravity anomalies (>= +100 mGal) is confined in the same triangular area. Going to the North the deep seismicity is more rare and the tensile regime decreases.

Both deep seismicity and tensile regime end reaching the N 41° parallel. Along this parallel the seismic activity seems almost absent, even through a lack of microseismic data, due to the geographic configuration of the seismic network, makes this conclusion partial. Anyway some indications of seismicity between Corsica and Sardinia Islands however exist (ING, 1990).

4. The Seismotectonic Belt

The geological/geophysical and seismological evidence supports the existence of tectonic and seismic activity along a belt across the Adriatic-Apennines systems. This evidence allows to define a more detailed geodynamic model of the area (Favali et al., 1993). The Plio-Pleistocenic westward sinking of the Adriatic microplate under the Apennines develops with differentiated rates and patterns. In the southern sector we can infer to a steeply dipping slab (about 70° as computed from Patacca and Scandone, 1989) and a relatively faster subduction with a slower eastward retreating of the trench. This is responsible in our opinion for the high seismic energy release with NW-SE trend and normal focal mechanisms. In the northern sector the sinking process has been hypothesized (Patacca and Scandone, 1989), but it is not very evident. The lower seismic and tectonic rates testify that this possible sinking process should anyhow develop with a smaller dipping and a lower rate. As a result, right-lateral strike-slip tectonics should occur in the Adriatic microplate along the above evidenced E-W belt. A present-day offset of 20-30 km across this belt can be desumed by the base of Pliocene contour lines and gives a first order appoximation of its entity (Bigi et al., 1989).

The Plio-Pleistocenic block tectonics observed in the central Apennines would represent the result of the accomodation of this offset, combined with its coeval thrusting in the eastern sector and with the extensional deformations along the Tyrrhenian margin. According to this model, it is the relative faster sinking of the southern sector that leads a larger extension in the central Tyrrhenian Sea, thus inducing an opposite strike-slip displacement in the Apennines-Tyrrhenian Sea sector. The central Tyrrhenian Sea itself is part of a semi-oceanic basin developed mostly since Messinian times (about 10 Ma ago). Its thin crust (down to 10 km) is generally related to an oceanization process that involved a continental lithosphere, after the main collision between Africa and Europe, in Early Cenozoic times, that caused the closing of the Tethys, the Upper Jurassic-Cretaceous oceanic basin (Ogniben et al., 1975; Bigi et al., 1989). A tectonic line located roughly along the N 41° parallel divides it from the northern Tyrrhenian basin that is characterized by a slightly older and more limited evolution (e.g., Selli et al., 1977; Wezel, 1985; Finetti and Del Ben, 1986; Sartori et al., 1989; Patacca et al., 1990). This reduced tectonic activity is testified also by its maximum depth (around 1 km).

Series of N-S to NNW-SSE normal listric and sub-vertical faults displace the basin (Bigi et al., 1989). Studies on the western margin in Corsica Island show that the northern basin underwent a delamination process, possibly through a symmetrical process (Julivet et al., 1991). Otherwise, evidence from the central basin shows that it suffered a rather asymmetrical extensional process, with the presence of sub-vertical faulting in its western sector and the presence of regional westverging listric and low angle normal faults (Sartori et al., 1989; Bigi et al., 1989). The northern Tyrrhenian basin was deeply affected by highly contaminated magmas, indicating either an intermediate thickness crust or a slow uprising of the magma itself that could reflect a relatively low tectonic rate (Serri, 1990; Turi et al., 1991). Greater water depths (3-4 km) characterize the central sector of the Tyrrhenian Sea. This fact, together with a minor crustal thickness, testifies that the oceanization process in this sector has almost reached completion. Along the eastern coastline, in the older part of the basin, a NW-SE trend becomes important (Zitellini et al., 1984). There, it is represented by relatively older faults and could testify of the earlier extensional direction in the area. Volcanism is also younger (Messinian to Recent) and shows peculiar sub-crustal petrological and geochemical features (Serri, 1990; Turi et al., 1991) which proves that the tectonic processes are relatively faster.

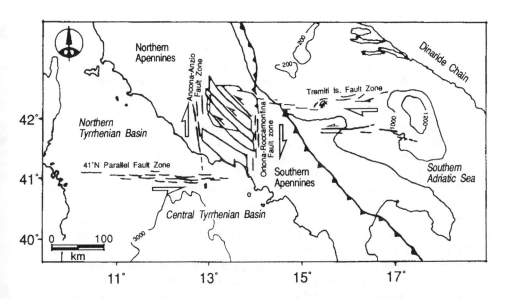

Figure 2. Model for the E-W seismotectonic belt. Relative maximum water depths are contoured; symbols as in Fig. 1.

The model is shown in Fig.2 with a sketch diagram of the Tyrrhenian Sea-Apennines-Adriatic Sea systems along the proposed E-W belt. The Sardinia-Corsica block, fragment of the European lithosphere, to the West does not seem to be involved in the kinematic process. The northern and central Tyrrhenian basins, the latter with a thinner crust, are divided roughly by the 41° discontinuity which acted in Plio-Pleistocenic times as a left-lateral transform fault (Sartori et al, 1989). The northern and southern Adriatic blocks are also divided, but by a right-lateral transform fault zone around 42° parallel. The deformed and uplifted central Apennines constitutes the transfer zone between the two longitudinal trasform zones. The opposite sense of motion in these two transforms is compatible with a faster hinge retreat and with a relative westward movement of the southern Adriatic block. This is interpreted by the dextral offset of the Adriatic foredeep at the Gargano latitudes (Fig.1 in Patacca and Scandone, 1989; Bigi et al., 1989). The relative slower eastward migration of the southern Adriatic block is justified admitting that the Adriatic lithosphere subduction is partly active through a moderate southwesternward movement. This partly compensates the migration, resulting in the observed right-lateral offset.

5. Conclusions

Several Authors (i.e. Patacca et al., 1990) state that the outest thrust front of the present-day central Apennines is sealed by Quaternary deposits and interpreted this evidence as the indication of a temporary interruption in the thrust tectonic processes. Again, Sartori et al. (1989) proposes a similar age as the ending time of main extensional tectonics in central Tyrrhenian basin. According to this evidence, a younger activity should be difficult to place. Yet, younger extensional related volcanism is present along the coastline together with evidence of coeval tectonic activity in the chain (among others, Alfonsi et al., 1991a, 1991b). Therefore, the basin extension-chain shortening process could well still be active in the central Apennines, with the extension concentrated along the eastern margin of the chain and the shortening included in the block tectonics event.

Geological/geophysical and seismological evidence indicates that the existing E-W complex deformational belt crossing the Adriatic Sea at the latitude of the Tremiti Islands and Gargano Promontory, finds its continuation in the Latium-Abruzzi Platform (LAP) domain and in the Tyrrhenian Sea along the N 41° parallel. This belt has a strike-slip displacement and a lithospheric significance dissecting the Adriatic microplate and separating two different seismotectonic provinces in the Apennines.

The uplifted LAP domain acts as a transfer zone, through a block rotation tectonics delimited by two N-S right-lateral strike-slip fault zones connected by a series of NW-SE left-lateral ones. In the southern edge of these two N-S lines Recent central volcanoes (Alban Hills and Roccamonfina) suggest a deep nature of these discontinuities.

The proposed model involves an opposite motion of the belt in the Tyrrhenian and Adriatic sides, showing left-lateral and right-lateral transform mechanisms respectively. This implies that the southern Adriatic lithosphere moves westward relative to the northern Adriatic one. The deep seismicity of the southern Tyrrhenian basin indicates a depth of the subduction process of over 500 km. If these depth are true also for our sector, then the westward wandering of the lithosphere plays a major role in the bending of the microplate.

North of the belt the Tyrrhenian Sea-Apennines-Adriatic Sea system evolved with less evidence of sinking processes. However, new seismic data indicate that the geodynamic process is confined within the first hundred km (Selvaggi and Amato, 1992). These depths seem to be too surficial to involve in the process the eastward relative wandering of the asthenosphere.

Acknowledgements

The authors wish to thank Daniela Riposati for drawing of the figures.

References

Alfonsi, L., Funiciello, R., Mattei, M., Girotti, O., Maiorani, A., Preite Martinez, M., Trudu, C. and Turi, B. (1991a) Structural and geochemical features of the Sabina strike-slip fault (central Apennines), Boll. Soc. Geol. Ital., 110, 217-230.

Alfonsi, L., Funiciello, R. and Mattei, M. (1991b) Strike-slip tectonics in the Sabina area, Boll. Soc. Geol. Ital., 110, 481-488.

Anderson, H. and Jackson, J. (1987) The deep seismicity of the Tyrrhenian Sea, Geophys. J. R. Astron. Soc., 91, 613-637.

Argnani, A., Favali, P., Frugoni, F., Gasperini, M., Ligi, M., Marani, M., Mattietti, G. and Mele, G. (1992) Foreland deformational pattern in the southern Adriatic Sea, in R. Funiciello and C. Laj (eds.), Proceedings 7° International School of Solid Earth Geophysics, Erice (TP) (in press).

Barberi, F., Gasparini, P., Innocenti, F. and Villari, L. (1973) Volcanism of the southern Tyrrhenian Sea and its geodynamic implications, J. Geophys. Res., 78, 5221-5232.

Basili, A. and Valensise, G. (1991) Contributo alla caratterizzazione della sismicità dell'area marsicano-fucense, in E. Boschi and M. Dragoni (eds.), Proceedings 2° Worshop "Aree sismogenetiche e rischio sismico in Italia", Erice (TP), 1986, pp.197-214.

Bigi, G., Castellarin, A., Catalano, R., Coli, M., Cosentino, D., Dal Piaz, G.V., Lentini, F., Parotto, M., Patacca, E., Praturlon, A., Salvini, F., Sartori, R., Scandone, P. and Vai, G.B. (1989) Synthetic structural-kinematic map of Italy (1:2,000,000), C.N.R. - P.F.G., Roma.

Caputo, M., Panza, G.F. and Postpischl, D. (1970) Deep structure of the Mediterranean basin, J. Geophys. Res., 75, 4919-4923.

Caputo, M., Panza, G.F. and Postpischl, D. (1972) New evidences about deep structure of the Lipari arc, Tectonophysics, 15, 219-231.

Cassano, E., Fichera, R. and Arisi Rota, F. (1986) Rilievo aeromagnetico d'Italia. Alcuni risultati interpretativi, Proceedings 5° Meeting of Gruppo Naz. Geofis. Terra Solida, Roma, pp.939-958.

Cerisola, R. and Montone, P. (1991) Analisi strutturale di un settore dei Monti Ausoni-Aurunci (Lazio, Italia centrale), Boll. Soc. Geol. Ital., 111, 449-457.

Channell, J.E.T. and Horváth, F. (1976) The African/Adriatic promontory as a palaeogeographical premise for Alpine orogeny and plate movements in the Carpatho-Balkan region, Tectonophysics, 35, 71-101.

Channell, J.E.T., D'Argenio, B. and Horváth, F. (1979) Adria, the African promontory, in Mesozoic Mediterranean paleogeography, Earth Sci. Rev., 15, 213-272.

Console, R., Di Giovambattista, R., Favali, P. and Smriglio, G. (1989) Lower Adriatic Sea seismic sequence (January 1986): spatial definition of the seismogenic structure, Tectonophysics, 166, 235-246.

Console, R., Di Giovambattista, R., Favali, P., Presgrave, B.W. and Smriglio, G. (1992) Seismicity of the Adriatic microplate, Tectonophysics, 217 (in press).

D'Argenio, B. and Horváth, F. (1984) Some remarks on the deformation history of Adria, from the Mesozoic to the Tertiary, Annales Geophysicae, 2, 143-146.

De' Dominicis, A. and Mazzoldi, G. (1987) Interpretazione geologico-strutturale del margine orientale della piattaforma Apula, Mem. Soc. Geol. Ital., 38, 163-176.

Doglioni, C. (1990) The global tectonic pattern, Journal of Geodynamics, 12, 21-38.

Favali, P., Mele, G. and Mattietti, G. (1990) Contribution to the study of the Apulian microplate geodynamics, Mem. Soc. Geol. Ital., 44, 71-80.

Favali, P., Funiciello, R., Mele, G., Mattietti, G. and Salvini, F. (1993) An active margin across the Adriatic Sea (central Mediterranean Sea), Tectonophysics, 219 (in press).

Finetti, I. (1982) Structure, stratigraphy and evolution of central Mediterranean, Boll. Geofis. Teor. Appl., 24, 247-312.

Finetti, I. and Del Ben, A. (1986) Geophysical study of the Tyrrhenian opening, Boll. Geofis. Teor. Appl., 38, 75-155.

Finetti, I. and Morelli, C. (1973) Geophysical exploration of the Mediterranean Sea, Boll. Geofis. Teor. Appl., 15, 263-341.

Finetti, I., Bricchi, G., Del Ben, A., Pipan, M. and Xuan, Z. (1987) Geophysical study of the Adria plate, Mem. Soc. Geol. Ital., 40, 335-344.

Funiciello, R., Montone, P., Salvini, F. and Tozzi, M. (1988) Caratteri strutturali del Promontorio del Gargano, Mem. Soc. Geol. Ital., 41 (in press).

Funiciello, R., Montone, P., Parotto, M., Salvini, F. and Tozzi, M. (1991) Geodynamical evolution of an intra-orogenic foreland, The Apulia case history (Italy), Boll. Soc. Geol. Ital., 110, 419-425.

Giardini, D. and Velonà, M. (1991) The deep seismicity of the Tyrrhenian Sea, Terra Nova, 3, 57-64.

Giorgetti, F. and Mosetti, F. (1969) General morphology of the Adriatic Sea, Boll. Geofis. Teor. Appl., 11, 49-56.

Istituto Nazionale di Geofisica (ING) (1990) Italian seismic catalogue (1450 b.C. - 1990 a.D.), Roma.

Julivet, L., Daniel, J.M. and Fournier, M. (1991) Geometry and kinematics of ductile extension in Alpine Corsica, Earth Planet. Sci. Let., 109, 278-291.

Malinverno, A. and Ryan, W.B.F. (1986) Extension of the Tyrrhenian Sea and shortening in the Apennines as result of arc migration driven by sinking lithosphere, Tectonics, 5, 227-245.

Mantovani, E., Babbucci, D. and Farsi, F. (1985) Tertiary evolution of the Mediterranean region, outstanding problems, Boll. Geofis. Teor. Appl., 26, 67-88.

McKenzie, D. (1972) Active tectonics of the Mediterranean region, Geophys. J. R. Astron. Soc., 30, 109-185.

Mongelli, F., Zito, G., Ciaranfi, N. and Pieri, P. (1989) Interpretation of heat flow density of the Apennine chain, Italy, Tectonophysics, 164, 267-280.

Montone, P. and Funiciello, R. (1989) Elementi di tettonica trascorrente alle Isole Tremiti (Puglia), Rend. Soc. Geol. Ital., 12, 7-12.

Montone, P. and Salvini, F. (1991) Evidences of strike-slip tectonics in the Apenninic chain near Tagliacozzo (L'Aquila), Abruzzi, central Italy, Boll. Soc. Geol. Ital., 110, 617-619.

Moretti, I. and Royden, L. (1988) Deflection, gravity anomalies and tectonics of doubly subducted continental lithosphere, Adriatic and Ionian Seas, Tectonics, 7, 875-893.

Ogniben, L., Parotto, M. and Praturlon, A. (eds.) (1975) Structural model of Italy, Quad. "La Ricerca Scientifica", C.N.R., 90, pp.496.

Patacca, E. and Scandone, P. (1989) Post-Tortonian mountain building in the Apennines. The role of the passive sinking of a relic lithospheric slab, in A. Boriani, M. Bonafede, G.B. Piccardo and G.B. Vai (eds.), The lithosphere in Italy, Rend. Acc. Naz. Lincei, Roma, 80, pp.157-176.

Patacca, E., Sartori, R. and Scandone, P. (1990) Tyrrhenian basin and Apenninic arcs, kinematic relations since Late Tortonian times, in "La Geologia italiana degli anni '90", 75° Meeting of the Italian Geological Society (invited paper), pp.102-107.

Postpischl, D. (ed.) (1985a) Atlas of isoseismal maps of Italian earthquakes, Quad. "La Ricerca Scientifica", C.N.R. - P.F.G., 114, vol. 2A, pp.164.

Postpischl, D. (ed.) (1985b) Catalogue of Italian earthquakes from 1000 to 1980, Quad. "La Ricerca Scientifica", C.N.R. - P.F.G., 114, vol. 2B, pp.239.

Royden, L., Patacca, E. and Scandone, P. (1987) Segmentation and configuration of subducted lithosphere in Italy: An important control on thrust-belt and foredeep-basin evolution, Geology, 15, 714-717.

Salvini, F. (1992) Tettonica a blocchi in settori crostali superficiali, modellizzazione ed esempi da dati strutturali in Appennino centrale, Studi Geol. Camerti, Spec. Pubbl. 1991/2, 237-248.

Salvini, F. and Tozzi, M. (1986) Evoluzione tettonica recente del margine tirrenico dell'Appennino Centrale in base a dati strutturali, implicazioni per l'evoluzione del Mare Tirreno, Mem. Soc. Geol. Ital., 36, 233-241.

Sartori, R. and ODP Leg 107 scientific staff (1989) Drillings of ODP Leg 107 in the Tyrrhenian Sea, Tentative basin evolution compared to deformations in the surrounding chains, in A. Boriani, M. Bonafede, G.B. Piccardo and G.B. Vai (eds.), The lithosphere in Italy, Rend. Acc. Naz. Lincei, Roma, 80, pp.139-156.

Selli, R., Lucchini, F., Rossi, P.L., Savelli, C. and Del Monte, M. (1977) Dati petrologici, petrochimici e radiometrici sui vulcani centro-tirrenici, Giornale Geologia, 42, 221-246.

Selvaggi, G. and Amato, A. (1992) Subcrustal earthquakes in the northern Apennines (Italy), evidence for a still active subduction? Geophys. Res. Let., vol. 19, no. 21, 2127-2130.

Serri, G. (1990) Neogene-Quaternary magmatism of the Tyrrhenian region, characterization of the magma sources and geodynamic implications, Mem. Soc. Geol. Ital., 44, 219-242.

Suhadolc, P. and Panza, G.F. (1989) Physical properties of the lithosphere-asthenosphere system in Europe from geophysical data, in A. Boriani, M. Bonafede, G.B. Piccardo and G.B. Vai (eds.), The lithosphere in Italy, Rend. Acc. Naz. Lincei, Roma, 80, pp.15-40.

Turi, B., Ferrara, G. and Taylor, H. (1991) Comparisons of $^{18}O/^{16}O$ and $^{87}Sr/^{86}Sr$ in volcanic rocks from the Pontine Islands, Ernici Mts. and Campania with other areas in Italy, in H. Taylor, G.R. O'Neill and I.R. Koplan (eds.), Debole isotopes geochemistry: a tribute to Samuel Epstein, Geochemical Soc., Spec. Pubbl. no 3, pp.307-324.

Zitellini, N., Marani, M. and Borsetti, A.M. (1984) Post-orogenic tectonic evolution of Palmarola and Ventotene basins (Pontine Archipelago), Mem. Soc. Geol. Ital., 27, 121-131.

Ward, S. and Valensise, G. (1989) Fault parameters and slip distribution of the 1915 Avezzano, Italy, earthquake derived from geodetic observations, Bull. Seism. Soc. Am., 79, 690-710.

Wezel, F.C. (1985) Structural features and basin tectonics of the Tyrrhenian Sea, in D.J. Stanley and F.C. Wezel (eds.), Geological evolution of the Mediterranean basin, Springer-Verlag, New York, pp.153-194.

STABLE FAULT SLIDING AND EARTHQUAKE NUCLEATION

M. DRAGONI
Department of Physics
University of Bologna
Viale Carlo Berti Pichat 8
40127 Bologna, Italy

ABSTRACT. The process of earthquake nucleation is studied assuming that faults are rupture surfaces on which sliding is controlled by friction. Earthquakes are assumed to arise through an instability of frictional sliding. Empirical slip laws indicate that, under constant ambient conditions, friction depends on time, slip rate and slip history. Regular stick-slip behaviour is induced by velocity weakening, a negative dependence of friction on slip rate. Velocity weakening is introduced into a model for a propagating Somigliana dislocation under slowly increasing ambient shear stress. The instability occurs when the rate at which friction decreases becomes greater than the rate at which the applied stress must increase to produce an advance of fault slip. The possibility that this condition is fulfilled depends on the velocity dependence and on the spatial distribution of friction on the fault. A critical nucleation width of the dislocation is associated with the instability and is controlled by the friction distribution, which determines the size of the initial slipping patch. Depending on the stress drop and the characteristic slip distance, the critical nucleation width may be greater for small earthquakes than for large earthquakes, with respect to the initial slipping patch.

1. Introduction

Earthquake nucleation is a physical process which is still scarcely understood. The properties of rocks and the ambient conditions prevailing in the shallow crust indicate that fault mechanics is dominated by frictional sliding on pre-existing rupture surfaces. Earthquakes are thought to arise through an instability of frictional sliding (Brace and Byerlee, 1966; Dieterich, 1974).

Stable frictional sliding is aseismic. The conditions under which seismic rather than aseismic slip occurs depend on the frictional stability transitions, which are strongly dependent upon lithology (Scholz, 1990). Faults that are currently aseismic are those in weakly consolidated sediments and ductile faults of various types, such as decollements. In general, seismic and aseismic slip may coexist on the same fault segment.

The first systematic observations of aseismic fault slip were made on the San Andreas Fault in central California (e.g. Scholz et al., 1969). Fault segments slipping aseismically have a high level of small earthquake activity, which however does not contribute significantly to slip. Scholz (1990) estimated that the seismic moment release of the creeping section of the San Andreas Fault accounts for less than 5% of its slip.

Direct observation of aseismic fault slip at the Earth's surface is not very common. However fault creep should be a widespread feature of tectonic plate boundaries. Since many years it has been observed (e.g. Davies and Brune, 1971; Jackson and McKenzie, 1988) that most of the relative plate displacement must occur aseismically. In fact seismic dislocations do not cover the whole area of plate boundaries, nor the slip associated with earthquakes is uniform over a dislocation surface (Archuleta, 1984; Hartzell and Heaton, 1986), while the long-term relative displacement along plate boundaries must be even for obvious mechanical reasons. Decollements, where slip occurs aseismically on weak ductile materials like salt, may on the other hand dominate the tectonics of some areas.

2. Frictional Slip and Instability

Laboratory experiments on the frictional sliding of rocks show that a dynamic instability can occur, resulting in a sudden slip and an associated stress drop. This phenomenon can occur repetitively: the instability is followed by a stationary period during which stress is recharged, followed by another instability. This behaviour is called *stick-slip* and was proposed by Brace and Byerlee (1966) as the mechanism of earthquakes. Earthquakes are recurring slip instabilities on pre-existing rupture surfaces.

In general, a dynamic slip instability occurs when the frictional strength decreases at a rate that exceeds the capability of the applied stress to follow (Stuart, 1981). The simplest picture of frictional sliding is that a static friction τ_s must be exceeded for slip to commence. Once slip has started, it is resisted by a dynamic friction τ_d which is smaller than τ_s. Experiments show however that τ_s and τ_d are not constant in time. If the surfaces are held in stationary contact for a time t, τ_s increases approximately as $\ln t$. During sliding, at constant slip rate V, τ_d is observed to depend on V, being lower for higher slip rates. Moreover the instability does not occur instantaneously at some threshold stress, but after an interval of accelerating slip, and takes place over a finite displacement, the

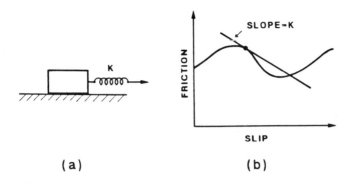

Fig. 1 - Schematic diagram illustrating the origin of frictional instability. (a) A heavy block sliding on a rough surface, where the driving force is transmitted to the block by a spring of rigidity K; (b) friction as a function of slip, where friction falls with displacement faster than the system can respond.

characteristic slip distance (Dieterich, 1979 a, b; Ruina, 1983; Rice and Ruina, 1983; Rice, 1983; Gu et al., 1984; Weeks and Tullis, 1985).

The frictional behaviour in which friction falls with slip is called *slip weakening* and may result in stick-slip (Fig. 1). However it does not provide a mechanism for the frictional resistance to regain its initial level and so does not lead to the regular stick-slip behaviour which is required to explain earthquakes. The basic phenomenon which leads to regular stick-slip is *velocity weakening*, a decrease of friction with increasing slip rate V. If V is an increasing function of time, slip weakening is just a consequence of velocity weakening.

If we consider a fault in the Earth's crust, lithological or geometrical changes along the fault may induce spatial changes in friction. It is commonly assumed that friction τ is locally proportional to the effective normal stress across the fault:

$$\tau = \kappa(\sigma_n - p) \qquad (1)$$

where κ is the coefficient of friction, σ_n is the applied normal stress and p is fluid pore pressure. Friction may vary in space as a consequence of changes in κ, σ_n and p. The nucleation, the propagation at finite speed and the final arrest of a fault dislocation are necessarily associated with inhomogeneities in friction and/or applied shear stress. Stability transitions may occur because of lithological changes along a fault, for instance when a dislocation front passes from consolidated to unconsolidated fault gouge at shallow depth. This may result in a friction fall (Marone and Scholz, 1988).

3. Empirical Slip Laws

Frictional sliding is a complex phenomenon, including the surmounting of obstacles and local processes of fracture or plastic flow (Paterson, 1978). Laboratory experiments on rocks of the kind initiated by Brace and Byerlee (1966) show that the form of frictional sliding depends mainly on ambient conditions, such as rock type, porosity and pore pressure, thickness of the gouge layer and, more importantly, temperature and pressure.

Under constant and uniform ambient conditions, fault strength depends on time, slip rate and slip history (Dieterich, 1978; Ruina, 1983; Weeks and Tullis, 1985). The experimental observations can be represented by constitutive relations including slip rate and history effects: the sliding history effects are usually represented by a state variable that evolves with displacement toward a steady-state value. A commonly employed constitutive law, yielding the coefficient of friction κ in terms of slip rate V and a state variable ψ, is (Ruina, 1983):

$$\kappa = \kappa_0 + \mathcal{A} \ln \frac{V}{V_0} + \mathcal{B}\psi \qquad (2)$$

where κ_0 is some basic friction, V_0 is a reference velocity, $\mathcal{A}$ and $\mathcal{B}$ are empirically determined constants. There are various forms of constitutive equations, which differ in detail, depending on the approximations employed, but all share the same approach and provide very similar representations of the data.

The constitutive equation is accompanied by an equation for the evolution of the state variable ψ:

$$\frac{d\psi}{dt} = -\frac{V}{\ell}(\psi + \ln \frac{V}{V_0}) \qquad (3)$$

where ℓ is the characteristic slip distance which is measured in the laboratory (e.g. Scholz, 1990). Equations (2) and (3) are similar to those proposed by Dieterich (1979a) for a wide range of slip rates. The second term in the r.h.s. of (2) describes the direct velocity effect, which is positive. The state variable includes a negative velocity effect, which evolves over the characteristic slip distance ℓ according to (3). A more general formulation may involve several state variables, each obeying to its own constitutive equation (Horowitz and Ruina, 1985; Tullis and Weeks, 1986); however the basic effects can be discussed using a single state variable and characteristic slip distance.

The state variable ψ represents the state of contact between the two sliding surfaces. For a stationary fault, it has been interpreted (Dieterich, 1979a; Dieterich and Conrad, 1984) as the average age of the contacts between the sliding surfaces. Actually the "stick stage" may include motion at very low rates: during this stage there is evidence that friction increases with increasing slip rate (Rabinowicz, 1958). Velocity weakening is interpreted as a consequence of the contact area decreasing with slip rate.

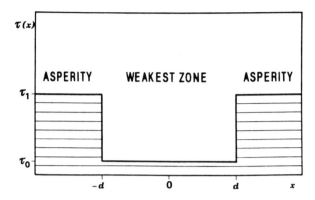

Fig. 2 - Static friction $\tau(x)$ around the weakest fault zone at time $t = 0$, where x is a coordinate along the fault. At $t = 0$ it is assumed that $\sigma = \tau_0$, where σ is the ambient shear stress, increasing with time, and the fault starts sliding.

4. Dislocation Model

We assume that faults are pre-existing rupture surfaces in the Earth's crust and that fault slip is controlled by friction. We consider a planar fault surface and describe the spatial dependence of friction by a piecewise function, that is as a combination of weak and strong patches. Such a function includes the effect of lithologic and geometric inhomogeneities of the fault as well as of all the other factors controlling the resistance to slip, like fluid pore pressure (Rice and Simons, 1976), fault gouge (Wang, 1984) and possible lithification processes (Angevine et al., 1982). Fault slip takes place in response to a gradually increasing ambient shear stress. If the ambient stress is uniform, slip starts in the weakest patch of the fault. A Somigliana dislocation is produced (Bilby and Eshelby, 1968).

To make the problem amenable to an analytical solution, a 2-D model is considered, where friction is variable only in one direction x on the fault plane. We assume that the weakest zone of the fault plane is $-d < x < d$, included between stronger zones (Fig. 2). The fault is stationary for $t < 0$. At $t = 0$ the static friction is

$$\tau(x) = \begin{cases} \tau_0, & |x| < d \\ \tau_1, & |x| > d \end{cases} \quad (4)$$

where $\tau_1 > \tau_0$. We assume that the ambient shear stress σ reaches the value τ_0 at $t = 0$: at that time the fault starts sliding and releasing a stress

$$\Delta\sigma(x,t) = \sigma(t) - \tau(x), \quad d < |x| < a(t) \quad (5)$$

where a is the dislocation half-width. Under equilibrium conditions, the increase of dislocation amplitude Δu and width $2a$ is controlled by the increase of ambient stress σ. The dislocation is confined by the surrounding higher-friction zones,

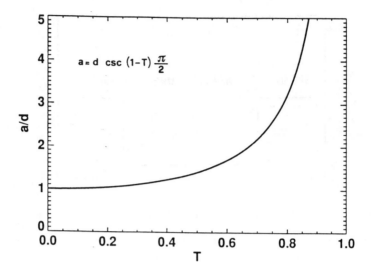

Fig. 3 - Dislocation half-width a as a function of the nondimensional stress parameter T, defined in (6). If the ambient shear stress σ is linearly increasing with time t, T is equivalent to t.

which play the role of asperities. In previous papers (Dragoni, 1990, 1992) it was shown that an aseismic dislocation can pass through an asperity, even if the ambient stress is lower than the asperity strength, due to the stress concentration produced by the dislocation around its front.

The solution technique of the Somigliana dislocation problem is summarized in the Appendix. The various quantities of the model are usefully expressed in terms of a nondimensional, time-dependent parameter

$$T = \frac{\sigma - \tau_0}{\tau_1 - \tau_0} \tag{6}$$

which is a measure of how much the ambient stress σ exceeds the friction level τ_0, relatively to the step $\tau_1 - \tau_0$. As σ increases from τ_0 to τ_1, the ratio T increases from 0 to 1. If σ is linearly increasing with time, T is equivalent to time and will be used in the following instead of t.

The half-width a of the propagating dislocation results (Dragoni, 1990)

$$a(T) = d \csc \varphi \tag{7}$$

where

$$\varphi = (1 - T)\frac{\pi}{2}, \quad 0 \leq T < 1 \tag{8}$$

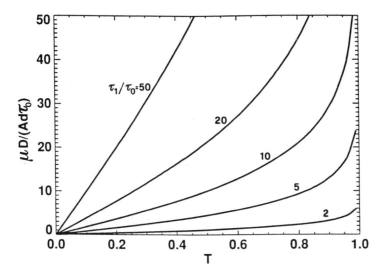

Fig. 4 - Fault slip $D \equiv \Delta u(0,t)$ as a function of the stress parameter T for different values of the ratio τ_1/τ_0.

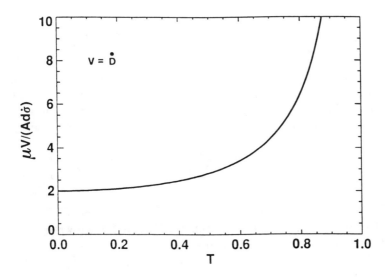

Fig. 5 - Fault slip rate $V \equiv \Delta \dot{u}(0,t)$ as a function of the stress parameter T.

A graph of a as a function of T is shown in Fig. 3. On real faults $a \to \infty$ is

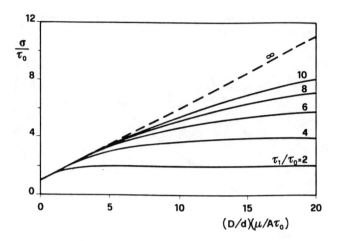

Fig. 6 - Ambient shear stress σ at $x = 0$ as a function of fault slip D, for different values of the ratio τ_1/τ_0.

never reached, since the propagating dislocation will encounter sooner or later other friction barriers slowing down the front motion.

At any time $t > 0$, the maximum fault slip is at $x = 0$ and is given by

$$\Delta u(0,t) \equiv D(T) = \frac{2A}{\pi\mu}(\tau_1 - \tau_0) \ a \sum_{n=1}^{\infty} \frac{\beta_n}{n} \sin \frac{n\pi}{2} \qquad (9)$$

which is obtained from (A.7), using (A.4)-(A.6). In (9) μ is the shear modulus, while A and $\beta_n(T)$ are given in the Appendix. A graph of $\mu D/(Ad\tau_0)$ as a function of T is shown in Fig. 4 for different values of the ratio τ_1/τ_0. The slip rate at $x = 0$ can be expressed as

$$\Delta \dot{u}(0,t) \equiv V(T) = \frac{A\dot{\sigma}}{\mu}\frac{a^2}{d} \sum_{n=1}^{\infty} \frac{\beta_n \cos\varphi - \delta_n \sin\varphi}{n} \sin \frac{n\pi}{2} \qquad (10)$$

where a dot denotes differentiation with respect to time and $\delta_n(T)$ is given in the Appendix. The nondimensional quantity $\mu V/(Ad\dot{\sigma})$ is shown in Fig. 5 as a function of T: $V(T)$ is independent of frictional resistances τ_0 and τ_1. For small values of T, V is fairly constant and approximately equal to its value at $T = 0$:

$$V_0 = \frac{2Ad\dot{\sigma}}{\mu} \qquad (11)$$

which is easily computed from (10).

The fault can be characterized by an effective stiffness (Dieterich, 1986), defined as the derivative of ambient stress with respect to fault slip. At fixed t, the minimum stiffness K_σ is found at $x = 0$, where the slip is maximum. Therefore the centre point $x = 0$ may be assumed to control the onset of instability for the entire fault patch and we define

$$K_\sigma \equiv \frac{d\sigma}{dD} \qquad (12)$$

A graph of σ as a function of D is shown in Fig. 6, whence it can be seen that $K_\sigma > 0$: K_σ represents the increment of ambient shear stress which is needed to produce a corresponding increment of fault slip. This slip hardening effect is produced as a consequence of the propagation through adjacent higher friction zones which resist the motion of dislocation fronts.

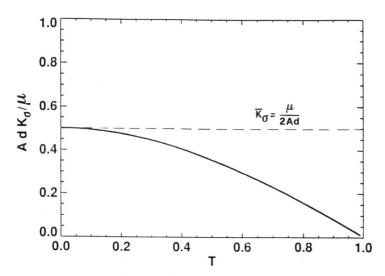

Fig. 7 - Effective fault stiffness K_σ at $x = 0$ as a function of T.

It can be seen that K_σ depends on τ_0 and τ_1 only through the nondimensional stress release T. Besides, it depends on the elastic constants and on the weak zone half-width d. A graph of the nondimensional quantity AdK_σ/μ as a function of T is shown in Fig. 7: K_σ is a bounded function of T and has a maximum

$$\bar{K}_\sigma = \frac{\mu}{2Ad} \qquad (13)$$

at $T = 0$. The existence of a maximum $\bar{K}_\sigma$ takes into account the possibility that fault slip starts at $t = 0$ on a zone with finite extension $2d$ and not just at a point.

5. Instability Conditions

As it stands, the model describes *stable sliding* and can well reproduce the early stage of fault slip originating from a weakly coupled zone. Slip is externally driven: the dislocation area enlarges as the shear stress at the dislocation edge increases due to increasing ambient stress. Now let us assume that slip is governed by equations (2) and (3). It can be seen in Fig. 5 that the slip rate is fairly constant or slowly increasing for values of T as large as 0.8, resulting in *steady-state slip*. During steady-state slip, it is assumed that the state variable ψ is constant in time (e.g. Ruina, 1983). Hence from (3)

$$\psi = -\ln \frac{V}{V_0} \tag{14}$$

where V_0 is given by (11). Using (14) in (2), we obtain the simplified Dieterich's law:

$$\kappa = \kappa_0 + (\mathcal{A} - \mathcal{B}) \ln \frac{V}{V_0} \tag{15}$$

If $\mathcal{A} - \mathcal{B} < 0$, the behaviour is velocity weakening and slip may become unstable (Rice and Ruina, 1983; Gu et al., 1984). In this case, friction is a decreasing function of slip rate. Multiplying (15) by $(\sigma_n - p)$, we obtain

$$\tau = \tau_0 \left(1 - \gamma \ln \frac{V}{V_0}\right) \tag{16}$$

where τ_0 is the value given in (4) and

$$\gamma = (\mathcal{B} - \mathcal{A}) \frac{\sigma_n - p}{\tau_0} \tag{17}$$

A graph of τ as a function of V is shown in Fig. 8 for different values of γ. Of course, due to a decrease in τ, the stress release $\Delta\sigma$ will increase faster in time with respect to the case of constant friction:

$$\Delta\dot{\sigma} = \dot{\sigma} + |\dot{\tau}| \tag{18}$$

This will in turn produce a higher V, since the fault needs to slide at a higher rate to balance the decreasing friction to the applied stress. There is a positive feedback. As a consequence, the stress parameter T will increase faster toward the value 1 than in the case of constant τ. From (16), velocity weakening is expressed by:

$$\frac{d\tau}{dV} = -\frac{\gamma \tau_0}{V} < 0 \tag{19}$$

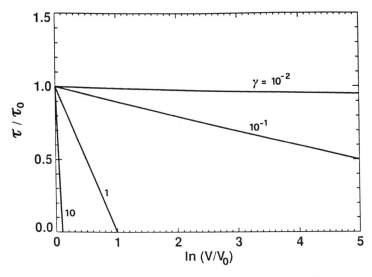

Fig. 8 - Steady-state friction τ as a function of slip rate V, for different values of γ. Large values of γ indicate a rapid decrease of friction with increasing slip rate.

Note that $|d\tau/dV|$ is a decreasing function of V and has a maximum at $V = V_0$. In Fig. 9, τ is plotted as a function of T, showing that $|d\tau/dT|$ is instead an increasing function of T. Correspondingly, we have slip weakening:

$$K_\tau \equiv \frac{d\tau}{dD} < 0 \qquad (20)$$

A graph of τ as a function of $\mu D/(Ad\tau_0)$ is shown in Fig. 10, for the case $\tau_1/\tau_0 = 10$: the slope $|K_\tau|$ is an increasing function of D, being 0 at $D = 0$. As a consequence, a given value of K_τ is reached after the fault has undergone a certain amount of slip: this is consistent with the observation of a characteristic slip distance. As a function of D, K_τ depends on γ and on the ratio τ_1/τ_0. Slip weakening is a necessary condition for instability. During slip weakening, sliding is stable as long as the rate $|K_\tau|$ at which τ decreases is less than the rate K_σ at which σ must increase to produce an advance of fault slip. On the contrary, instability occurs if the friction decrease is overwhelming: hence the condition for instability is (e.g. Stuart, 1981)

$$|K_\tau| > K_\sigma \qquad (21)$$

or

$$R \equiv \frac{|K_\tau|}{K_\sigma} > 1 \qquad (22)$$

It can be easily seen that

$$R = \frac{1}{\tau_1 - \tau_0} \left| \frac{d\tau}{dT} \right| \qquad (23)$$

In Fig.s 11 a-c, R is plotted as a function of T for different values of τ_1/τ_0 and γ. As a general rule, instability is favoured by small values of τ_1/τ_0 and large values of γ. If γ is large, the friction decrease is fast and $|K_\tau|$ increases rapidly: the instability condition is reached at small values of T. Let T_c be the critical value of T at which $R = 1$. For smaller values of γ, T_c increases, but it may happen that friction vanishes before R reaches 1: in this case there is no instability. If the initial slipping patch is very weak with respect to adjacent patches (e.g. $\tau_1/\tau_0 > 10$), the instability takes place only when friction has a strong velocity dependence ($\gamma \gg 10$). Lower values of γ may lead to instability when τ_1/τ_0 is relatively small. The abrupt interruptions of the curves for R in Fig. 11 are due to the fact that friction falls eventually to zero with increasing slip rate and no further decrease is possible.

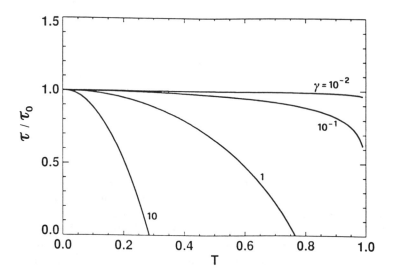

Fig. 9 - Friction τ at $x = 0$ as a function of T, for different values of γ.

If the instability condition (22) is fulfilled, this occurs when a critical value T_c of T is overcome. As a consequence, there is a minimum dislocation width for unstable fault slip. The growth of the slipping patch up to the point of instability is called *nucleation*. Accordingly, a critical half-width a_c, also called *nucleation width*, can be introduced and defined as

$$a_c \equiv a(T_c) \qquad (24)$$

The minimum value for a_c is d, the half-width of the weakest zone. In fact $a_c \to d$, when $\gamma \to \infty$. At fixed τ_1/τ_0, if γ is large instability may be reached, but $a_c \approx d$. If γ is smaller, instability is more difficult to reach, but if it is reached a_c may be sensibly larger than d.

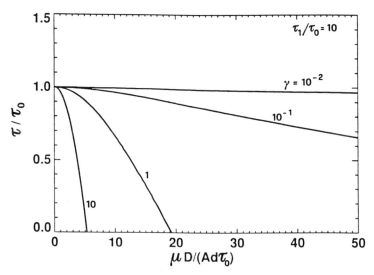

Fig. 10 - Friction τ at $x = 0$ as a function of fault slip D, for different values of γ. The case $\tau_1/\tau_0 = 10$ is considered.

Further constraints come from estimates of the characteristic slip distance ℓ. Direct measurements of ℓ for prepared laboratory faults fall in the range 1 to 50 μm, indicating that ℓ is sensitive to surface roughness and gouge particle dimensions (Dieterich, 1986). Since natural faults are much rougher and have larger and more heterogeneous gouge particles than laboratory faults, values for ℓ extrapolated for natural faults are substantially greater, probably up to several mm. However, stability studies indicate that ℓ cannot exceed a value of about 100 mm, or the fault sliding will become stable (Tse and Rice, 1986; Stuart, 1988).

In our model $\ell \equiv D(T_c)$. Assuming an upper limit for ℓ (e.g. $\ell < 10$ cm) sets a lower limit on the possible values of γ leading to instability. This limit depends on the size d of the initial slipping patch. With reference to Fig. 10, if the initial slipping patch is large (e.g. $d > 1$ km), large values of γ are required (e.g. $\gamma \gg 10$). If instead the initial slipping patch is small ($d < 1$ km), γ can be smaller.

In modelling slip weakening (e.g. Dieterich, 1986) it has been often assumed that the friction decrease is $\Delta\tau$ and takes place over a characteristic slip distance ℓ, so that

$$|K_\tau| \approx \frac{\Delta\tau}{\ell} \tag{25}$$

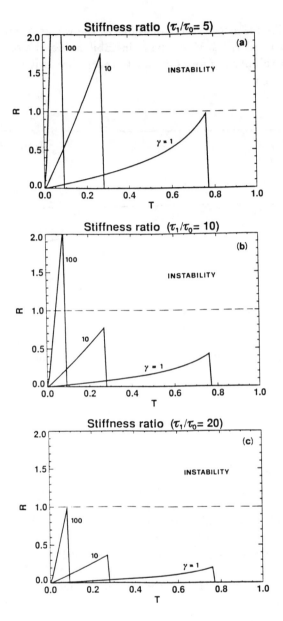

Fig. 11 - Ratio R as as function of T for different values of τ_1/τ_0 and γ. Instability is reached when the curves for R overpass the dashed line corresponding to $R = 1$.

Hence high values of γ correspond to large stress drops and/or low values of ℓ.

A calculation of ℓ based on fault topography (Scholz, 1990) shows that ℓ should decrease with depth, ranging from 1 to 10 mm at seismogenic depths. Accordingly, γ may be sensibly greater for large earthquakes originating at the base of the seismogenic layer, than for smaller earthquakes. It follows that, for large earthquakes, the nucleation width is close to the width of the weakest fault zone, where slip initiated: since the motion of dislocation fronts is very slow during the initial stable sliding, the instability condition is reached before the dislocation area has sensibly enlarged.

6. Conclusions

The introduction of an empirical constitutive law for friction into a model for stable frictional sliding under equilibrium conditions gives an insight into the conditions for stability transitions on a fault. Under a uniform ambient shear stress, slowly increasing with time, sliding starts in the weakest fault patch, being resisted by adjacent asperities. In its early stage, slip is stable and aseismic at a fairly constant rate. As the slip rate increases, friction decreases, resulting in velocity weakening. In general, the instability occurs when the rate at which friction decreases becomes greater than the rate at which the applied stress must increase to produce an advance of fault slip. The possibility that this condition is fulfilled depends on the velocity dependence and on the spatial distribution of friction on the fault. A critical nucleation width of the dislocation is associated with the instability.

The results indicate the crucial role played by friction inhomogeneity on a fault. The nucleation width is in fact controlled by the friction distribution, which determines the size of the initial slipping patch. The critical nucleation patch coincides with the initial slipping patch when large stress drops and low values of the characteristic slip distance are considered, and when the initial patch is large. These features may be common to great earthquakes originated at the base of the seismogenic layer. Smaller earthquakes are plausibly characterized by smaller initial patches, stress drops and, if shallower, by greater characteristic slip distances: in this case the critical nucleation patch may result much larger than the initial patch.

7. Appendix

Let us consider a fault plane embedded in an infinite elastic medium with shear modulus μ and Poisson modulus ν. Consider the x direction along the fault plane and let $\Delta\sigma(x,t)$ be the shear stress gradually released by a dislocation process according to the equilibrium equation (e.g. Bilby and Eshelby, 1968)

$$\Delta\sigma(x,t) - \frac{\mu b}{2\pi A} \int_{-a}^{a} \frac{\mathcal{D}(\xi,t)}{x-\xi} d\xi = 0, \quad |x| < a \qquad (A.1)$$

where $\mathcal{D}$ is a distribution of infinitesimal dislocations, each having Burgers' vector $b\mathcal{D}d\xi$, and $A = 1$ or $1 - \nu$ for antiplane or inplane deformation, respectively.

The Somigliana dislocation problem can be solved analytically by employing a solution technique based on Chebyshev polynomials (Erdogan et al., 1973; Bonafede et al., 1985). A detailed solution is given in Dragoni (1990). Here only a few points are recalled. After making the change of variable

$$y = x/a \qquad (A.2)$$

where a is the dislocation half-width, the stress release $\Delta\sigma(ay,t)$ on the dislocation surface ($|y| \leq 1$) can be expanded into Chebyshev polynomials of the second kind $U_n(y)$:

$$\Delta\sigma(ay,t) = \frac{\mu}{2A} \sum_{n=1}^{\infty} \alpha_n(t) U_{n-1}(y) \qquad (A.3)$$

where

$$U_{n-1}(y) = \frac{\sin n\theta}{\sin \theta} \qquad (A.4)$$

with $\theta = \arccos y$. The evaluation of the Chebyshev coefficients with $\Delta\sigma$ given by (5) yields

$$\alpha_n(t) = \frac{2A}{\pi\mu}(\tau_1 - \tau_0)\beta_n(t) \qquad (A.5)$$

where τ_1 and τ_0 are defined in (4) and

$$\beta_n(t) = \begin{cases} \sin 2\varphi, & n = 1 \\ 2 \sin \frac{n\pi}{2} \left[\frac{\sin(n-1)\varphi}{n-1} + \frac{\sin(n+1)\varphi}{n+1} \right], & n > 1 \end{cases} \qquad (A.6)$$

where φ is defined in (8). The slip amplitude Δu is then computed according to

$$\Delta u(ay,t) = a(t)\sqrt{1-y^2} \sum_{n=1}^{\infty} \frac{\alpha_n(t)}{n} U_{n-1}(y) \qquad (A.7)$$

where $|y| \leq 1$ and $a(t)$ is given by (7). The slip rate $\Delta\dot{u}$ can be computed by differentiation of the series in (A.7) term by term. Its value at $x = 0$ has been given in (10), where

$$\delta_n(t) = 2 \begin{cases} \cos 2\varphi, & n = 1 \\ 2 \sin \frac{n\pi}{2} \cos n\varphi \cos \varphi, & n > 1 \end{cases} \qquad (A.8)$$

No stress singularities are present in the solution.

References

Angevine C.L., Turcotte D.L. and Furnish M.D. (1982). Pressure solution lithification as a mechanism for the stick-slip behavior of faults, *Tectonics*, **1**, 151-160.

Archuleta R.J. (1984). A faulting model for the 1979 Imperial Valley earthquake, *J. Geophys. Res.*, **89**, 4559-4585.

Bilby B.A. and Eshelby J.D. (1968). Dislocations and the theory of fracture, in *Fracture, An Advanced Treatise*, vol. 1, pp. 99-182, ed. Liebowitz, H., Academic Press, New York.

Bonafede M., Dragoni M. and Boschi E. (1985). Quasi-static crack models and the frictional stress threshold criterion for slip arrest, *Geophys. J. R. astr. Soc.*, **83**, 615-637.

Brace W.F. and Byerlee J.D. (1966). Stick-slip as a mechanism for earthquakes, *Science*, **153**, 990-992.

Davies G.F. and Brune J.N. (1971). Regional and global fault slip rates from seismicity, *Nature Phys. Sci.*, **229**, 101-107.

Dieterich J.H. (1974). Earthquake mechanisms and modeling, *Annu. Rev. Earth Planet. Sci.*, **2**, 275-301.

Dieterich J.H. (1978). Time-dependent friction and the mechanism of stick-slip, *Pure Appl. Geophys.*, **116**, 790-806.

Dieterich J.H. (1979a). Modeling of rock friction, 1. Experimental results and constitutive equations, *J. Geophys. Res.*, **84**, 2161-2168.

Dieterich J.H. (1979b). Modeling of rock friction, 2. Simulation of preseismic slip, *J. Geophys. Res.*, **84**, 2169-2175.

Dieterich J.H. (1986). A model for the nucleation of earthquake slip. In *Earthquake Source Mechanics*, eds. Das S., Boatwright J. and Scholz C. H., Amer. Geophys. Union, Washington: 25-35.

Dieterich J.H. and Conrad G. (1984). Effect of humidity on time- and velocity-dependent friction in rocks, *J. Geophys. Res.*, **89**, 4196-4202.

Dragoni M. (1990). A model of interseismic fault slip in the presence of asperities, *Geophys. J. Int.*, **101**, 147-156.

Dragoni M. (1992). A dislocation model of aseismic fault slip under nonuniform friction, *Terra Nova*, **4**, 501-508.

Erdogan F., Gupta G.D. and Cook T.S. (1973). Numerical solution of singular integral equations, in *Mechanics of Fracture*, vol. 1, chapt. 7, ed. Sih, G.C.,

Noordhoff, Leyden.

Gu J.C., Rice J.R., Ruina A.L. and Tse S.T. (1984). Slip motion and stability of a single degree of freedom elastic system with rate and state dependent friction, *J. Mech. Phys. Sol.*, **32**, 167-196.

Hartzell S.H. and Heaton T.H. (1986). Rupture history of the 1984 Morgan Hill, California, earthquake from the inversion of strong motion records, *Bull. Seism. Soc. Am.*, **76**, 649-674.

Horowitz F. and Ruina A. (1985). Frictional slip patterns generated in a spatially homogeneous elastic fault model, *Eos*, **66**, 1069.

Jackson J. and McKenzie D. (1988). The relationship between plate motions and seismic moment tensors, and the rates of active deformation in the Mediterranean and Middle East, *Geophys. J.*, **93**, 45-73.

Marone C. and Scholz C. (1988). The depth of seismic faulting and the upper transition from stable to unstable slip regimes, *Geophys. Res. Lett.*, **15**, 621-624.

Paterson M. S. (1978). *Experimental Rock Deformation - The Brittle Field*, Springer-Verlag, Berlin.

Rabinowicz E. (1958). The intrinsic variables affecting the stick-slip process, *Proc. Phys. Soc. London*, **71**, 668-675.

Rice J.R. (1983). Constitutive relations for fault slip and earthquake instabilities, *Pure Appl. Geophys.*, **121**, 443-475.

Rice J.R. and Ruina A.L. (1983). Stability of steady frictional slipping, *J. Appl. Mech.*, **105**, 343-349.

Rice J. R. and Simons D. A. (1976). The stabilization of spreading shear faults by coupled deformation-diffusion effects in fluid-infiltrated porous materials, *J. Geophys. Res.*, **81**, 5322-5334.

Ruina A. (1983). Slip instability and state variable friction laws, *J. Geophys. Res.*, **88**, 10359-10370.

Scholz C.H. (1990). *The Mechanics of Earthquakes and Faulting*, Cambridge University Press, Cambridge.

Scholz C.H., Wyss M. and Smith S.W. (1969). Seismic and aseismic slip on the San Andreas fault, *J. Geophys. Res.*, **74**, 2049-2069.

Stuart W.D. (1981). Stiffness method for anticipating earthquakes. *Bull. Seism. Soc. Am.*, **71**: 363-370.

Stuart W.D. (1988). Forecast model for the great earthquakes at the Nankai trough subduction zone, *Pure Appl. Geophys.*, **126**, 619-642.

Tse S. and Rice J. (1986). Crustal earthquake instability in relation to the depth variation of frictional slip properties, *J. Geophys. Res.*, **91**, 9452-9472.

Tullis T.E. and Weeks J.D. (1986). Constitutive behavior and stability of frictional sliding of granite, *Pure Appl. Geophys.*, **124**, 383-414.

Wang C.-Y. (1984). On the constitution of the San Andreas Fault Zone in Central California, *J. Geophys. Res.*, **89**, 5858-5866.

Weeks J.D. and Tullis T.E. (1985). Frictional sliding in dolomite: a variation in constitutive behavior, *J. Geophys Res.*, **90**, 7821-7826.

PALEOMAGNETISM IN THE MEDITERRANEUM FROM SPAIN TO THE AEGEAN:
a review of data relevant to Cenozoic movements

E. MÁRTON
*Eötvös Loránd Geophysical Institute
1145 Budapest, Columbus u. 17-23.
Hungary*

ABSTRACT. Paleomagnetic observations relevant to the tectonics of the Alpine-Mediterranean zone are of very different quality. By introducing the concept of effective precision, the first part of this review demonstrates how such data can safely be applied to regional tectonics; then guidelines are suggested for fast check on data and also on the reliability of interpretations.
The second part deals with paleomagnetic evidences of displacements. It addresses the main issues in tectonically interpreted paleomagnetism and assesses models in the light of effective precision and in the context of all available relevant paleomagnetic observations. The regions highlighted are Iberia with deformed margins, the Sardinia-Corsica block, the Central Mediterraneum, the Northern Calcareous Alps, the South Pannonian unit, the Balkanids, the Hellenids and the Central Aegean area.

1. Introduction

The initial interest in modern paleomagnetic studies (in the early 1950s) was essentially geomagnetic. However, the method was soon applied also to geological problems and provided the first clear evidence for continental drift (Runcorn, 1956).

Since the mid 1960s, there has been a tremendous expansion in the number of centres studying paleomagnetism. Valuable information rapidly accumulated about the behaviour of the Earth's magnetic field in geologic times as well as about the displacements of continents. Both geomagnetic and geologic branches of the method provided essential data to the formulation of the plate tectonic theory.

One important consequence for paleomagnetism of the general acceptance of plate tectonics was a shift of the interest towards mobile belts. Studies have been extremely active in the Alpine-Mediterranean zone, where plate tectonic reconstruction of the classical definition (i.e. based on the oceanic magnetic anomalies) was not directly applicable.

Mobile belts were a new challenge, for movements in them are not only fast and small-scale compared to the displacements of large plates, but the ancient remanence of the rocks is often overprinted. Therefore, the resolution of the method had to be improved. Indeed, increasingly sophisticated equipment and methods started to appear by the mid 1970s. From this time on paleomagnetism has developed very fast.

Observations available today from the Alpine-Mediterranean zone are the products of different periods and also of a large number of non-standardized research centres. In making a tectonic analysis of such data, it is necessary to determine first what reliance can be placed on each of them. Unfortunately, there is no general agreement concerning the real value in tectonic interpretation of the data.

The quality of paleomagnetic data was first judged by criteria for minimum reliability, appropriate to the period in which they were published (e.g. Irving, 1964, McElhinny, 1973) and later by classification systems (e.g. Briden and Duff, 1981, Márton, E., 1986, Westphal, 1989).

Classification systems permit to distribute paleomagnetic data between several categories. Nevertheless, in tectonic interpretations they are hardly ever respected: it is only paleomagnetic mean directions (or poles) together with respective confidence intervals and the age of the rock that enter the game. However it is often forgotten that the size of the confidence interval strongly depends on the way it was estimated. There is a general tendency in paleomagnetism to produce as small confidence interval as possible. As a result, it quite often becames irrealistic.

Paleomagnetism is expected to answer questions concerning the movements of different areas in geologic times and with respect to each other. In these matters, statistical criteria are considered as decisive, tacitly assuming that errors are reflected in and expressed by statistical parameters. Yet, the application of statistical criteria in paleomagnetism is far from rigorous and consistent. Data from stable and mobile areas are treated differently.

In mobile belts, most authors automatically postulate relative movement when "coeval" paleomagnetic mean directions - together with their respective confidence intervals - stand apart. In less obvious cases, conclusion concerning the lack or presence of relative motion is reached after some further statistical analysis. On the contrary, rigidity in paleomagnetic sense of areas known as stable is never questioned, though coeval paleomagnetic poles from different parts of massifs or continents quite often significantly differ (e.g. Eurasia in the Neogene, Bohemian Massif in the late Paleozoic, see Krs, 1968, 1982).

In view of this situation the author of the present review thinks that operating mechanically with statistical precision may lead to misinterpretation. Since the statistical aspect of the quality of paleomagnetic observations is not yet fully appreciated, the first part of the review will deal with this problem. The reader is referred to textbooks in the matters of appropriate cleaning, component analysis identification of the NRM (natural remanent magnetization) components, the magnetic mineralogy, field and consistency tests etc, where these undeniably very important points about paleomagnetic observations are exhaustively discussed e.g. by McElhinny (1973), Tarling (1983), O'Reilly (1984), Butler (1992).

2. Statistical (apparent) precision and effective precision of paleomagnetic data

2.1. STANDARD PROCEDURE OF OBTAINING PALEOMAGNETIC DATA

Statistical precision of paleomagnetic data is expressed by statistical parameters (A95, Table 1 and $k=(N-1)/(N-R)$). They are normally accepted at their face value and cited e.g. in data bases. Effective precision has so far not been defined. In order to understand why such concept is needed, the standard procedure of obtaining data will be briefly outlined.

The basic element in paleomagnetism is the sample, fully oriented in situ in the field. Normally, several samples are collected from an outcrop (called site or locality) and several outcrops are sampled from an area (Fig.1). Specimens cut from the sample are subjected to measurement and cleaning in the laboratory. As a result, characteristic remanence is isolated. Directions of magnetization are usually expressed in polar coordinates (declination, inclination and total intensity). Individual directions are combined to give a mean direction.

Table 1. Calculation of mean paleomagnetic direction and the radius of the confidence circle

$$tgD = \sum_{i=1}^{N} m_i \Big/ \sum_{i=1}^{N} l_i \qquad sinI = \sum_{i=1}^{N} n_i \Big/ R$$

$$cosA_{95} = 1 - \frac{N-R}{R}(20^{1/(N-1)} - 1)$$

$$R^2 = (\sum_{i=1}^{N} l_i)^2 + (\sum_{i=1}^{N} m_i)^2 + (\sum_{i=1}^{N} n_i)^2$$

m_i, l_i, n_i, components of the remanence in the coordinate system of the sample or present system or tectonic system; N number of samples

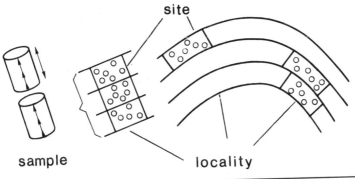

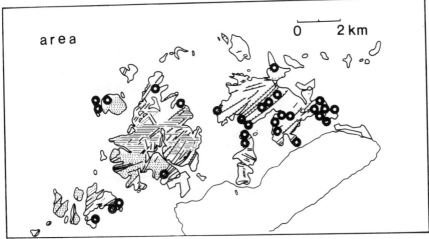

Fig.1. Levels of data combination. 1 Sample: independently oriented core or hand samples; 2 Site: a sedimentary layer, a dyke or a lava flow; 3 Locality: assembly of sites closely related in age and space; 4 Area: localities or sites from a tectonic unit closely related in age.

Orthodox data combination starts from specimen level and proceeds towards area level using at each level the formulas of Table 1. Calculations take place in the appropriate coordinate systems. It means that above sample level we calculate paleomagnetic directions, both in the present or field system (core orientation data incorporated), and also in the tectonic system. In the latter, tilt and plunge angles are also incorporated. Thus, the sampled rock unit is restored to the position at the time of the acquisition of the remanence.

Paleomagnetic elements are automatically referred to the present geographic system. However, displacements must often be seen in relation to much less precisely defined coordinate systems (e.g. microplates in relation to large plates or to each other, Fig.2).

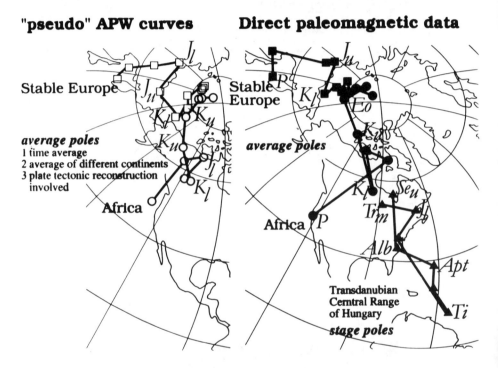

Fig.2. Apparent polar wander (APW) curves for stable Europe and Africa as paleomagnetic reference systems (data by Westphal et al., 1986) and an example of the APW curve for a tectonic unit of the Mediterraneum (Transdanubian Central Range, data by Márton and Márton, 1983, Márton E., 1987) as reference system. α_{95} for the "pseudo" APW curves are between 5-16° for the direct APW: 3-28° for the Transdanubian Central Range: 7-18°. The number of "independent" observations entering statistical calculations is that of the areas or localities.
key: P Permian; J_e early, J_u late Jurassic; K_e early, K_u late Cretaceous; Eo Eocene; Tr_m mid Triassic; Ti Tithonian; Apt Aptian; Alb Albian; Se_u late Senonian.

When data combination follows the pyramidal pattern recommended by Irving (1964), one may expect the mean paleomagnetic direction defined at area level (but not below!) to be free of serious bias, and the confidence interval realistic. Unfortunately, the method reduces drastically the number of "samples" as the specimen-sample-site-locality-area levels are reached, one after the other. Since statistical evaluation needs a large number of data, paleomagnetists very often depart from the orthodox way in order to obtain "precise" figures. The arbitrary treatment became common practice because of the usually limited number of suitable outcrops (of the same age!) within a tectonic unit and the tremendous effort needed to produce independent paleomagnetic directions, sufficient in number, perfectly cleaned and thoroughly analyzed. However, the impact of non-orthodox statistical procedure on the real value (effective precision) has just started to be analyzed (Márton, E., 1990b).

2.2. DATA COMBINATION AND EFFECTIVE PRECISION

It is common knowledge among paleomagnetists that paleomagnetic vectors, despite of careful field and laboratory processing, may depart from the ancient field direction. There are basically three factors responsible for this situation: 1. magnetic overprint, 2.local tectonics, 3.magnetic anisotropy.

The orthodox way of data processing minimizes their combined effect; at the same time prevents overinterpretation (i.e. postulation of relative movements, when the differences in directions are due to other factors) by defining maximum size for the confidence interval. Paleomagnetic data thus obtained are as precise as they look i.e. statistical and effective precisions are close.

While departures from the orthodox way always reduce the radius of the confidence circle (Fig.3), they do not lead to serious bias in the mean direction, unless some localities or sites are represented by large, others by small number of samples. In such cases more weight is given to sites or localities with large number of samples, while there is no guarantee that those are the best (i.e. free from overprint, perfect tectonic correction, anisotropy effect insignificant).

Unfortunately, existing data bases contain no information about the manner of data

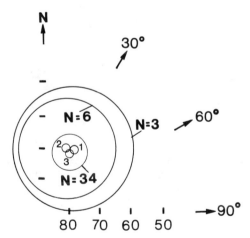

Fig.3. Reduction of the radius of confidence circle by departing from orthodox data combination. Example from the Mecsek Mts Hungary: 34 samples, collected at 6 sites and 3 localities from Anisian limestone. Due to the even distribution of samples between sites and localities the position of the mean direction is not influenced considerably.

combination. Therefore, without consulting original publications it is impossible to know how the two kinds of precisions are related.

Improvement in statistical precision may also be achieved by data rejection. Compilations usually indicate the number of collected as well as that of the evaluated samples. Rejection causes no serious discrepancy between statistical and effective precision when the proportion of the rejected samples is small compared to the "good ones". In the opposite case, however, it is important to check the reason: others than weak magnetization imply much lower effective than statistical precision.

2.3. OVERPRINT AND EFFECTIVE PRECISION

There are paleomagnetic methods that deal correctly with minor overprints obvious from the smeared distribution of paleomagnetic vectors (Khramov and Shollo, 1967, Halls, 1976, McFadden and McElhinny, 1988). They are, however, limited to cases, where the studied rock units are tilted with different strikes.

Overprints do not always cause scatter at site or locality level. Such overprints may be partial or complete. Red sediments and slowly cooling igneous bodies are suspect for the first (Fig.4a) and all kinds of rocks exhibiting remanence direction close to the direction of the present Earth's magnetic field at the sampling site in the field system (Fig.4b), for the second. Undetected overprints are one of the main dangers in paleomagnetism, for they may lead to much lower effective than statistical precision.

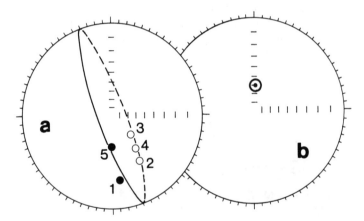

Fig.4. Unremovable overprint. Site mean directions, in field system, on stereographic projection. Partial overprint in a sheeted dyke, 45m thick (a) due to slow cooling through polarity reversals, Bükk Mts, Hungary; Dots: lower, circles: upper hemisphere. 1-5 are distributed along a great circle. Complete overprint (b). Footwall in a bauxite pit, Transdanubian Range, Hungary. Red dolomite, age Triassic. Mean direction (dot) with A95.

The best way to minimize the danger from undetected overprint is to base an area mean paleomagnetic direction on geographically distributed localities or sites and on different rock types, for it is highly unlikely that the proportion of the overprint to the original remanence is constant over a large area and in different lithologies.

2.4. LOCAL TECTONICS AND EFFECTIVE PRECISION

Simple tilt correction restores the position of the paleomagnetic vector precisely in tectonically quiet areas. Complex tectonics requires a thorough analysis of the local conditions, followed by the quantification of a full tectonic correction at each site or locality. In all cases, when simple tilt correction around a horizontal axis is applied, instead of full correction, declination will be affected. Yet the error is dependent on both dip and plunge magnitude (Fig.5).

Moreover, the order in which tilt and plunge corrections are applied also affects declination, though inclination is fully restored (Márton, P., 1977). Consequently, it is impossible to know if

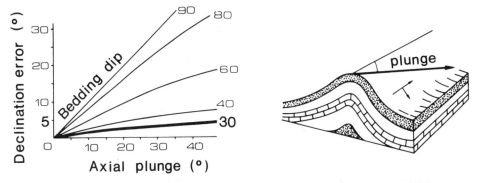

Fig.5. Apparent rotation due to uncorrected-for plunge. Modified after Tarling, 1983.

individual tectonic corrections are perfect or faulty. Thus, the effective precision in tectonically highly complicated areas is inevitably lower than statistical, unless supported by positive fold test, based on geographically distributed locality or site means (e.g. Cirilli et al., 1984).

2.5. THE INFLUENCE OF ANISOTROPY ON EFFECTIVE PRECISION

Metamorphic rocks are usually carefully studied for the effect of anisotropy on the paleomagnetic vector (e.g. Heller, 1973, Kligfield et al., 1983, Márton E. et al., 1987). Otherwise it is normally dismissed as insignificant below 10 percent. Unfortunately, there are examples to prove the

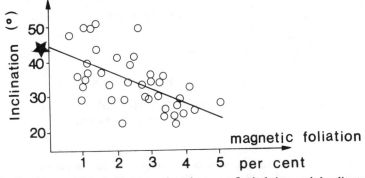

Fig.6. Inclination correlates with magnetic anisotropy. Ignimbrites and rhyolite tuffs. Bükk Mts. Hungary. Mean paleomagnetic inclination is 35°, 10° less than inclination at zero anisotropy.

contrary (Fig.6). Magnetic "schistosity" when coincides with bedding planes may deflect inclinations towards the horizontal. Clays, marls with high clay content, sheeted dykes etc. may be suspected for some error in inclination even on a regional scale, for tectonic correction will systematize minimum susceptibility axes.

2.6. THE TIME FACTOR

In regional tectonic applications of paleomagnetism time has an important influence on effective precision. On one hand, effective precision will be much lower than statistical, unless time represented by the studied rocks is long enough to average out secular variation (Fig.7).

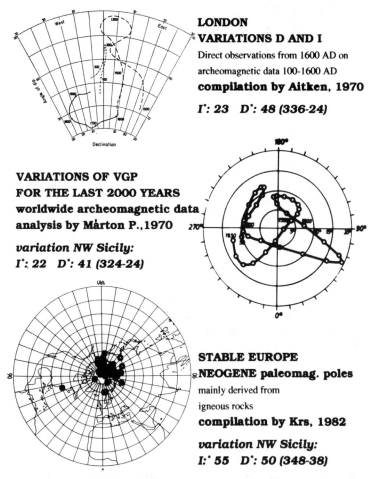

Fig.7. Secular variation in historic times (top left and middle right) and uncertainties expected in declination and inclination in stable regions due to complex sources of error (bottom left). I° and D° are maximum variations in inclination and declination, respectively; numbers in bracket are the extreme values of declination, that may appear as CW or CCW "rotations".

In sediments, the magnetization within an individual sample may well tend to satisfy such requirement. Fast cooling lava flows, thin dykes, however, often preserve a record of past secular variation; thus, they need to be sampled more extensively than sediments. In fact, it is only an overall mean direction based on a large number of site mean directions, obtained on distinct igneous bodies, which may be regarded as free from bias due to secular variation.

An other aspect of time is that paleomagnetic poles must be coeval in comparisons between tectonic units. When stage (even better substage) poles are compared, this condition is fulfilled. Such comparisons, however, are not very common for two reasons. Either the age of the source rocks in one or the other, or in both tectonic units is not known precisely, or stage poles are available, but not of the same stage. To get around the lack of precisely dated poles, it is common practice to average data for long time intervals (e.g. early Cretaceous, late Cretaceous). Unfortunately, the same poles are used to demonstrate small relative movements, forgetting that age differences may manifest themselves in statistically different pole positions. In such cases the relationship of the effective to statistical precision will depend on the speed of the apparent polar wander: in times of fast APW (Fig.2, Cretaceous for Africa and the Transdanubian Central Range) loose timing causes major uncertainties, while in quiet periods effective precision may even coincide with statistical.

2.7. GENERAL ASSESSMENT OF THE PALEOMAGNETIC DATA IN TECTONIC INTERPRETATION

As it was demonstrated, in regional tectonic interpretation effective precision is well estimated by statistical when:
a) the direction (pole) is based on geographically distributed sites or localities and the number entering statistical calculation is that of the localities or sites (geographically distributed!)
b) the observation represents sufficiently long and reasonably short time
c) deflection from the ancient field is unlikely
d) the age of the magnetization is convincingly tied to the age of the rock or to some important younger geologic event.

Short of one or more of the above criteria, any interpretation in terms of regional tectonics based on the comparison of a pair of "coeval" directions or poles must be regarded as tentative.

Users of paleomagnetic data may not always afford to reach back to original papers where concerning points a-d sufficient information may be found. Moreover, the information is not contained in explicit form, but the reader is left to form opinion from documentations of the laboratory procedures, the analysis of the NRM components, the carrier of the magnetization, field and consistency tests applied etc. about the quality of the observation.

Luckily, it is possible to suggest a few guidelines for fast and effective check on data as well as on the reliability of interpretation.

1. Interpretations often make use of the fact that the two angular elements of a paleomagnetic vector do not reflect events of the same age. Thus the elements are often treated separately: declinations may be plotted on maps to indicate rotations, inclinations appear recalculated as paleolatitudes (Table 2). However, in the interpretation of the declination differences as tectonic rotations, the inclinations cannot be ignored. Within a tectonic unit of the dimension of a microplate inclinations of the "same age" must be close. When they are very different - which is quite often the case - one has to suspect error in the data even if the statistical parameters belonging to each observation seem to be excellent. Check on the inclinations is a very efficient way to form a judgement on "paleomagnetic proofs" e.g. for secondary deformations of arcs. However, in comparisons between tectonic units of different origin (like a Southern Tethyan and

a Nothern Tethyan one, see later Fig.17-18), the inclinations are not expected to agree.

2. Concerning declinations, the tectonic interpretation is far from simple. Any declination may be composite. For instance, a declination measured on a Permian rock may be the manifestation of a single rotation at any time after the acquisition of the remanence; on the other hand, it can be the net effect of consecutive rotations, in opposite senses. Therefore, declination has to be resolved into components of different ages. This is done by monitoring younger rotations i.e. by defining longer segments of apparent polar wander curves. It is important to know, however, if rotations observed on younger rocks reflect the movement of detached rock masses or are representative of the overall displacements of thicker crustal or litospheric fragments.

Table 2. Interpretation of paleomagnetic observation in terms of tectonics

Inclination ⇒ LATITUDE of the source area
AT THE TIME OF MAGNETIZATION
Declination ⇒ measures the NET ROTATION
AFTER THE TIME OF MAGNETIZATION

Tectonic interpretation uses
declination or inclination or
both paleomagnetic elements
in the form of a mean direction
with statistical parameters
or
paleomagnetic pole
derived from the elements

3. Time sequences of paleomagnetic observations (APW curves) may also help to constrain the age of the magnetization. Naturally, the lower limit of the age is that of the stratigraphic or isotope age of rock. The upper limit is always a matter of consideration. Tests (fold, conglomerate, reversal) are useful in this respect. Nevertheless, reliability is enhanced by demonstrating that "old" directions are different from younger ones.

4. There is a special case, when APW curves or field test are not needed to suspect that the magnetization is younger than the age of the source rock. It is when the paleomagnetic direction in the present system (before tectonic correction) is aligned with the present field direction. Such directions, recalculated in tectonic system, may depart significantly from the present elements. Yet, it is hazardous to regard them as true indications of tectonic rotation.

5. The most important point about APW curves (time sequences) is that they can express systematic differences. In view of the fact that paleomagnetic data with coinciding effective and statistical precisions are still few and those existing possess fairly large A95, relative rotations may be regarded as proven only when the APW curves of tectonic units are systematically displaced. In such situation the rotation angle may also be confidently estimated.

Systematic differences in APW curves arise when there are systematic differences in paleodeclinations or paleoinclinations or both. Thus, in order to demonstrate relative movements between tectonic units it is not necessary to operate with APW curves: it is sufficient to illustrate that such differences do exist in either paleodeclinations or paleoinclinations. The advantage of such illustration is that their graphic representation is much more expressive than that of the APW curves.

6. Most tectonically oriented paleomagnetic papers on the Mediterranean region compare observations from the mobile belt to either stable Europe, or Africa, or both. In such comparison only large and systematic differences can be safely relied on. The reason is that while the general features of the APW curves for both large plates are fairly well established (Fig.2), individual poles are characterized by low statistical and even lower effective (mainly because of loose timing) precision. Moreover, they lack details because of averaging processes and rejection of data that do not fit the general trend. Owing to the much lower effective precision of African and European poles, small relative movements with respect to the large plates cannot be postulated paleomagnetically, despite of the sometimes high effective precision of Mediterranean data.

7. A special aspect of the effective precision of African poles is the disagreement on the length of the late Jurassic - early Cretaceous segment of the African APW curve (fast APW for Gondwana continents see Schult et al., 1981, Schmidt and Embleton, 1982, Márton and Márton, 1983, Valencio et al., 1983, moderate APW: e.g. Lowrie, 1986, Westphal et al., 1986). Depending on the standpoint of an author in this matter, the same large rotations may be interpreted as signs of movement independent of Africa or on the contrary, evidence for coordinated movement.

3. CENOZOIC TECTONIC MOVEMENTS IN THE ALPINE-MEDITERRANEAN ZONE AS REFLECTED IN PALEOMAGNETIC OBSERVATIONS

Paleomagnetic observations available from the Alpine-Mediterranean zone have been interpreted by the authors of the data as well as other scientists. Some models, relying on paleomagnetic observations, are known to meet general approval (e.g. Van der Voo, 1969). In other cases conflicting ideas seem to gain support from the same set of data (e.g. VandenBerg, 1979, Channell et al., 1979). Regardless of the situation, this review reaches back to original publications. Conclusion and models will be assessed in the light of effective precision and in the context of all available relevant paleomagnetic data.

In general, the review analyses trends and tendencies. The data involved are all characterized by their authors with good or satisfactory statistical precision (A95 is typically less than 15°). However, these parameters will not be cited for they are simply not comparable due to the problem discussed in 2.2.

Although the main goal is Cenozoic history, Mesozoic and sometimes even older data will also be discussed for two reasons. Firstly, the declination rotation observed on pre-Cenozoic rocks is quite often partly or fully due to young tectonic movements. Secondly, the most convincing evidence for relative movements is the systematic separation of coeval poles, paleodeclinations or paleoinclinations; the best way to estimate the angle of a Cenozoic rotation is to match as complete as possible segments of APW or declination curves (comprising data older than the rotation), since it is exactly the young movement that "splits" off the original curve into identical, but displaced curves.

However, Cenozoic data will not be discussed when they are clearly not relevant to the displacements of their basement (e.g. Dela Pierre et al., 1992) for it is beyond the scope of this review to deal with local tectonics.

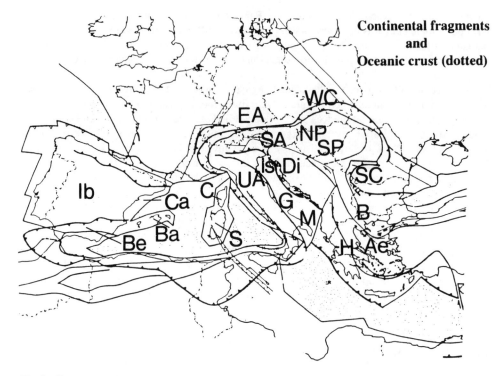

Fig.8. Continental fragments and oceanic crust in the Mediterranean area. Modified after Gealey (1988). Key: Ib Iberian microplate; Be-Ba Betics-Balearics; Ca Catalanian Coastal Range; C Corsica; S Sardinia; G Gargano; M Murge; Is Istria; UA Umbrian Apennines; Di Dinarids; SA Southern Alps; EA Eastern Alps; NP North Pannonian Unit; SP South Pannonian Unit; WC West Carpathians; SC South Carpathians; B Balkanids; H Hellenids; Ae Central Aegean area.

3.1. IBERIA

Of the several continental fragments of the Alpine-Mediterranean zone (Fig.8) Iberia was one of the earliest studied with the purpose of demonstrating mobility. Van der Voo (1969) concluded from his own results that "stable" Iberia must have rotated CCW by about 35° with respect to stable Europe. The angle corresponded to what was required by the opening of the Bay of Biscay. He also demonstrated that the rotation terminated before the end of Cretaceous.

The conclusion that "stable" Iberia rotated CCW during the Cretaceous have been confirmed by later studies. However, the agreement between the angle of rotation as postulated by Van der Voo on paleomagnetic grounds and the angle required to open the Bay of Biscay must be regarded as coincidental today (Fig.9).

Compared to the present orientation, Permo-Carboniferous, late Triassic - early Jurassic, early-mid Cretaceous do show CCW deviation. On the other hand, Triassic (a single locality, since data from the Pyrenées must be disregarded) and late Jurassic declinations are close to zero, in harmony with observations from the Catalanian Coastal Range (Fig.9, full circles). If we rely on statistical precision we must conclude that the rotations of Iberia were complex even before the

Cretaceous. There is an other possibility. We may regard some differences in declination as consequences of low effective precision (Triassic through Oxfordian). The mean declination computed for the Permian-Oxfordian time interval is still suggestive of a rotation (in relation to stable Europe) required to open the Bay of Biscay.

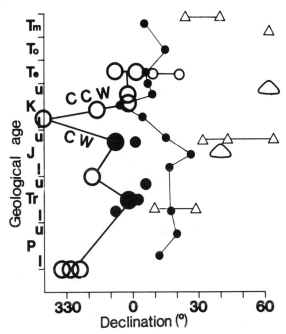

Fig.9. Declination change with time for stable Iberia (large hollow and full circles: Van der Voo, 1969, Schott et al., 1981, Galdeano et al., 1989), the Catalanian Coastal Range (small hollow and full circles: Parés et al., 1992; circles are full when paleomagnetic results of matching age are available from both stable Iberia and the Catalanian Coastal Range), Betics (large triangle, Platzman and Lowrie, 1992), Balearics (small triangle, Parés et al., 1992) compared with declinations expected for Madrid in stable European system (dots, Westphal et al., 1986).

Unfortunately the late Jurassic - early Cretaceous seems to be full of events. CW rotation after the Oxfordian is needed to achieve declination less than 320°. The break-up of the South Portuguese, Lusitanian and Iberian Basins may have accompanied this rotation. Then Iberia starts a rotation in the opposite sense which ends with the Albian, when according to plate tectonic models, the Bay of Biscay just starts to open (Fig.10). The angle of the late Jurassic - early Cretaceous rotations is large enough to be significant, even if the effective precision of the data is surely somewhat lower than the statistical (A95: 4-10°, after data rejection).

It is true that the relative positions of Iberia and stable Europe seem to have changed after the mid-Cretaceous. However, this took place in the Cenozoic, when stable Europe rotated sightly in the CW sense while Iberia have remained fixed.

The authors of the Cretaceous paleomagnetic data (Galdeano et al., 1989) postulated independent Iberia during the early Cretaceous. Since the African reference framework is poorly constrained for the late Jurassic - early Cretaceous, it is equally possible to see Iberia as moving

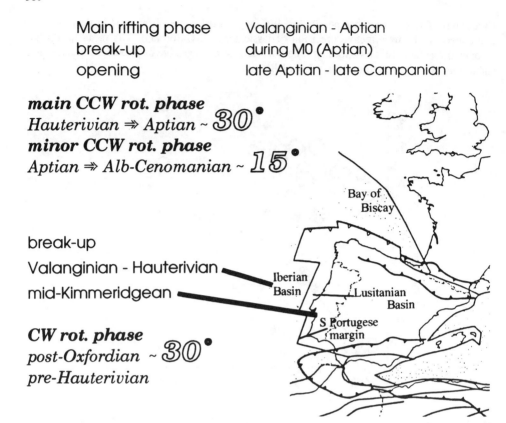

Fig.10. Important events in the displacement history of Iberia as indicated by plate tectonics and paleomagnetism.

in coordination with Africa (Fig.11) and with continental fragments of southern Tethyan origin (much higher effective precision than any African data).

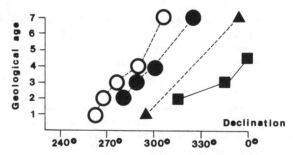

Fig.11. Declination trends during the Cretaceous for stable Iberia (square), for Africa (triangle), for the Umbrian Apennines (full circle) and for the Transdanubian Central Range (circle). Declinations are recalculated for Madrid. 1.Tithonian, 2.Neocomian, 3.Aptian, 4.Albian, 5.Cenomanian, 6.Turonian, 7.Senonian.

Concerning the internal state of the microplate, it seems to be quite rigid in paleomagnetic sense (Fig.12).

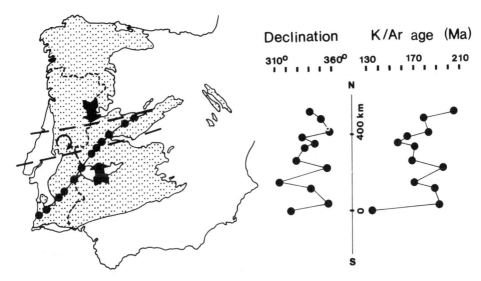

Fig.12. Rotation predicted for the Spanish Central System (Vegas et al., 1990) and paleomagnetic sampling sites (dots) along the Plasencia-dyke (left); declination and K/Ar age variation along the Plasencia-dyke (Schott et al., 1981). Mean declination outside of the system: D=337° I=48° A95=10° and within the system: D=346° I=46° A95=9° do not differ.

Of the deformed margins, the Catalanian Coastal Range does not seem to differ from the rest of the microplate while in the Betics and the Balearics large CW rotations were observed (Fig.9).

Models explaining the enormous difference between the Iberian microplate and its present southern margin are highly speculative (Osete et al., 1988, Freeman et al., 1989, Allerton et al., 1991, Platzman and Lowrie, 1992); their pre-folding relative position is also an open question: to some, the deformed southern margin had been moving in coordination with Iberia prior to its main deformation phases (Platzman and Lowrie, 1992), to others (Villalain et al., 1991) there is no evidence for it.

The rotations resulting in an overall declination difference of about 60° between "stable" Iberia and the Betics-Balearics are attributed to mid-late Miocene tectonic events (constrained only for the Balearics). An interesting aspect of the observations is that an additional rotation required to open the Valencia trough is not evident in data from the Betics and the Balearics, respectively.

3.2. CORSICA-SARDINIA

It is generally accepted that the Corsica-Sardinia unit rotated away from the southern coast of France in the Miocene. On the other hand there is a difference in opinion about the manner of rotation. To some, the Corsica-Sardinia unit behaved as a single rigid block (e.g. Fig.8), to others, Sardinia rotated about 60° CCW, while Corsica only 30° (Westphal et al., 1976).

On the present evidence, paleomagnetism supports the first version (Vigliotti et al., 1990 and Fig. 13). Combined paleomagnetic and radiometric studies dated the movement at 19.5 Ma (Montigny et al., 1981) and also estimated the angle (30°, in the CCW sense).

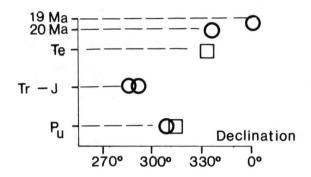

Fig.13. Paleomagnetic declinations for Sardinia (circle) and Corsica (square). Studied ages: late Permian (P_u: Vigliotti at al, 1990), Triassic-Jurassic (Tr-J: Horner-Lowrie, 1981), Eocene (Te, secondary remanence: Vigliotti-Kent, 1990) and around 19.5 Ma (Manzoni et al., 1980, Montigny et al., 1981).

Older than Miocene data from the two islands are also indicative of CCW net rotation (Fig.13). Comparison of the angles observed on rocks of different ages reveals that prior to the CCW rotation of about 19.5 Ma, rotation in the same sense, but of an even larger angle must have taken place (Jurassic-Cretaceous declinations for Sardinia show an additional 50°). The involvement of Corsica in such rotation is indirectly evidenced by coinciding Permian declinations.

Whatever age we assign to the motion (between Jurassic and early Miocene), which led to the declination deviation in excess of that of the Miocene, we can not find evidence of such movement for stable Europe (Fig.9). Thus, we must consider the possibility of Sardinia-Corsica moving independently of stable Europe in the late Mesozoic or the Paleogene.

3.3. COUNTERCLOCKWISE ROTATED PALEOMAGNETIC MEGAUNIT OF THE CENTRAL MEDITERRANEUM

CCW rotation of about 30° during the Miocene is not unique to the Sardinia-Corsica block. Over a large area paleomagnetic indications point to similar Cenozoic mobility. Regionally consistent paleomagnetic directions relevant to Cenozoic rotations are of two kinds. One group is of late Senonian - Cenozoic age. These are directly showing the sense and angle of Cenozoic rotation (Fig. 14). To an other group belong observations on Cenomanian - Turonian (Murge: Márton, E. and Nardi, in prep.; Gargano: VandenBerg, 1983; Dinarids, down to Split: Márton, E. and Veljović, 1987, Márton, E. et al., 1990b, Márton, E. and Milićević, see Fig. 20; Abruzzi: Márton, P. and D'Andrea, 1992). Regardless of their "autochthony (e.g. Transdanubian Central Range or Gargano-Murge) or "allochthony" (Outer West Carpathians, Abruzzi), with the exception of the Abruzzi with extremely complex local tectonics, they all exhibit CCW rotation that is similar in angle for rocks of the same age.

When average poles are calculated from data representing the sampling areas indicated by symbols in Fig.14 for Senonian - Eocene and Oligocene - early Miocene times, we find that the poles are departing from poles of similar age for Africa and stable Europe in the CCW sense by about 30°. Surprisingly, the size of the confidence circle belonging to the average poles matches

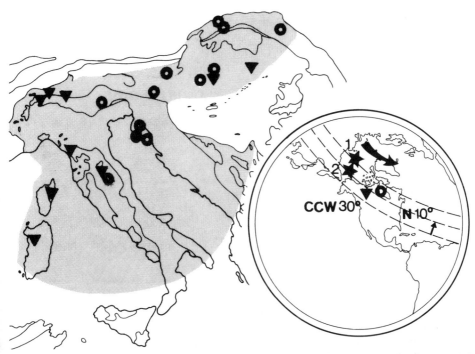

Fig.14. Definition of a paleomagnetic megaunit in the Central Mediterraneum. The Megaunit (left side, shaded area) is characterized by about 30°CCW overall rotation with respect both to stable Europe and Africa (right side) during the Miocene. Key on the map: source areas of latest Cretaceous+Eocene (circle) and Oligocene+Early Miocene (triangle) observations showing the sense and angle of rotations directly; key, right side: combined stable European and African pole (star) for the Neogene (1) and the Paleogene (2);triangle and circle poles based on area mean poles of the shaded area. Small circles (dashed lines) indicate poles of equal distance from the Central Mediterraneum. (Compilation by Márton, E. and Mauritsch, 1990, with the addition of data by Heller, 1973; Heller et al., 1989; Vigliotti and Kent, 1990).

that of stable Europe for the Neogene and Paleogene (Table 3). This indicates that the rigidity in the Cenozoic of the shaded area indicated in Fig.14 (in paleomagnetic sense) matches that of stable areas.

Eocene data from Salento area (Tozzi et al., 1988) and Gargano (Tozzi et al., 1989) were disregarded during calculations. The reason is that the mean paleomagnetic directions for both areas are close to the present field direction before tectonic correction. In view of this situation, full remagnetization is not unlikely, despite of the carrier being magnetite and the occurence of reversed polarity (see for examples: Morris and Robertson, in press, Márton, E. and Nardi, 1991).

Although in the S-SE part of the area, tentatively outlined in Fig.14, direct evidence for regionally consistent Miocene rotations is lacking, it is possible to regard the shaded area as a paleomagnetic megaunit in the Central Mediterraneum. The reason is that areas with inferred Cenozoic CCW rotations are connected to those with direct evidence for it via Cretaceous poles.

Table 3. Degree of coordination within the CCW rotated megaunit of the Central Mediterranean compared with that of stable Europe. A95 (heavy figures) based on number of area means (N).

for *CCW rotated megaunit*	for *stable Europe* (compilation by Krs, 1982)
Oligocene +early Miocene **4°** (N=12)	Neogene **4.5°** (N=41)
latest Cretaceous +Eocene **7°** (N=8)	Paleogene **5.3°** (N=31)

This megaunit must have moved in a microplate-like manner, independently from both stable Europe and Africa, during the Miocene.

The term "African affinity" is very often used in connection with CCW rotated paleomagnetic directions (e.g. Heller at al., 1989) in this region. It is important to remember that such observations are basically due to Cenozoic rotation and not necessarily signify coordinated movements with Africa.

The paleomagnetic poles shown in Fig.14 characterize the overall movement of the megaunit in relation to stable Europe and Africa. While the area cannot be conceived as "absolutely rigid", only a few paleomagnetic differences may be regarded as really significant. Comparison of minimum two pairs of "coeval" poles reveal, for instance, that the Abruzzi (Márton, P., and D'Andrea, 1992), the Southern Alps (VandenBerg and Zijderveld, 1982), the Dinaric fold belt (Márton, E. et al., 1990b), the Umbrian Apennines and the Transdanubian Central Range (Cirilli at al., 1984) moved slightly with respect to the "stable Adriatic microplate" (Márton, E. and Veljović, 1983, VandenBerg, 1983, Márton, E. and Nardi, 1991). The rotation angle in these cases is rather poorly defined. Moreover, it is difficult to say if existing differences are inherited, i.e. due to pre-Miocene relative movements or they were generated in the process of the Miocene rotation. It is very likely that both are true.

Concerning the possibilities of paleomagnetism in demonstrating small relative movements we can mention the example of the Umbrian Apennines and the Transdanubian Central Range; for both well-defined Cretaceous - Paleogene APW curves are available, thus a rotation of 15° between them is well documented (Fig.15).

The rotation of the paleomagnetic megaunit characterized by westerly declinations must have ended by the late Miocene, since in coherently behaving areas (e.g. Umbrian-Marches region,

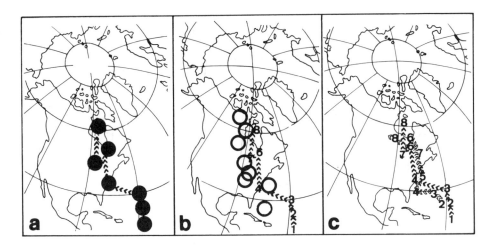

Fig.15. Cenozoic relative rotation between the Transdanubian Central Range (full circles, a) and the Umbrian Apennines (circles, b): Cretaceous-Paleogene APW curves are of the same character, but displaced. They can be brought into coincidence by a CCW rotation of 15° of the latter around a pole of rotation situated in the gulf of Taranto (c).

Sardinia, NE Hungary) declinations measured on late Miocene rocks do not depart significantly from the present north (Dela Pierre et al., 1992, Montigny et al., 1981, Márton and Márton, 1991).

In Italy, there are several places from where rotations were reported on younger than Miocene rocks. However, those observations are not relevant to the overall movement of the peninsula north of Calabria since they represent detached rock masses that quite often exhibit widely different paleodeclinations within a small area (e.g. Dela Pierre et al., 1992).

3.4. THE CALABRIAN ARC

Incoronato and Nardi (1987) reviewed paleomagnetic data from the Southern Apennines and Sicily. They concluded that except for six localities (2 Cretaceous in the Southern Apennines and 4 Jurassic in NW Sicily) the data were useless for tectonic interpretation. However, they agreed with the previous interpretation of data from the Calabrian arc (Channell et al., 1980) that suggested close correlation between paleomagnetic declinations and the curvature of the arc. Though the timing of the deformation was not constrained paleomagnetically, most authors dealing with the Peri-Tyrrhenian area (e.g. Channell et al., 1980, Nairn et al., 1985) assumed that it started in the Miocene.

More recently published data are of Plio-Pleistocene ages (Aïfa et al., 1988, Sagnotti, 1992). Thus, they can not be used together with Mesozoic ones to fill the large gap between the Trapanese platform and the northern part of the Southern Apennines, the source areas of the Mesozoic data. Therefore, we must be content to conclude that while paleomagnetic data (after rejecting non-consistent ones from the Apennines) are in harmony with expectations in a bending model, they are far too few to constrain the bending.

3.5. NORTHERN CALCAREOUS ALPS

North of the CCW rotated paleomagnetic megaunit of the Central Mediterraneum, the nappe system of the Northern Calcareous Alps (NCA) is thrusted over the stable European margin. In the NCA, east of the western margin of the Tauern window, late Senonian-Paleocene Gosau formation exhibits near-zero declinations (Mauritsch and Becke, 1987), i.e. no evidence of overall rotation with respect to stable Europe in the Cenozoic. Triassic-Jurassic is clockwise rotated (Mauritsch and Becke, 1987, Heer, 1982, Channell, 1991), i.e. the rotation pattern is just the opposite of the "African pattern" (Fig.16). All these observations suggest that, if the NCA was ever moving in coordination with Africa, it must have separated from other tectonic units of southern Tethyan origin before the late Cretaceous.

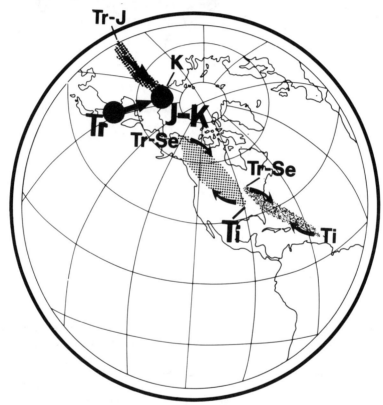

Fig.16. Rotation pattern, as indicated by paleomagnetic poles, of the eastern part of the Northern Calcareous Alps (top left) which joins stable Europe (poles shown by full circles) by the end of Cretaceous, compared with the Mesozoic pattern for Africa (centre) and that for the Transdanubian Central Range (bottom right). Curved arrows indicate the trend of the APW which corresponds to CW rotation during the Jurassic and CCW rotation during the Cretaceous. The fields indicated by shading are the areas that contain most of the Mesozoic poles for Africa and the Transdanubian Central Range, respectively. Key: Tr, Triassic; J, Jurassic; Ti, Tithonian; K, Cretaceous; Se, Senonian.

3.6. THE SOUTH PANNONIAN UNIT

The South Pannonian unit (Fig.17) is one of the most problematic part of the Alpine-Mediterranean zone. A number of geological as well as paleomagnetic characteristics (note the different character of the declination trend in the Transdanubian Central Range and the South Pannonian unit, respectively, Fig.18) indicate that prior to the Cenozoic (probably Neogene) the two units were not moving together. Moreover, paleomagnetic observations from the South Pannonian unit (Márton, E. 1986, Pătraşcu and Panaiotu 1991, Pătraşcu et al., 1990), as well as from the South Carpathians (Pătraşcu et al., 1992, also a weak indication from the Northeast Carpathians, Bazhenov and Burtman, 1980) suggest that they all participated in a large Cenozoic CW rotation (Fig.17).

Fig.17. Paleomagnetic indications (smaller arrow heads) and evidences (larger arrow heads) for Cenozoic clockwise rotation from the South Pannonian unit and the Carpathians. Data by Bazhenov and Burtman (1980), Márton, E. (1986), Pătraşcu et al. (1990), Pătraşcu and Panaiotu (1991) and Pătraşcu et al. (1992). Age of the source rocks: Cretaceous through early Miocene. North of the tectonic line, indicated by heavy line, tectonic units rotated CCW in the Miocene (angle of rotation is shown by a large arrow, data by Márton, E., 1986).

The contrast on either side of the Mid-Pannonian Mobile Belt is so striking that we must identify the belt as plate boundary, active in the Cenozoic, even if most authors outside Hungary fail to recognize its primary importance (e.g. Westphal et al., 1986).

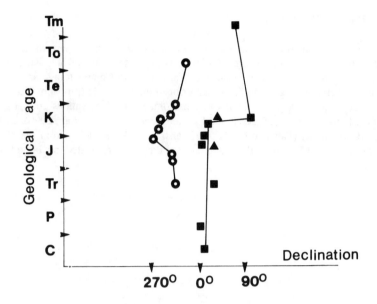

Fig.18. Declination for the Transdanubian Central Range (dots) and for the South Pannonian unit (Mecsek Mts: squares, Villány Hills: triangles) as a function of geologic age (Márton, E., 1990a). The former shows African pattern, i.e. clockwise rotation in the late Jurassic, compensated by CCW rotation in the early Cretaceous, the latter the lack of it. Key: C, Carboniferous; P, Permian; Tr, Triassic; J, Jurassic; K, Cretaceous; Te, Eocene; To, Oligocene; Tm, Miocene.

3.7. BALKANIDS AND HELLENIDS

Paleomagnetic data (Carboniferous through Oligocene) from Bulgaria do not show clear evidence for relative movements with respect to the present orientation or stable Europe (Márton, E., 1988). In contrast, paleomagnetic observations indicate high mobility in the Hellenids, and in the Balkanids of Greece.

Despite of the large number of modern studies, the tectonic interpretation of paleomagnetic data from Greece is not easy due to extreme tectonic complexity. The difficulties may be best illustrated by the example of the Hellenic arc.

According to Kissel and Laj (1988) paleomagnetic data constrain the deformation of the arc in the following way. Before 16 Ma, the arc was almost rectilinear with an E-W trend. The curvature has been acquired during two distinct phases: during the first (16-12.5 Ma) deformation was continuous along the arc, as shown by CW declination rotation in the west, CCW declination rotation in the east, and no rotation in the south; the second phase (0-6 Ma), a CW rigid block rotation, is restricted to the western segment.

The attractive model proposed by Kissel and Laj (1988) has some weak points. Concerning the first phase the problem is that the constraint put on the continuous deformation model by paleomagnetic data is very poor. To illustrate this, indicators for the postulated rotations (see Fig.5 of Kissel and Laj, 1988) are split into two groups according to their relevance to each of the deformation phases (Fig. 19). When data bearing on the second phase are omitted, it becomes evident that the model is based on interpolation.

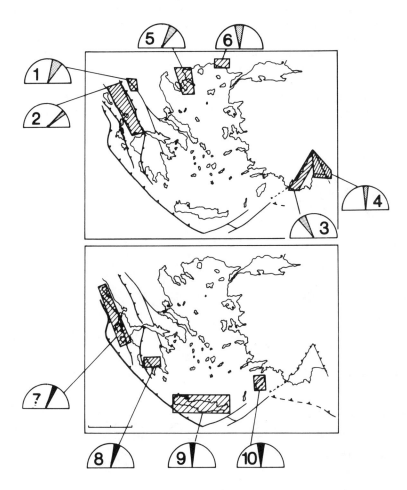

Fig.19. Paleomagnetic indicators for postulated rotations at different parts of the Aegean deformed in two phases (Kissel and Laj, 1988). Top: indicators for rotation relevant to the 1st phase (16-12.5 Ma). Bottom: those for the 2nd phase (0-6 Ma). Data by 1: Kissel and Laj (1988), 2: Kissel et al. (1984, 1985), Horner and Freeman (1983), 3: Kissel and Poisson (1987), 4: Kissel and Poisson (1986), 5: Kondopoulou and Westphal (1986), 6: Kissel et al. (1986), 7-10: Laj et al. (1982), Valente et al. (1982).

Concerning the second phase, it is convincingly demonstrated that the eastern and southern segments are not affected by rotations. The western segment also seems to be sufficiently represented to justify the conclusion of its moderate CW rotation. However, when the model of two-phase deformation for the paleomagnetically well-represented western segment is tested against all available observations, we discover that the rotation angle does not always change, either in space, or in time as predicted by the model.

The model predicts a CW rotation of about 25° for each of the deformation phases for the NW part of the Hellenids. On the mainland e.g. late Cretaceous observations are consistent with the prediction. At the same time they convincingly demonstrate about 90° CW rotation of the Western Hellenids with respect to the Adriatic region (compare coeval data from the Hellenids and the area NW of it). The problem is, however, that CW rotation of about 50° is observed not only on pre-Tortonian but also on Tortonian rocks, which are supposed to show about 25° (Fig. 20). If there are differences, they are between the different zones of the Hellenids (Ionian and Pindos) suggested by systematic, but very small displacement of the respective declination curves (Mauritsch, pers. comm.).

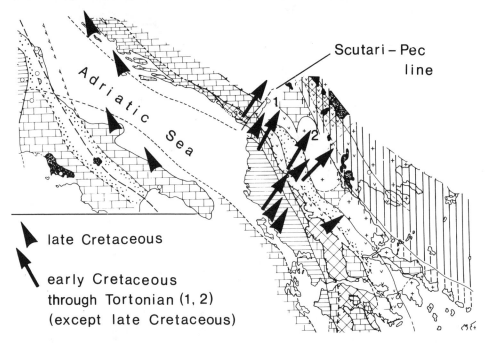

Fig.20. Paleomagnetically indicated rotations around the Scutary-Peć tectonic zone. Late Cretaceous declinations are consistently CCW rotated at both sides of the Adriatic See, while CW rotated in the Hellenids. Declinations observed on rocks of different ages are uniform in the Hellenids. References: VandenBerg, 1983; Márton, E. - Nardi, 1991; Márton, E. - Milićević, unpublished; Horner, 1983; Márton, E. et al., 1990a; Speranza et al., 1992; Mauritsch, unpublished.

The next problem is that the "autochthonous" Paxos zone (Paxos island) does not exhibit post Eocene declination rotation (Fig.21). Further complications are encountered in Corfu island. This island belongs to the Ionian zone, and as in the NW part of the Hellenids on the mainland, rocks of different ages, relevant to both postulated phases of deformation were studied from here. The maximum angle of declination rotation observed on Tortonian and younger rocks is about 25°, in agreement with the model. However, in Eocene, Cretaceous and Jurassic rocks, also studied in Corfu (Horner, 1983), there is no clear evidence for an older rotation phase (Fig.21). In contrast on Lefkas island, the angle of CW rotation is 90° on Eocene and Cretaceous rocks, also belonging to the Ionian zone.

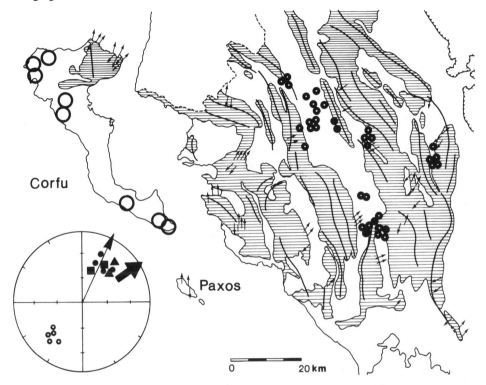

Fig.21. Paleomagnetic picture of part of the Paxos and Ionian zones. Tiny arrows on the map are declinations by Horner (1983) observed on Paleocene-Eocene. Dots on the map are Oligocene sampling sites of Kissel et al. (1985), large circles on Corfu island are the sampling sites for Tortonian and younger sediments. The stereographic net at the bottom left corner compares average Paleocene - through Oligocene declinations (heavy arrow) from the mainland (small dots) with individual site mean directions from Corfu island plus Jurassic (squares) and Cretaceous (triangles) directions from the same island. Max. declination deviation calculated from observations on the Tortonian and younger sediments from Corfu are shown in light arrow.

One would assume that such an enormous rotation is not really pausible, except for two reasons. One is that the same angle was found by independent studies (Horner, 1983, Márton E. et al., 1990a). The second, that large CW rotation is not unique to the island of Lefkas. It also characterizes the Mesozoic (including late Cretaceous) of Central Greece, and even western Aegean (Fig.22) that were interpreted by Morris (1992) as related to a shear zone connecting the North Anatolian Transform Fault to the Hellenic Trough. It stands to reason to regard the paleomagnetic observations from Lefkas island as the missing link.

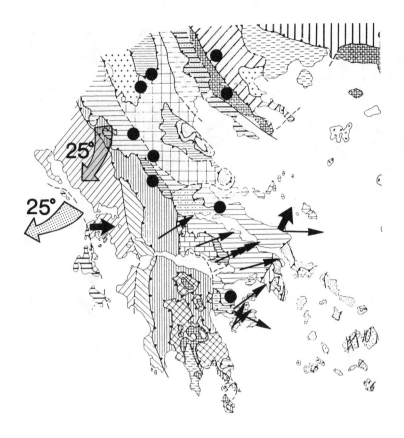

Fig.22. Clockwise declination rotations that may be relevant to movements along the North Anatolian fault (arrows, light: Mesozoic, heavy: Cenozoic observations) and locations exhibiting counterclockwise rotations (full circles) in Greece. References: Horner, 1983, Pavlides et al., 1988, Surmont, 1989, Márton, E. et al., 1990a, Morris, 1992, Edel et al. (in press).

As we have seen, the variations in the angle of CW rotations does not seem to support any kind of unifying model. The picture is mosaic-like. To complicate the situation further, indications from CCW rotations were also reported from several areas (Fig.22).

In view of the fact that paleomagnetic control of the Hellenids is still unsatisfactory for such a complex area, there is a need for more, intensive and extensive studies, especially on Cretaceous and younger ages. Fortunately, a large proportion of the existing data is of good quality. Therefore, future interpretations can easily accommodate them.

3.8. THE CENTRAL AEGEAN AND WESTERN ANATOLIA

Paleomagnetic observations from this area were interpreted in terms of block rotations connected to strike-slip movements along the North Anatolian fault and due to regional extension since the Tortonian (Kissel and Laj, 1988). Excepting Evia (3) and the Bergama-Izmir area (10), where sediments were also studied, the directions interpreted in terms of relative rotations are based on

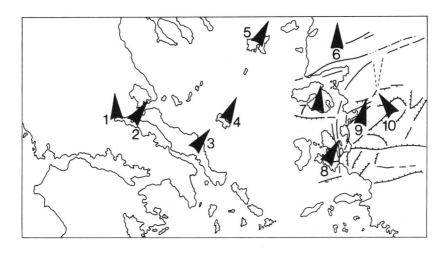

Fig.23. Paleomagnetic results from the internal regions of the Aegean domain. Observations on late Cenozoic rocks; all extrusive volcanics, except 2 (sediments) and 10 (sediments and volcanics mix). References: Kissel and Laj, 1988, Kissel et al., 1989, Westphal and Kondopoulou, 1989.

extrusive igneous rocks. Thus effective precision is expected to be much lower than statistical. The most serious objection to a purely tectonic interpretation of the individual mean declinations (Fig.23: 1-10) is non-averaged out secular variation (see Fig.7), sometimes high portion of data rejected, or treatment of non-conform sites in a deliberate way (Table 4).

Table 4. Summary of additional information relevant to each of the mean declinations of Fig. 23.

	age (Ma)	number of sites used	rejected	mean inclination	scatter in inclination
1	3	4	-	59°	54-62°
2	2	2	-	51°	45-59°
3	13	6	1	44°	36-53°
4	15	4	1*	45°	38-55°
5	17-21	10	?	48°	?
6	10-20	4	-	67°	61-70°
7	17	17	9	54°	20-69°
8	17-22	8	-	52°	16-85°
9**	17-18	3	?	39°	29-48°
10**	7-18	10	?	54°	15-71°
			OVERALL MEAN INCLINATION (3-10):		39-67°=28°

* with mean: D=289° I=38° A95=6°
** unclear reason for defining two groups

4. A generalized overview of the study area in the Cenozoic

Between stable Europe, slightly rotating CW and Africa, just about terminating its CCW rotation that had started with the early Cretaceous, continental fragments as well as most of the less stable areas move in a consistent manner (Fig. 24) in the Cenozoic.

The overall position of three areas, stable Iberia (1), the Northern Calcareous Alps (2) and different tectonic units in Bulgaria (3) does not seem to change relative to the present north (the last one must have moved with stable Europe since the Carboniferous). Everywhere else, significant, sometimes large declination rotations are observed.

The western segment of the study area comprises two main regions: stable Iberia with folded margins and the CCW rotated paleomagnetic megaunit of the Central Mediterraneum. In the light of Cenozoic and older paleomagnetic data we can envisage both as moving coordination with Africa after the late Jurassic opening in the Tethys. However, their Cenozoic history differs. Iberia got obstacled by the end of Cretaceous and the change in relative position with respect to stable Europe in the Cenozoic is due to slight CW rotation of the latter. At the southern margin of Iberia, in the Betic-Balearic zone (4), the main period of mobility is of Miocene age (large CW rotations). The same is true of the CCW rotated paleomagnetic megaunit of the Central

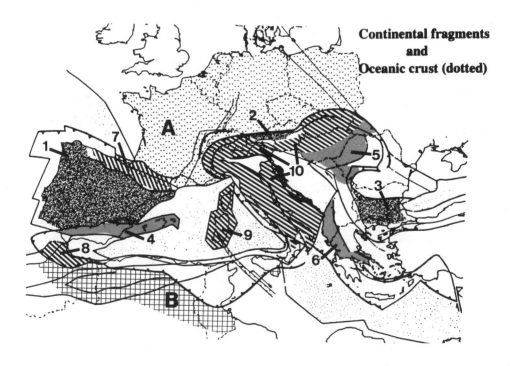

Fig.24. Generalized overview showing areas of different paleomagnetic characteristics in the Cenozoic. Key: lack of overall rotation: 1, 2, 3; slight and large CW rotations, respectively: A and 4, 5, 6; slight and large CCW rotations: B and (7, 8), 9, 10.

Mediterraneum (9, 10) that rotated by about 30° in a microplate-like manner (high coordination, but not rigid in the strict sense).

Paleomagnetic evidence shows that by the end of Cretaceous the Northern Calcareous Alps was already in the same overall position as it is today, but prior to its emplacement, it had rotated CW by large angle, probably in the late Cretaceous.

CCW and CW rotated regions are in contact at the Mid-Pannonian Mobile Belt and probably also at the Scutari-Peć line. CW rotated are the South Pannonian unit, with the Southern Carpathians (5) and the Western Hellenids, plus Balkanids in Greece (6). Both exhibit large rotation angles, but the timing of movements is different: in the first area the rotations ended with the Miocene, while in the second it is continuing till our days.

5. Acknowledgements

It is a pleasure to acknowledge the help with discussions and materials of J.B. Edel, J. Galindo-Zaldivar, A. Jabaloy, D. Kondopoulou, H.J. Mauritsch, A. Morris, C. Panaiotu, J.M. Parés and A.H.F. Robertson. Many thanks also go to R. Bordás, G. Imre and A. Lugosi for technical support.

Special thanks are due to E. Mantovani for having encouraged this work. The manuscript benefited from the comments of an anonymous reviewer.

6. References

Aitken, M.J. (1970) 'Dating by archaeomagnetic and thermoluminescent methods', Phil. Trans. Roy. Soc. London, A269, 77-88.

Aïfa, T., Feinberg, H., Pozzi, J.-P. (1988) 'Pliocene-Pleistocene evolution of the Tyrrhenian arc: paleomagnetic determination of uplift and rotational deformation', Earth Planet. Sci. Lett., 87, 438-452.

Allerton, S., Platt, J.P., McClelland, E., Lonergan, L. (1991) 'Block rotations in the eastern Betic Cordillera, southern Spain', IUGG XX. General Assembly, Vienna, abstracts, 86.

Bazhenov, M.L., Burtman, V.S. (1980) 'About the nature of the northern part of the Carpathians (in Russian)', Dokladi Academii nauki USSR, 255, 681-685.

Briden, J.C., Duff, B.A. (1981) 'Pre-Carboniferous paleomagnetism of Europe north of the Alpine Orogenic Belt', in M.W. McElhinny, D.A. Valencio (eds.), Paleo-reconstruction of the Continents, pp. 137-150.

Butler, R.F. (1992) Palaeomagnetism. Magnetic domains to geologic terranes. Blackwell Scientific Publications, Oxford.

Channell, J.E.T. (1991) 'New paleomagnetic data from the Southern Alps', IUGG XX. General Assembly, Vienna, Abstracts, p. 87.

Channell, J.E.T., D'Argenio, B., Horváth, F. (1979) 'Adria, the African Promontory, in Mesozoic Mediterranean Palaeogeography', Earth Science Reviews, 15, 213-292.

Channell, J.E.T., Catalano, R., D'Argenio, B. (1980) ' Paleomagnetism and deformation of the Mesozoic continental margin in Sicily, Tectonophysics, 58, 391-407.

Cirilli, S., Márton, P., Vigliotti, L. (1984) 'Implications of a combined biostratigraphic and palaeomagnetic study of the Umbrian Maiolica Formation', Earth Planet. Sci. Lett., 69, 203-214.

Dela Pierre, F., Ghisetti, F., Lanza, R., Vezzani, L. (1992) 'Paleomagnetic and structural evidence of Neogene tectonic rotation of the Gran Sasso range (central Apennines, Italy)', Tectonophysics, 215, 335-348.

Edel, J.B., Kondopoulou, D., Pavlides, S., Westphal, M. (in press) 'Paleomagnetic evidence for a large counterclockwise rotation of northern Greece prior to the Tertiary clockwise rotation', Geodinamica Acta.

Freeman, R., Sàbat, F., Lowrie, W., Fontboté, J.-M. (1989) 'Paleomagnetic results from Mallorca (Balearic islands, Spain)', Tectonics, 8, 591-608.

Galdeano, A., Moreau, M.G., Pozzi, J.P., Berthou, P.Y., Malod, J.A. (1989) 'New paleomagnetic results from Cretaceous sediments near Lisboa (Portugal) and implications for the rotation of Iberia', Earth Planet. Sci. Lett., 92, 95-106.

Gealey, W.K. (1988) 'Plate tectonic evolution of the Mediterranean-Middle East region', in Scortese, C.R. and Sager, W.W. (eds.), Mesozoic and Cenozoic Plate Reconstructions, Tectonophysics, 155, 285-306.

Halls, H.C. (1976) 'A least-squares method to find a remanence direction from converging remagnetization circles', Geophys. J. R. Astron. Soc., 45, 297-304.

Heer, L., (1982) 'Paläomagnetische Testuntersuchungen in der Nördlichen Kalkalpen im Gebiet zwischen Golling und Kössen', MSc Thesis, Technical University, Munich.

Heller, F. (1973) 'Magnetic anisotropy of granitic rocks of the Bergell Massif (Switzerland)', Earth Planet. Sci. Lett., 20, 180-188.

Heller, F., Lowrie, W., Hirt, A.M. (1989) 'A review of paleomagnetic and magnetic anisotropy results from the Alps', Geol. Soc. Spec. Publ., 45, 399-420.

Horner, F.J. (1983) 'Paleomagnetism of carbonates in the Southern Tethys. Implications for the polarity of the Earth's magnetic field in the early Jurassic and the tectonics in the Ionian zone of Greece (in German)', PhD Thesis, Zurich.

Horner, F., Freeman, R. (1983) 'Palaeomagnetic evidence from pelagic limestones for clockwise rotation of the Ionian zone, Western Greece', Tectonophysics, 98, 11-27.

Horner, F., Lowrie, W. (1981) 'Paleomagnetic evidence from Mesozoic carbonate rocks for the rotation of Sardinia', J. Geophys., 49, 11-19.

Incoronato, A., Nardi, G. (1987) 'Tertiary rotations in Southern Apennines and western Sicily', Rend. Soc. Geol. It., 9, 131-136.

Irving, E., (1964) Palaeomagnetism and its application to geological and geophysical problems, John Wiley and sons, New York/London/Sidney.

Khramov, A.N., Shollo, L.E. (1967) Palaeomagnetism (in Russian), Niedra, Leningrad.

Kissel, C., Laj, C. (1988) 'The Tertiary geodynamic evolution of the Aegean arc: a paleomagnetic reconstruction', Tectonophysics, 146, 183-201.

Kissel, C., Poisson, A. (1986) 'Etude paléomagnetique préliminaire de formations néogènes du bassin d'Antalya (Taurides occidentales, Turquie)', C. R. Acad. Sci. Paris, 302, 711-716.

Kissel, C., Poisson, A. (1987) 'Etude paléomagnetique des formations cénozoïque des Bey Daglari (Taurides occidentales)', C. R. Acad. Sci. Paris, 304, 343-348.

Kissel, C., Jamet, M., Laj, C. (1984) 'Palaeomagnetic evidence of Miocene and Pliocene rotational deformations of the Aegean Area', in J.E. Dixon, A.H.F. Robertson, (eds.), The Geological Evolution of the Eastern Mediterranean, Spec. publ. Geol. Soc., London, 17, 669-679.

Kissel, C., Laj, C., Müller, C. (1985) 'Tertiary geodinamical evolution of northwestern Greece: paleomagnetic results', Earth Planet. Sci. Lett., 72, 190-204.

Kissel, C., Kondopoulou, D., Laj, C., Papadopoulos, P. (1986) 'New paleomagnetic data from Oligocene formations of Northern Aegea', Geophys. Res. Lett., 13, 1039-1042.

Kissel, C., Laj, C., Poisson, A., Simeakis, K. (1989) 'A pattern of block rotations in Central Agea', in C. Kissel, C. Laj, (eds), Paleomagnetic Rotations and Continental Deformation, NATO ASI Series, Kluwer Academic Publishers, Dordrecht/Boston/London, pp. 115-129.

Kligfield, R., Lowrie, W., Hirt, A., Siddans, A.W.B. (1983) 'Effect of progressive deformation on remanent magnetization of Permian redbeds from the Alpes Maritime (France)', Tectonophysics, 97, 59-85.

Kondopolou, D., Westphal, M. (1986) 'Paleomagnetism of the Tertiary intrusives from Chalkidiki (Northern Greece)', J. Geophys., 59, 62-66.

Krs, M. (1968) 'Rheological Aspects of Palaeomagnetism? Review of data from Bohemian Massif', XIII. Int. Geol. Congress, 5, 87-96.

Krs, M., (1982) 'Implication of Statistical Evaluation of Phanerozoic Palaeomagnetic Data (Eurasia, Africa)', Rozpravy Ceskoslovenské Akademie Ved, Rada Matematickych a Prírodních

Ved, 92, 1-86.

Laj, C., Jamet, M., Sorel, D., Valente, J.P. (1982) 'First paleomagnetic results from Mio-Pliocene series of the Hellenic sedimentary arc', Tectonophysics, 86, 45-67.

Lowrie, W. (1986) 'Paleomagnetism and the Adriatic Promontory: a reappraisal', Tectonics, 5, 797-807.

Manzoni, M., Marini, A., Vigliotti, L. (1980) 'Dislocazioni tettoniche dedotte dalle direzioni di magnetizzazione primaria del distretto vulcanico di Sarroch (Sardegna)', Boll. Geof. Teor. Appl., 22, 139-152.

Márton, E. (1981) 'Tectonic implications of paleomagnetic data for the Carpatho-Pannonian region', Earth Evolution Sciences, 3-4, 257-264.

Márton, E. (1986) 'Paleomagnetism of igneous rocks from the Velence Hills and Mecsek Mountains', Geophysical Transactions of ELGI, 32, 83-145.

Márton, E. (1987) 'Paleomagnetism and tectonics in the Mediterranean region', J. Geodyn., 7, 33-57.

Márton, E., (1988) 'Paleomagnetism - An overview of the Central Mediterranean', in M. Rakús, J. Dercourt, A.E.M. Nairn, (eds.), Evolution of the northern margin of Tethys: the results of IGCP Project 198, Mémoires de la Societé Géologique de France, Paris, Nouvelle Série, 154, 223-244.

Márton, E. (1990a) 'Kinematics of the principal tectonic units of Hungary from paleomagnetic observations', Acta Geod. Geoph. Mont. Hung., 25, 387-397.

Márton, E. (1990b) 'Palaeomagnetic data base: the problem of precision', IAGA 'New Trends in Geomagnetism II. (Rock magnetism, palaeomagnetism and database usage' conference, Bechyně, South Bohemia, abstract.

Márton, E., Márton, P. (1983) 'A refined apparent polar wander curve for the Transdanubian Central Mountains and its bearing on the Mediterranean tectonic history', Tectonophysics, 98, 43-57.

Márton, E., Márton, P. (1991) 'Tertiary rotations in NE- Hungary', IUGG XX. General Assembly, Vienna, Abstracts, p. 95.

Márton, E., Nardi, G. (1991) 'Is it still evidenced paleomagnetically that Apulia region has been part of Africa? First results from Murge', IUGG XX. General Assembly, Vienna, Abstracts, p. 87.

Márton, E., Nardi, G. (in prep.) 'Cretaceous paleomagnetic results from Murge (Apulia): tectonic implications', (submitted to Geophys. J. Int.).

Márton, E., Mauritsch, H. J. (1990) 'Structural applications and discussion of a paleomagnetic post-Paleozoic database for the Central Mediterranean', Phys. Earth. Planet. Int., 62, 46-59.

Márton, E., Veljović, D. (1983) 'Paleomagnetism of the Istria peninsula, Yugoslavia', Tectonophysics, 91, 73-87.

Márton, E., Veljović, D. (1987) 'Palaeomagnetism of Cretaceous carbonates from the northwestern part of the Dinaric fold belt', Tectonophysics, 134, 331-338.

Márton, E., Mauritsch, H.J., Pahr, A. (1987) 'Paläomagnetische Untersuchungen in der Rechnitzer Fenstergruppe', Mitt. österr. geol. Ges., 80, 185-225.

Márton, E., Papanikolaou, D.J., Lekkas, E. (1990a) 'Paleomagnetic results from the Pindos, Paxos and Ionian zones of Greece', Phys. Earth Planet. Int., 62, 60-69.

Márton, E., Veljovic, D., Milicevic, V. (1990b) 'Paleomagnetism of the Kvarner islands, Yugoslavia', Phys. Earth Planet. Int., 62, 70-81.

Márton, P. (1970) 'Secular variation of the geomagnetic virtual dipole during the last 2000 years as inferred from the spherical harmonic analysis of the available archeomagnetic data', PAGEOPH, 81, 163-176.

Márton, P. (1977) 'On the principles of the tectonic implications of paleomagnetism (in Hungarian)', Magyar Geofizika, 18, 161-165.

Márton, P., D'Andrea, M. (1992) 'Paleomagnetically inferred tectonic rotations of the Abruzzi and northwestern Umbria', Tectonophysics, 202, 43-53.

Mauritsch, H.J., Becke, M. (1987) 'Paleomagnetic investigations in the Eastern Alps and the Southern Border Zone', in H.W. Flügel, and P. Faupl, (eds.), Geodynamics of the Eastern Alps. Vienna, Deuticke, pp. 282-309.

McElhinny, M. W. (1973) Paleomagnetism and plate tectonics, University Press, Cambridge.

McFadden, P.L., McElhinny, M.W. (1988) 'The combined analysis of remagnetization circles and direct observations in palaeomagnetism', Earth Planet. Sci. Lett., 87, 161-172.

Montigny, R., Edel, J.B., Thuizat, R. (1981) 'Oligo-Miocene rotation of Sardinia: K-Ar ages and paleomagnetic data of Tertiary volcanics', Earth Planet. Sci. Lett., 54, 261-271.

Morris, A. (1992) 'Large clockwise tectonic rotation of Central and Southern Greece deduced from palaeomagnetic analysis of Mesozoic carbonates', EGS XVII. General Assembly, Edinburgh, abstracts, p. C28.

Morris, A., Robertson, A.H.F. (in press) 'Miocene remagnetisation of Mesozoic Antalya Complex units of Isparta Angle, SW Turkey', Tectonophysics.

Nairn, A.E.M., Nardi, G., Gregor, C.B., Incoronato, A., (1985) 'Coherence of the Trapanese units during tectonic emplacement in western Sicily', Boll. Soc. Geol. It., 104, 267-272.

O'Reilly, W. (1984) Rock and mineral magnetism, Blackie and Son, Glasgow.

Osete, M.L., Freeman, R., Vegas, R. (1988) 'Preliminary palaeomagnetic results from the Subbetic Zone (Betic Cordillera, southern Spain): kinematic and structural implications', Phys. Earth Planet. Inter., 52, 283-300.

Parés, J.M., Freeman, R., Roca, E. (1992) 'Neogene structural development in the Valencia trough margins from paleomagnetic data', Tectonophysics, 203, 111-124.

Pătrașcu, St., Panaiotu, C. (1991) 'Timing of rotational motion of Apuseni Mts. (Romania): paleomagnetic data from Tertiary magmatic rocks', IUGG XX. General Assembly, Vienna, Abstracts, p. 95.

Pătrașcu, St., Bleahu, M., Panaiotu, C. (1990) 'Tectonic implications of paleomagnetic research into Upper Cretaceous magmatic rocks in the Apuseni Mountains, Romania', Tectonophysics, 180, 309-322.

Pătrașcu, St., Bleahu, M., Panaiotu, C., Panaiotu, C.E. (1992) 'The paleomagnetism of the Upper Cretaceous magmatic rocks in the Banat area of the South Carpathians: tectonic implications', Tectonophysics, 213, 341-352.

Pavlides, S.B., Kondopoulou, D.P., Kilias, A.A., Westphal, M. (1988) 'Complex rotational deformations in the Serbo-Macedonian massif (north Greece): structural and paleomagnetic evidence', Tectonophysics, 145, 329-335.

Platzman, E., Lowrie, W. (1992) 'Paleomagnetic evidence for rotation of the Iberian Peninsula and the external Betic Cordillera, Southern Spain', Earth Planet. Sci. Lett., 108, 45-60.

Runcorn, S.K. (1956) 'Paleomagnetic comparison between Europe and North America', Proc. Canad. Assoc. Geol. 8, 77-85.

Sagnotti, L. (1992) 'Paleomagnetic evidence for a pleistocene counterclockwise rotation of the Sant'Arcangelo Basin, southern Italy', Geophys. Res. Lett., 19/2, 135-138.

Schmidt, P.W., Embleton, B.J.J. (1982) 'Comments on "Palaeomagnetism of Upper Cretaceous Volcanics and Nubian Sandstones of Wadi Natash, SE Egypt and Implications for the Polar Wander Path for Africa in the Mesozoic" (by Schult, A., Hussain, A.G. and Soffel, H.C., 1981. J. Geophys., 50, 16-22.)', J. Geophys., 51, 150-151.

Schott, J-J., Montigny, R., Thuizat, R. (1981) 'Paleomagnetism and potassium-argon age of the Messejana Dike (Portugal and Spain): angular limitation to the rotation of the Iberian Peninsula since the Middle Jurassic', Earth Planet. Sci. Lett., 53, 457-470.

Schult, A., Hussain, A.G., Soffel, H.C. (1981) 'Palaeomagnetism of Upper Cretaceous volcanics and Nubian sandstones of Wadi Natash, S-E Egypt and implications on the polar wander path for Africa in the Mesozoic', J. Geophys., 50, 16-22.

Speranza, F., Kissel, C., Islami, I., Hyseni, A., Laj, C. (1992) 'First paleomagnetic evidence for rotation of the Ionian zone of Albania', Geophys. Res. Lett., 19/7, 697-700.

Surmont, J. (1989) 'Paléomagnetisme dans le Hellénides internes: analyse des aimantations superposées par la méthode des cercles de réaimantation', Can. J. Earth Sci., 26, 2479-2494.

Tarling, D.H. (1983) Palaeomagnetism. Principles and Application in Geology, Geophysics and Archeology, Chapman and Hall, London.

Tozzi, M., Kissel, C., Funicello, R., Laj, C., Parotto, M. (1988) 'A clockwise rotation of Southern Apulia?', Geophys. Res. Lett., 15, 681-684.

Tozzi, M., Funicello, R., Parotto, M. (1989) 'Paleomagnetismo ed evoluzione geodinamica del settore settentrionale dell'avampaese apulo', Atti 8° Convegno Naz. GNGTS, Roma.

Valencio, D.A., Vilas, J.F., Pacca, I.G. (1983) 'The significance of the palaeomagnetism of Jurassic-Cretaceous rocks from South America: predrift movements, hairpins and magnetostratigraphy', Geophys. J. R. astr. Soc., 73, 135-151.

Valente, J.P., Laj, C., Sorel, D., Roy, S., Valet, J.P. (1982) 'Paleomagnetic results from Mio-Pliocene marine series in Crete', Earth Planet. Sci. Lett., 57, 159-172.

VandenBerg, J., (1979) 'Paleomagnetism and the changing configuration of the Western Mediterranean area in the Mesozoic and early Cenozoic eras', Geologica Ultraiectina, 20, 147-153.

VandenBerg, J. (1983) 'Reappraisal of paleomagnetic data from Gargano (South Italy)', Tectonophysics, 98, 29-41.

VandenBerg, I., Zijderveld, J.D.A. (1982) 'Paleomagnetism in the Mediterranean area', Geodynamics series, 7. Alpine-Mediterranean Geodynamics, Final report of working group 3. Amer.Geophys.Union/Geol.Soc.Amer., pp. 83-112.

Van der Voo, R. (1969) 'Paleomagnetic evidence for the rotation of the Iberian Peninsula', Tectonophysics, 7, 5-56.

Vegas, R., Vázquez, J.T., Suriñach, E., Marcos, A. (1990) 'Model of distributed deformation, block rotations and crustal thickening for the formation of the Spanish Central System', Tectonophysics, 184, 367-378.

Vigliotti, L., Kent, D.V. (1990) 'Paleomagnetic results of Tertiary sediments from Corsica: evidence of post Eocene rotation', Phys. Earth Planet. Int., 62, 97-108.

Vigliotti, L., Alvarez, W., McWilliams, M. (1990) 'No relative rotation detected between Corsica and Sardinia', Earth Planet. Sci. Lett., 98, 313-318.

Villalain, J.J., Osete, M.L., Vegas, R., Garcia-Dueñas, V., Heller, F. (1991) 'New palaeomagnetic results in the Western Subbetics, Betic Cordilleras (Southern Spain)', IUGG XX. General Assembly, Vienna, Abstracts, p. 86.

Westphal, M. (1989) 'The Strasbourg paleomagnetic data base', Geophys. J. 97, 361-363.

Westphal, M., Kondopoulou, D. (1989) 'Paleomagnetism of Miocene volcanics from Lemnos island: implications for block rotations in the vicinity of the North Aegean Trough', International workshop on active and recent strike-slip tectonics, Florence, Abstracts, p. 34.

Westphal, W., Orsini, J., Vellutini, P. (1976) 'Le microcontinent Corso-Sarde, sa position initiale: donné paléomagnétiques et raccords geologiques', Tectonophysics, 30, 141-157.

Westphal, M., Bazhenov, M., Lauer, J.P., Pechersky, M., Sibuet, J.C., (1986) 'Paleomagnetic implications on the evolution of the Tethys belt from the Atlantic ocean to the Pamir since the Triassic', Tectonophysics, 123, 37-82.

USE OF THE PALEOMAGNETIC DATABASES FOR GEODYNAMICAL STUDIES: SOME EXAMPLES FROM THE MEDITERRANEAN REGION

G. SCALERA, P. FAVALI and F. FLORINDO
Istituto Nazionale di Geofisica
Via di Vigna Murata 605, 00143, Roma, Italy

ABSTRACT. The publication of Global Paleomagnetic Database allows an easier use of the paleomagnetic data, which are fundamental tools to constrain regional and global geodynamic models. The Stable Africa and Stable Europe, the Arabian plate and finally the Mediterranean region paleomagnetic data, over Mesozoic-Cenozoic, have been critically re–examined using suitable computer programs developed at the *Istituto Nazionale di Geofisica*. The data have been selected on the basis of space, time and quality filtering with the convinction that the higher quality data are more useful for geodynamic studies, than averaging larger datasets without any filtering of original studies. Besides the extraction software, a mapping program to plot sampling localities and paleopoles with associate confidence ellipses, using cartographic projections, enhances the understanding of the mutual relations among groups of paleopole belonging to different areas. The African and European Apparent Polar Wander Paths (APWPs) has been computed, using only higher quality African and European data respectively, and compared with some already published APWPs. Some applications to regional tectonics of the Mediterranean region have been done. In particular Italy and Istria have been considered.

1. Introduction

The paleopole data were used since fifthies (e.g., Irving, 1958) to strongly support the global mobilistic theory against the fissism. These data, which give information about latitudinal position of the sites through the geological time, can have also paleogeodetic importance. The continue methodological improvement of the paleomagnetic measurements has allowed the use of paleomagnetism also in regional geodynamical studies. For all these reasons many researchers collected paleopole data, for their personal use (e.g., Perrin et al., 1988; Piper, 1988) even if their choose was biased by some subjective criteria.

Several regional databases were then compiled with the purpose to overcome the dispersion of the data sources and the subjectivity of the choice. During the eighties, five regional paleomagnetic databases (DBs) – the natural evolution of the periodic lists of paleopoles published by Irving and McElhinny on the "Geophysical Journal of the Royal Astronomical Society" from sixties to eighties – were compiled by Khramov (Russian Federation), Pesonen (Scandinavia), Irving (Canada), Lunyendyk and Butler (USA) and Westphal (Europe). These Authors previosly defined common criteria to be able to merge the regional DBs in one worldwide database. Finally this global DB (Global Paleomagnetic Database – GPDB) was coordinated and published by McElhinny and Lock (1990a; 1990b) and Lock and McElhinny (1991), and contains global data of about 5000 paleopoles, collected until 1988 for the world and 1989 for Russian Federation. The next four years (1989–1992) have been collected by Van der Voo and should be merged to the GPDB in 1993. This GPDB synthesize all the information concerning the paleopole parameters and their quality stored in a complex file structure. Nevertheless the contents of the GPDB are very heterogeneous collecting data that has been produced by the Authors with completely different purposes (global geodynamics, block rotations, polarity inversions, magnetostratigraphy, etc...).

The publication of the GPDB has allowed an easier use of the paleomagnetic data, which are fundamental tools to costrain geodynamical models. It is unavoidable to take into account this large amount of data expecially in this study, which purpose is to give some examples at different scale as contribution to the comprehension of the Mediterranean geodynamics.

TIME WINDOWS		INC. (Ma)
0 - 5	Pliocene - Quaternary	5
5 - 15	Middle - Late Miocene	10
15 - 25	Early Miocene	10
25 - 35	Oligocene	10
35 - 45	Middle - Late Eocene	10
45 - 55	Early - Middle Eocene	10
55 - 65	Paleocene	10
65 - 100	Late Cretaceous	30
100 - 140	Early Cretaceous	30
140 - 175	Middle-Late Jurassic	30
175 - 210	Early Jurassic	30
210 - 245	Triassic	30

Table 1 Time windows in Ma used in this study and with the corrispondent names of epochs. Age uncertanties in Ma (INC) are associated to these time windows.

QUALITY FACTORS	Q = A	Q = B	Q = C	Q = D
SITES	≥ 8	≥ 6	≥ 4	< 4
SAMPLES	≥ 8	≥ 6	≥ 4	< 4
Dp - Dm	≤15	≤15	≤20	>20
LABORATORY AND ANALYTICAL PROCEDURES*	≥ 3	≥ 2	≥ 2	< 2

* : 0 = No demagnetization;
   1 = Only pilot demagnetization on some samples;
   2 = All samples treated, blanket treatment only;
   3 = Stereonets with J/J_0, or vector plots provided (*Zijderveld, 1967*);
   4 = *Principal component analysis* (PCA) (*Kischvink, 1980*) plus vector or stereoplots and J/J_0;
   5 = PCA plus vector plots plus multiple treatments that are successfully isolating vectors (e.g. AF and Thermal, or Thermal and Chemical etc).

Table 2 The 4 quality classes of data defined in this study.

2. Geological–Tectonic Framework

The geodynamic evolution of the Mediterranean area (part of the Tethys belt) is mainly caused by the complex interaction between the Euroasiatic and African plates. It is then necessary to consider the relationships of the Mediterranean area with Eurasia, Africa and the middle oceanic ridges of Atlantic and Indian Oceans. The area that we have considered in this study geographically includes the African continent (excluding Madagascar isle), the Europe until the Urals and part of Asia (Minor Asia, Caucasus and Arabian peninsula). In this zone traces of all orogenies can be found, either the two Paleozoic events (Caledonian 570–370 Ma; Variscan–Hercynian 370–220 Ma), or the Alpine Meso–Cenozoic orogeny (Early Mesozoic Alpine 220–65 Ma; Middle Cenozoic Alpine 65–20 Ma; Upper Cenozoic Alpine 20 Ma to present) (Bally et al., 1985; Bilal and Van Eysinga, 1987).

The Caledonian orogeny played in Africa a role only in the Mauritanides. In Europe it interested areas such as Scandinavia, Scotland, Wales and Ardenne. Besides the Caledonian orogenic belts, wide zones constituted by craton basins located mainly in continental pre-Mesozoic lithosphere, and by shields deformed by Precambrian orogenic episodes. The most of Africa is composed by cratons, such as western African, Congo, Tanzanian and Kalahari

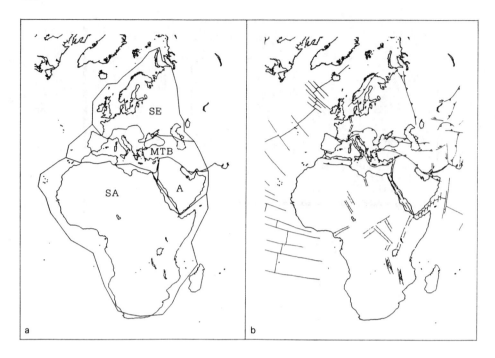

Fig. 1 Map of: a) the four poligonals (SA=Stable Africa; SE=Stable Europe; A=Arabia; MTB=Mobile Tethys Belt); b) sketch of the tectonic setting. The poligonals have been traced referring to the following maps and papers: Merla et al., 1973; Stocklin and Nabavi, 1973; UNESCO, 1976; Martin et al., 1978; Bartov, 1979; BRGM, 1980; AA.VV., 1981; Nairn et al., 1981; Boccaletti and Dainelli, 1982; Boccaletti et al., 1985; Salmon et al., 1988; Condie, 1989; Kampunzu and Lubala, 1991; Kruczyk et al., 1992.

cratons with other old deformed areas like Katanga. In Europe it is possible to find old deformed zones like Karelian system and Fennosarmatian shield, and wide craton basins (e.g., North Sea Basin and Russian–Ukrainan Basin). The Variscan–Hercynian orogeny strongly marked in Africa the Morocco and the southern African Cape, in Europe the Central Massif, the Variscides and the Urals.

The Meso–Cenozoic Alpine orogeny interested in Africa the Maghrebides, north of the Saharian platform, and the Cape Range in the southern tip of the continent. The rift-valleys of East Africa, which are divided in western and eastern branches, were formed during Neogene and are associated to volcanic activity. Other volcanic activity took place in Ahaggar and Tibesti during Tertiary and Quaternary. In Europe the Alpine orogeny deformed the mountain belts of Iberian peninsula, Pyrenees, Alps, Apennines, Dinarides, Hellenides, Carpathians, Crimean peninsula, Pontides, Taurides, Zagros, Greater and Lesser Caucasus, Elzburg and Kopet–Dag (at the East of Caspian Sea). Besides these belts, some recent oceanized or not-oceanized basins are recognized, like in Alboran Sea, Tyrrhenian Sea, Adria basin, Aegean Sea, Pannonian basin and Black Sea.

The complex interaction between Africa and Europe has determined in the zone of Alpine orogeny (the so-called Tethys belt) a fragmentation in several microplates in mutual movement (McKenzie, 1972; Kissel and Laj, 1989). Therefore it is necessary for geodynamic approaches to separate the areas of Africa and Europe interested by older orogenies (defined stable) from the Tethys belt (defined mobile). In this work we have studied only a limited portion of the mobile Tethys belt from Iberia to Caucasus.

3. Filtering

The paleopoles contained in the GPDB have to be grouped according to the geography, geology and tectonics of the sites, the age of the primary magnetization, and the reliability of the data. A computer program of extraction allows a quick selection of suitable set of data. This software developed by some of the Authors (Sagnotti et al., 1993) can select the data in space, time, quality, and from a maximum of 20 criteria. A cartographic map projection program allows the automatic plot of the poles (Scalera, 1990) immediately evaluating the pole distributions and the declinations.

3.1 SPATIAL FILTERING

Following the geologic-tectonic framework in which the reasons of the distinction between stable and mobile areas were explained, we grouped the paleopoles on the basis of their geographic coordinates. We have drawn initially three polygonals: Stable Africa (SA), Stable Europe (SE), Mobile Tethys Belt (MTB). The limits of these polygonals were realized using many tectonic maps and publications (see references in Fig.1).

Beside the just described polygonals a fourth one was considered, including the Arabian plate (A). Albeit the Arabian geodynamics is of great interest, too scarce data are contained in this last poligonal, preventing its analysis in this study.

3.2 TIME FILTERING

The data spatially divided have been selected also following the age of magnetization of the rocks, according to the time windows of Table 1.

The width of the time windows has to be calibrated in agreement to the following reasons:

i) the decreasing amount of data for unit of time (time density) as coming back to the geologic time;

ii) the need to divide the time in periods and epochs according to the Geological Time (Bilal and Van Eysinga, 1987).

The first window (5 Ma) enclose Pliocene and Quaternary, then six windows (10 Ma each) cover from Miocene to Paleocene, the five last windows (35 Ma each, except the Early Cretaceous) from Cretaceous to Triassic. Also the maximum amplitudes of the range

STABLE AFRICA

Time Windows (M.a)	Number of data after filtering			
	Q = A	Q = B	Q = C	Q = D
0 - 5	--	3	6	19
5 - 15	--	3	1	3
15 - 25	--	--	--	5
25 - 35	--	--	1	2
35 - 45	1	--	--	--
45 - 55	--	--	--	--
55 - 65	--	--	--	--
65 - 100	2	3	1	3
100 - 140	--	1	1	3
140 - 175	--	1	--	3
175 - 210	--	5	3	4
210 - 245	--	--	2	4

STABLE EUROPE

Time Windows (M.a)	Number of data after filtering			
	Q = A	Q = B	Q = C	Q = D
0 - 5	2	6	3	71
5 - 15	--	2	2	10
15 - 25	--	1	--	5
25 - 35	--	--	1	--
35 - 45	--	--	--	--
45 - 55	--	--	2	7
55 - 65	--	3	3	8
65 - 100	1	--	1	4
100 - 140	--	--	--	5
140 - 175	--	4	--	6
175 - 210	--	2	2	3
210 - 245	--	4	1	50

ALPINE TETHYS BELT

Time Windows (M.a)	Number of data after filtering			
	Q = A	Q = B	Q = C	Q = D
0 - 5	1	11	4	150
5 - 15	3	3	6	45
15 - 25	3	--	--	9
25 - 35	--	1	--	6
35 - 45	1	4	3	5
45 - 55	--	2	1	6
55 - 65	--	--	--	13
65 - 100	2	12	1	103
100 - 140	--	5	1	43
140 - 175	1	11	4	96
175 - 210	--	3	1	33
210 - 245	--	--	7	30

ITALY

Time Windows (M.a)	Number of data after filtering			
	Q = A	Q = B	Q = C	Q = D
0 - 5	1	2	--	10
5 - 15	--	1	--	4
15 - 25	--	--	--	2
25 - 35	--	1	--	1
35 - 45	--	4	1	--
45 - 55	--	2	1	1
55 - 65	--	--	--	2
65 - 100	1	6	1	31
100 - 140	--	2	--	2
140 - 175	--	--	1	5
175 - 210	--	--	--	3
210 - 245	--	--	--	5

Table 3 Number of data divided in the 4 data quality classes and in the 12 time windows for: a) Stable Africa; b) Stable Europe; c) Mobile Tethys Belt; d) Italian Peninsula.

uncertainty of the magnetization ages is function of the geological time and never overcome the time size of the windows (Table 1).

3.3 QUALITY FILTERING

Besides this space-time filtering, different reasons suggest to operate a quality filtering of the data because of the need to evaluate lack of remagnetizations, quality of age determination and structural control.

For istance – among others – the data, kept for magnetostratigraphic purposes and contained in the GPDB, are tipically referred to only few localities and are not distributed along tectonic structures. They, therefore, cannot give a statistically significant pole. Old studies do not fulfill some modern criteria of laboratory and analytical procedures (see Table 2). Some old data have an insufficient number of sites and samples to allow a significant Fisher statistics (Fisher, 1953) and mean pole computation. Therefore it is necessary for geodynamic studies to select carefully paleopoles, and in literature different Authors tried to define quality criteria.

Van der Voo and French (1974) considered reliable data only those characterized by successful cleaning of any possible secondary component of magnetization and based on a considerable number of samples. Moreover, a well determined age of the rocks (supposing that the magnetization has about the same age), and finally structural controls were also requested.

May and Butler (1986) used more restrictive criteria. These criteria required at least 10 sites per pole to consider stable primary component of magnetization. Furthermore, the Authors prescribed Fisher precision parameter of Virtual Geomagnetic Pole (VGP) dispersion (K) between 20 and 150, and α_{95} confidence interval less than 15. The maximum age uncertainty is accepted 10 Ma and an appropriate understanding of necessary structural corrections is requested.

Van der Voo (1990) proposed 7 equally-weighted reliability criteria: (1) well-determined age of the rocks, (2) minimum number of data and statistical precision, (3) demagnetization, (4) field tests, (5) tectonic coherence, (6) reversal, and (7) no suspicion of remagnetization. The assigned quality index (from 0 to 7) equals the satisfied number of criteria.

Li and Powell (1991) assigned to the data 5 quality classes (A, B, C, D and E). Paleomagnetic poles of A and B classes had a good age and paleohorizontal control. They were considered as "key-poles". The C class poles generally have paleohorizontal control, but lack tight age control, or they are statistically less reliable. The D and E class poles are considered too unreliable to be used.

Besse and Courtillot (1991) assumed the following minimum reliability criteria: at least 6 sites per pole and 6 samples per site, α_{95} confidence interval less than 15, evidence for successful alternate field and/or thermal demagnetization and maximum date uncertainty of 15 Ma. They tried to spot evidences of remagnetization, to check quality of age determination and to assess tectonic structures, including fold test. For APWP construction they also excluded data from mobile zones.

We have selected the paleopoles implementing some criteria already proposed by Besse and Courtillot (B&C) (1991). Four different quality classes have been defined only with the first two fulfilling modern criteria (Table 2).

The A quality class has more restrictive criteria than the B&C ones. We defined A class paleopoles as "key-poles". The minimum number of sites and samples was choosen as 8. The minimum value for both Dp and Dm – the angular semi-amplitude of the axes of the 95% confidence ellipse – was choosen less than or equal to 15. The laboratory and analytical procedures are required greater then or equal to 3. The complete list of the laboratory and analytical procedures is shown in Table 2.

The B class was choosen with very similar criteria like the B&C ones. The only difference is the time windowing of the age of the primary magnetization, which in our case is variable versus the geologic age (see Table 1). This choice has narrower time ranges than those of the B&C criteria since Quaternary until Paleocene, while for the remaining ages (Cretaceous–Triassic) the time ranges are equal to B&C ones. The minimum value for Dp and Dm is still choosen less than or equal to 15. The laboratory and analytical procedures are required greater then or equal to 2 (Table 2).

An "intermediate" C class has been defined with 4 for the number of sites and samples,

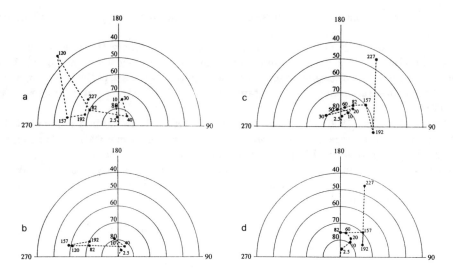

Fig. 2 Apparent Polar Wander Paths (APWPs) for Stable Africa and Stable Europe computed from the present up to 245 Ma; a) Stable Africa APWP with A, B and C quality classes of data; b) Stable Africa APWP with A and B quality classes of data; c) Stable Europe APWP with A, B and C quality classes of data; d) Stable Europe APWP with A and B quality classes of data; The numbers near the dots indicate the central values of each time window.

slightly less than the minimum which can be inferred from Fisher statistics. In this case the minimum value for Dp and Dm was choosen less than or equal to 20. This C class has been used to verify the stability and/or instability of the APWP, and to indicate their possible trends.

A last D class encloses all the remaining data with the ranges of uncertainty of magnetization ages as prescribed in Table 1. However these, joined to other data with greater time uncertainties, cannot have practical use in geodynamic studies.

Finally we have checked that our set of A and B quality class data resulted mostly furnished with at least one field test (e.g. fold test, reversal test, conglomerate test ...), which are important in assess the quality of data, in particular to test if the NRM is the true primary magnetization.

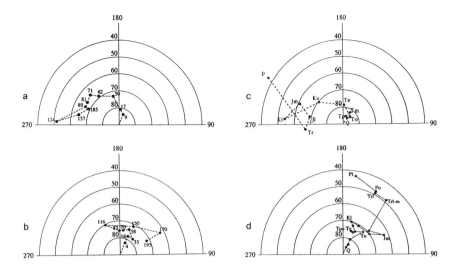

Fig. 3 Polar wander curves for Africa and Eurasia, re-examined by other Authors; a) African APWP and b) European APWP by Besse and Courtillot (1991); c) African APWP and d) European APWP by Westphal et al. (1986).

4. African and European APWPs

In our study we used data belonging to the first two quality classes (Table 2). We adopted also the C class data for the APWP of Africa and Europe only with the aim to verify the results. It was very surprising that most of the data (about 80%) does not fall in the A and B quality classes, useful for geodynamic studies (Table 3ac). Stable Europe (SE) and Stable Africa (SA) have data less than Mobile Tethys Belt (MTB). Finally Arabia has too few data that we did not considered at all.

First of all we computed the VGPs of SA and SE for the non-empty time windows of Table 1. We plotted also their APWP (Figs. 2a,b and 2c,d respectively) using the Fisher's mean pole if possible, or the "key-pole" or the B class pole when only one pole is representative of the considered time period.

The African APWP with A and B quality data (Fig.2b) show a little recent loop between 2.5 and 40 Ma, and a southward trend which terminates in a cusp near 157 Ma. Then the path come back toward the present pole. If the path is constructed using also the C quality data (Fig.2a), the general shape of the African APWP is preserved with differences of few degree except for a large discrepancy at 120 Ma due to a spurious data (Hailwood, 1974).

The European APWP only with A and B quality data (Fig.2d) show a regular progression between 2.5 and 82 Ma – these numbers are the central values of the time windows – and two cusps at 82 and 192 Ma, which can be interpreted as geological episodes of reorganization of the lithospheric plate boundaries and resulting driving forces, although different interpretations are possible. The same trend is shown using also the C class data (Fig.2c), but the uncertainty increases over 15° expecially for the older ages later than 20 Ma.

Our results are close to those published by different Authors (Fig.3a–d). Our difference with other sources is limited to 10°–15° for both Europe and Africa. This is, certainly, a confirmation of a substantial similarity among the paths and of their utility for paleogeographical reconstructions at global scale. On the other hand, it is a measure of the limit that we cannot overcome, if geodynamic considerations are done on the basis of paleopole data at more regional scale.

5. Applications to regional geodynamics

The paleomagnetic poles making up an APWP of a major plate also can be used as "reference pole" to detect motions (vertical-axis rotations and latitudinal transport) of crustal blocks. Essentially two basic methods exist to analyze these motions: the direction–space and pole–space approaches (Butler, 1992). We adopt in this study the second criterium. In this approach, which involves analysis of spherical triangles, the comparison between the reference pole of the stable part of the plate and the observed pole , determined for the considered crustal block, is performed for each geological period. The poleward transport P is:

$$P = p_o - p_r$$

where p_o represents the great–circle distance in degrees between the sampling site and the observed pole, and p_r is the great–circle distance between the sampling site and the reference pole. The value of P is positive if the block has moved toward the reference pole. The rotation R around a vertical axis in degrees is:

$$R = \cos^{-1} \frac{(\cos s - \cos p_o \cos p_r)}{\sin p_o \sin p_r}$$

where s is the great circle distance between the observed and reference poles.

A comparison among the Tethyan paleopoles and the African and European APWP is performed choosing 4 time windows, where a sufficient number of data is present. In these examples (Figs. 4–7) the links between paleopole distribution and the tectonic–structural evolution of the sites is shown.

Early Jurassic

In figure 4 two opposite big rotations can be noted. The two European sites are located in the central part of the northern calcareous Alps (Austria), and from the transdanubian

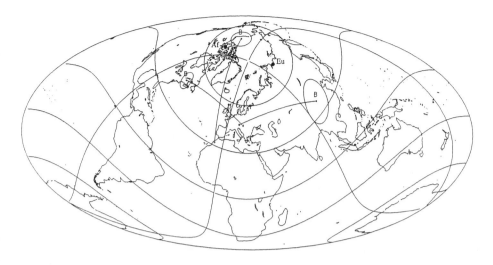

Fig. 4 Mobile Tethys Belt: Sampling localities and positions of their poles with confidence ellipses for Early Jurassic (175–210 Ma). In this and following (Figs.5–7) figures Eu and Af represent the coeval mean poles for Europe and Africa respectively.

central mountains (Hungary). The first paleomagnetic result comes from samples of Liassic limestones (Mauritsch and Frisch, 1978). The latter result came from a complete sequence of Mesozoic marine sediments (Marton and Marton, 1981). The value of the rotations of the two localities has been re-evaluated with respect to the Early Jurassic European mean pole in about $R_{Europe}^{North.Alps} \pm \triangle R = 41.4 \pm 15.5$ clockwise in the first case, and in $R_{Europe}^{Transd.Mount.} \pm \triangle R = 77.8 \pm 11.3$ counterclockwise in the last one.

Middle–Upper Jurassic

In figure 5 the paleopoles coming from the European mobile belts show a low dispersion distribution besides the European mean pole. In the case of the Caucasian paleopoles it is possible to note the typical pattern with an elongated and arcuated distribution. This pattern is an index of mutual rotations of different amount of little blocks (MacDonald, 1980). Among the African data the Tunisian pole presents a clockwise rotation with respect to African mean pole, even if the Authors considered the site belong to the stable and undisturbed eastern platform (Nairn et al., 1981). This evidence can be interpreted as a

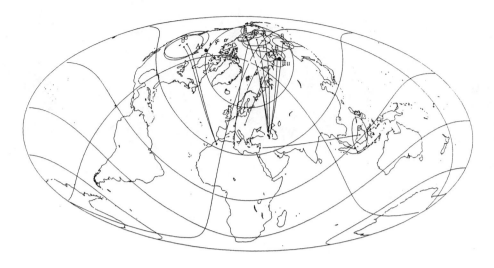

Fig. 5 Mobile Tethys Belt: Sampling localities and positions of their poles with confidence ellipses for Middle–Upper Jurassic (140–175 Ma).

coherent movement of this sector of the platform, which is close to the tectonized area of the N–S Bou Kornine fault system. Another anomalous result in this time window is the paleomagnetic investigation on Upper Jurassic volcanic rocks from Lebanon, near the Jordan–Dead sea transform fault system (Gregor et al., 1974). The results show a very strong rotation relative to the African mean pole. It can be interpreted as block rotation kinematics ascribed to the evolution of the Jordan–Dead Sea left–lateral transform fault. This evolution is recent starting about 30 Ma ago (Westphal et al., 1986).

Upper Cretaceous

This time window have a relatively large number of data (Fig.6). Also in this case an elongated and arcuated distribution can be noted for the whole data set. Big rotations of about $R^{Troodos}_{Europe} \pm \triangle R = 85.1 \pm 7.9$ and $R^{Troodos}_{Africa} \pm \triangle R = 65.9 \pm 11.7$ of the Troodos microplate (Cyprus) can be easily seen. According to Clube et al. (1985) this rotation occurred relatively quickly since the Upper Cretaceous to Upper Eocene. Regional geological consideration support the rotation of a small fragment of oceanic crust (Troodos ophiolite) that was stranded adjacent to an active continental margin.

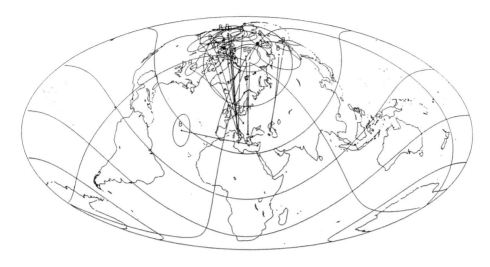

Fig. 6 Mobile Tethys Belt: Sampling localities and positions of their poles with confidence ellipses for Upper Cretaceous (65–100 Ma).

Early Miocene

This time window has only two poles coming from Izmir region in western Anatolia (Fig.7). The paleomagnetic investigations on Miocene tuff and lacustrine marly limestones and lava flows from this region show an anticlockwise rotation of about $R^{Izmir}_{Europe} \pm \triangle R = 41.8 \pm 10.5$. The results obtained in this area, together with others from Aegean domain, suggest that the Early Miocene arc was almost rectilinear trending E–W, and that the curvature was acquired by opposed rotational deformation at each terminations (Kissel et al., 1986).

Finally we have outlined geodynamical peculiarities at more regional and local scale. In particular Italy and Istria have been taken into account, evaluating their possible rotations.

5.1 ITALY

Several paleomagnetic studies were carried out in the last 15 years on the Italian peninsula (e.g., Channell et al., 1978; 1980; Horner and Lowrie, 1981; Tauxe et al., 1983; Besse et al., 1984; Sagnotti, 1992; Mattei et al., 1992). Furthermore many authors published critical reviews of the existing data (e.g. Vandenberg and Zijderveld, 1982; Lowrie, 1986). In these

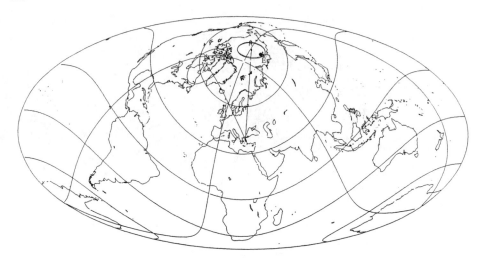

Fig. 7 Mobile Tethys Belt: Sampling localities and positions of their poles with confidence ellipses for Early Miocene (15–25 Ma).

papers great counterclockwise rotations have been determined for Italy since Early Cretaceous. But up to now the data are concentrated in space and in time and they are very poor for Tertiary and Quaternary (Table 3d). The complete lists of Italian poles extracted from GPDB (1979-1988), their referenced Authors, and the map of the their sampling sites, are in press in Sagnotti et al. (1993). The northern Apennines, the central Italy up to Campania, Puglia and Messina Strait are particularly lacking of sampling sites. Higher densities of data are present in central–eastern Alps, Umbrian–Marche Apennines, Sardinia and Sicily. This lack and concentration of data do not allow to detail with a sufficiently narrow time window the evolution of the main deformation phases and the movements of the different Italian geological structures.

Figure 8a,b shows the distribution of the Italian paleopoles in Upper Cretaceous (65–100 Ma). The data come from different sites along the entire peninsula. All the A and B data (Fig.8a) present minor or major counterclockwise declination rotations with respect to the European coeval pole, with the unique A quality pole very close to the centre of the distribution. If no quality filtering is applied (Fig.8b) the dispersion increases, but the counterclockwise rotations with respect to the Upper Cretaceous European pole are statistically confirmed.

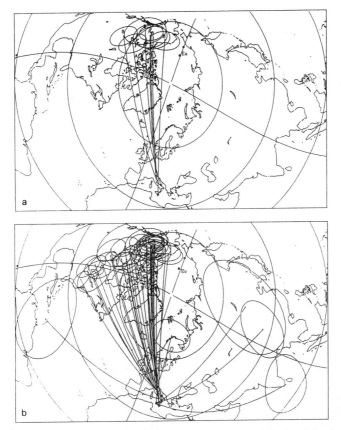

Fig. 8 Upper Cretaceous (65–100 Ma) Italian paleopoles. Different lower limits of the data quality are shown in a) and b) with the aim to put in evidence the effect of the quality filtering. a) A and B quality classes; b) any filtering applied.

On the other hand, the wide time span of the window (35 Ma), in which a finite lenght segment of the Cretaceous APWP must be present, has to be considered. The different sampling sites and related paleopole positions could be so distant in time as the entire time window and, therefore, they should be representative of different lapse of geological time. Furthermore, the different parts of the Italian Peninsula could have different rotation story leading to a concomitant dispersion cause. For these reasons, we cannot compute an "absolute" rotation of the peninsula but we only give – as indicative of the magnitude order – the value of the rotation R of the A "key–pole" with respect to the European and African coeval poles, and the paleo–latitudinal transports P.

$$R^{Italy}_{Europe} \pm \triangle R = -19.9 \pm 7.5$$

$$P^{Italy}_{Europe} \pm \triangle P = 7.9 \pm 6.7$$

$$R^{Italy}_{Africa} \pm \triangle R = 1.3 \pm 12.4$$

$$P^{Italy}_{Africa} \pm \triangle P = 11.1 \pm 10.5$$

The A and B quality poles (Fig.8a) show a distribution around the African pole. Albeit the poles from central Italy (Umbria) and Gargano show a more accentuated rotation - a fact which indicates a different behaviour of different geological structures - this distribution could be considered as a clue of the influence of African plate movements on the rotations of different Italian geological domains. Moreover the limitate amount of the rotation (about 20°) can be considered as a restrictive boundary condition to some extreme paleogeographical reconstructions, in which the complete closing of the Tyrrhenian and Alboran Seas requires nearly 90° rotation, which should be justified by particular models of block movements (Kissel and Laj, 1989). Notwithstanding these limitations many Authors have estrapolated a counterclockwise rotation to the entire peninsula since Cretaceous until present.

5.2 ISTRIA

We have also analyzed the paleomagnetic results for Istria peninsula (Yugoslavia) using always the pole-space approach and its formalism. The purpose is to detect any possible post-Upper Cretaceous relative movement of this region with respect to the Stable Africa and the Stable Europe.

We have choosen this area due to its role in the Mediterranean evolution. The Istria peninsula, the Gargano, the southern Alps and the Iblean platform are emerged portions of the Adriatic region, foreland of the Apennines and Dinarides belts. The Adriatic geodynamic evolution mainly influenced the tectonic setting of the Western Mediterranean area. The samples collected by Marton and Veljovic (1983) yield a mean Cretaceous pole (lat. 53.0, long. 275.0 and α_{95} =4.8) which indicates a counterclockwise rotation of about 30° with respect to the coeval African pole (Irving, 1977) in post-Mesozoic time.

Lowrie (1986), using the same data, evaluated about 17° counterclockwise rotation - with 11° of associated standard deviation - of Istria with respect to the coeval reference pole obtained from the African APWP by Van der Voo and French (1974).

Following Lowrie (1986) we have averaged the paleomagnetic results for Istria collected by Marton and Veljovic (1983) for the Upper Cretaceous to give mean directions and pole positions. The Upper Cretaceous mean pole (lat. 59.0, long. 279.0 and α_{95} =15.0) has been compared with our coeval African (lat. 72.4, long. 249.9 and α_{95} =10.8) and European (lat. 75.8, long. 181.1 and α_{95} =3.8) reference poles. The results confirm a post-Cretaceous counterclockwise rotation of Istria

$$R^{Istria}_{Africa} \pm \triangle R = 21.3 \pm 18.$$

$$R^{Istria}_{Europe} \pm \triangle R = 42.5 \pm 15.2$$

with very large confidence intervals. Moreover the values of the confidence intervals of the latitudinal transports

$$P^{Istria}_{Africa} \pm \triangle P = -1.4 \pm 14.8$$

$$P^{Istria}_{Europe} \pm \triangle P = -4.0 \pm 12.4$$

prevent any speculations about this kind of mutual movement between the involved plates.

6. Conclusions

The availability of the GPDB has opened new possibilities in geodynamic studies. The main advantage is the fast selection of suitable set of data by means of computer extraction programs. The selection has been made with space, time and quality filtering. However the automatic handling of data does not exclude the necessity to read the original papers, to be sure to evaluate, without ambiguities, the true reliability of the data and to avoid errors of trascription in the database.

We have re-examined some peculiar aspects of the Mediterranean geodynamics. A camparison has been performed among the European and African APWP just published by several Authors and our APWP computed using only the first two or three higher quality classes of data. An uncertainty of $10°-15°$ is found among the different APWP, even if the general trend is clearly confirmed. The value of uncertainty is assumed as a probable limit that we cannot overcome to ascertain the mutual displacements of the plates. The same uncertainty magnitude is found to compute rotations and latitudinal transports for Italy and Istria with respect to coeval European and African reference poles. However future improvements of the paleomagnetic methods and techniques and consequent increase of the quality of the data cannot be excluded, and they can lead to better constrain the results and narrow confidence intervals.

The Upper Cretaceous Italian data indicate a counterclockwise rotation (computed by means of the quality A datum) of about $20°$ with respect to the coeval European pole. This amount is very far from the values published in other paleogeographic reconstructions. This could suggest that the kinematics of opening of Alboran and Tyrrhenian Seas cannot be accounted for a simple rotation of Italy as a rigid block (Vandenberg et al.,1978), but it can be better fitted by a fragmentation of Italy in more little blocks characterized by different rotation patterns (Mattei et al., 1992; Favali et al., 1993, and this volume), as it can be seen in the dispersion of high quality poles coming from sites located along the entire peninsula.

Istria in the same time window shows a strong rotation with respect to the European pole (about $42°$), confirming its evolution as indipendent body in more connection with the African geological history. Both Italy and Istria do not have resolvable latitudinal transports due to their large confidence intervals.

The African and European APWPs have been computed using African and European data without the use of relative motions models to transfer all data in a common reference frame. Despite that, using the GPDB, the number of high quality data used is comparable or greater than the the previous papers. In this study we have used data from only two major plates. A truly global study will require data from at least all the surrounding plates.

References

AA.VV. (1981) *Tectonic map of Italy*, 1:1500000, C.N.R.–P.F.G., Pubbl. no.269, G.E.O., Roma.

Argnani, A., Favali, P., Frugoni, F., Gasperini, M., Ligi, M., Marani, M., Mattietti, G. and Mele, G. (1992) *Foreland deformational pattern in the Southern Adriatic Sea*, in R. Funiciello and C. Laj (eds.), Proceedings 7° International School of Solid Earth Geophysics, Erice (TP) (in press).

Bally, A.W., Catalano, R. and Oldow, J. (1985) *Elementi di Tettonica Regionale*, Pitagora Editrice, Bologna, pp.276.

Bartov, Y. (1979) *Geological map of Israel*, 1:500000, Geol. Survey of Israel.

Besse, J. and Courtillot, V. (1991) *Revised and Synthetic Apparent Polar Wander Paths of the African, Eurasian, North America and Indian Plates, and True Polar Wander Since 200 Ma*, J. Geophys. Res., **96**, 4029–4050.

Besse, J., Pozzi, J.P., Mascle, G. and Feinberg, H. (1984) *Paleomagnetic study of Sicily: consequences for the deformation of Italian and African margins over the last 100 Ma*, Earth Planet. Sci. Lett., **67**, 377–390.

Bilal, V.H. and Van Eysinga, F.W.B. (1987) *Geological Time Table, Fourth revised enlarged and updated edition*, Elsevier, Amsterdam.

Butler, R.F. (1992) *Paleomagnetism: magnetic domains to geologic terranes*, Blackwell Sci. Publ., Boston, pp.319.

Boccaletti, M. and Dainelli, P. (1982) *Il Sistema Regmatico Neogenico–Quaternario nell'area Mediterranea: esempio di deformazione plastico/rigida postcollisionale*, Mem. Soc. Geol. Ital., **24**.

Boccaletti, M., Conedera, C., Dainelli, P. and Gočev, P. (1985) *Tectonic map of the western Mediterranean area*, 1:2500000, C.N.R., Pubbl. no.158.

BRMG, Bureau de Recherches Geologiques et Minieres (1980) *Carte geologique de la France et de la marge continentale*, 1:1500000, France.

Channell, J.E.T., Lowrie, W., Medizza, F. and Alvarez, W. (1978) *Paleomagnetism and tectonics in Umbria, Italy*, Earth Planet. Sci. Lett., **39**, 199–210.

Channell, J.E.T., Catalano, R. and D'Argenio, B. (1980) *Paleomagnetism of the Mesozoic continental margin in Sicily*, Tectonophysics, **61**, 391–407.

Clube, T.M.M., Creer, K.M. and Robertson, A.H.F. (1985) *Paleorotation of the Troodos microplate, Cyprus*, Nature, **317**, 523–525.

Condie, K.C. (1989) *Tectonic map of the Earth*, 1:36000000, Williams & Heintz Map Corp.

Favali, P., Funiciello, R., Mele, G., Mattietti, G. and Salvini, F. (1993) *An active margin across the Adriatic Sea (central Mediterranean Sea)*, Tectonophysics, **219** (in press).

Favali, P., Funiciello, R. and Salvini, F. (this volume) *Geological and seismological evidence of strike-slip displacement along the E–W Adriatic–Central Apennines belt*.

Fisher, R.A. (1953) Dispersion on a sphere, Proc. R. Astron. Soc., **A217**, 295–305.

Gregor, C.B., Mertzman, S., Nairn, A.E.M. and Negendank, J. (1974) *The paleomagnetism of some Mesozoic and Cenozoic volcanic rocks from the Lebanon*, Tectonophysics, **21**, 375–395.

Hailwood, E.A. (1974) *Paleomagnetism of the Msissi Norite (Morocco) and the Paleozoic reconstruction of Gondwanaland*, Earth Planet. Sci. Lett., **23**, 376–386.

Horner, F. and Lowrie, W. (1981) *Paleomagnetic evidence from Mesozoic carbonate rocks for the rotation of Sardinia*, J. Geophys., **49**, 11–19.

Irving, E. (1958) *Rock magnetism: a new approach to the problems of polar wandering and continental drift*, in S.W. Carey (ed.), Continental Drift, a Symposium, Hobart, 24–57.

Irving, E. (1977) *DRift of the major continental blocks since the Devonian* Nature, **270**, 304–309.

Kampunzu, A.B. and Lubala, R.T. (1991) *Magmatism in extensional structural setting.* Springer-Verlag, Berlin, pp.620.

Kirschvink, J.L. (1980) *The least-square line and plane and the analysis of paleomagnetic data*, Geophys. J. R. Astron. Soc., **62**, 699–718.

Kissel, C., Laj, C. Poisson, A., Savascin, Y., Simeakis, K. and Mercier, J.L. (1986) *Paleomagnetic evidence for Neogene rotational deformations in the Aegean domain*, Tectonics, **5**, 783–795.

Kissel, C. and Laj, C. (eds.) (1989) *Paleomagnetic rotations and continental deformation*, NATO ASI Series, **vol.254**, pp.516.

Kruczyk, J., Kadzialko-Hofmokl, M., Lefeld, J., Pagac, P. and Tunyi, I. (1992) *Paleomagnetism of Jurassic sediments as evidence for oroclinal bending of the Inner West Carpathians*, Tectonophysics, **206**, 315–324.

Li, Z.X. and Powell C.McA. (1991) *Late Proterozoic to Early Paleozoic Paleomagnetism and the formation of Gondwanaland*, Proceedings of the Gondwana-8° meeting, Hobart (in press).

Lock, J. and McElhinny, M.W. (1991) *The Global Paleomagnetic Database*, Surveys in Geophysics, **12**, 317–491.

Lowrie, W. (1986) *Paleomagnetism and the Adriatic promontory: a reappraisal*, Tectonics, **5**, 797–807.

MacDonald, W.D. (1980) *Net tectonic rotation, apparent tectonic rotation, and the structural tilt correction in paleomagnetic studies*, J. Geophys. Res., **85**, 3659–3669.

Martin, D.L., Nairn, A.E.M., Noltimier, H.C., Petty, M.H. and Schmitt, T.J. (1978) *Paleozoic and Mesozoic paleomagnetic results from Morocco*, Tectonophysics, **44**, 91–114.

Marton, E. and Marton, P. (1981) *Mesozoic paleomagnetism of the Transdanubiam Central Mountains and its tectonic implications*, Tectonophysics, **72**, 129–140.

Marton, E. and Veljovic, D. (1983) *Paleomagnetism of the Istria Peninsula, Yugoslavia.* Tectonophysics, **91**, 73–87.

Mattei, M., Funiciello, R., Kissel, C. and Laj, C. (1992) *Rotazione di blocchi crostali neogenici nell'Appen-nino cent-rale: ana-lisi paleo-magne-tiche e di ani-so-tropia della suscettività magnetica (AMS)*, Studi Geol. Camerti, Spec. Pubbl. 1991/2, 221–229.

Mauritsch, H.J. and Frisch, W. (1978) *Paleomagnetic data from the central part of the northern calcareous Alps, Austria*, J. Geophys., **44**, 623–637.

May, S.R. and Butler, R.F. (1986) *North-American Jurassic apparent polar wander: implications for plate motions, paleogeography and Cordilleran tectonics*, J. Geophys. Res., **91**, 11519–11544.

McElhinny, M.W. and Lock, J. (1990a) *Global Paleomagnetic Database Project*, Phys. Earth Planet. Int., **63**, 1–6.

McElhinny, M.W. and Lock, J. (1990b) *IAGA global paleomagnetic database*, Geophys. J.

Int., **101**, 763-766.
McKenzie, D. (1972) *Active tectonics of the Mediterranean region*, Geophys. J. R. Astron. Soc., **30**, 109-185.
Merla, G., Abbate, E., Canuti, P., Sagri, M. and Tacconi, P. (1973) *Carta geologica dell'Etiopia e della Somalia*, 1:2000000, C.N.R.
Nairn, A.E.M., Schmitt, T.J. and Smithwick, M.E. (1981) *A paleomagnetic study of the Upper Mesozoic succession in the northern Tunisia*, Geophys. J.R. Astron. Soc., **65**, 1-18.
Perrin, M., Elston, D.P. and Moussine-Pouchkine, A. (1988) *Paleomagnetism of Proterozoic and Cambrian strata, Adrar de Mauritanie, cratonic West Africa*, J. Geophys. Res., **93**, 2159-2178.
Piper, J.D.A. (1988) *Paleomagnetic database*, Open Univ. Press, Milton Keynes, pp.264.
Sagnotti, L., (1992) *Paleomagnetic evidence for a pleistocene counterclockwise rotation of the Sant'Arcangelo basin, Southern Italy*. Geophys. Res. Lett., **19**, 135-138.
Sagnotti, L., Scalera, G. and Florindo, F. (1993) *Automatic management of the global paleomagnetic database (ASCII version)*, Geoinformatica (in press).
Salmon, E., Edel, J.B., Pique, A. and Westphal, M. (1988) *Possible origins of Permian remagnetizations in Devonian and Carboniferous limestones from the Moroccan Anti-Atlas (Tafilalet) and Meseta*, Phys. Earth Planet. Int., **52**, 339-351.
Scalera, G. (1990) *Palaepoles on an expanding Earth: a comparison between synthetic and real data sets*, Phys. Earth Planet. Int., **62**, 126-140.
Stocklin, J. and Nabavi, M.H. (1973) *Tectonic map of Iran*, 1:2500000, Geol. Survey of Iran.
Tauxe, L., Opdyke, N.D., Pasini, G. and Elmi (1983) *Age of the Plio-Pleistocene boundary in the Vrica section, Southern Italy*, Nature, **304**, 125-129.
UNESCO (1976) *Geological World Atlas*, Commission for the Geological map of the World (C.G.M.W.).
Vandenberg, J., Klootwijk, C.T. and Wonders, A.A.H. (1978) *Late Mesozoic and Cenozoic movements of the Italian peninsula: further paleomagnetic data from the Umbrian sequence*, Geol. Soc. Am. Bull., **89**, 133-150.
Vandenberg, J. and Zijderveld, J.D.A. (1982) *Paleomagnetism in the Mediterranean area*, in H. Berckhemer and K. Hsü (eds.), Alpine-Mediterranean Geodynamics, Am. Geophys. Union - Geol. Soc. Am., Geodynamics Series, **7**, 83-112.
Van der Voo, R. (1990) *Phanerozoic paleomagnetic poles from Europe and North America and comparison with continental reconstructions*, Rev. Geophys., **28**, 167-206.
Van der Voo, R. and French, R.B. (1974) *Apparent polar wandering for the Atlantic-bordering continents: Late Carboniferous to Eocene*, Earth Sci. Rev., 10, 99-119.
Zijderveld, J.D.A. (1967) *A.C. demagnetization of rocks: analysis of result*, in Methods in Paleomagnetism, Elsevier, New York, 254-286.
Westphal, M., Bazhenov, M.L., Lauer, J.P., Pechersky, D.M. and Sibuet, J. (1986) *Paleomagnetic implications of the evolution of the Tethys belt from the Atlantic Ocean to the Pamirs since the Triassic*, Tectonophysics, **123**, 37-82.